Telecommunications Engineering

Edited by
Owen Thompson

Larsen & Keller
www.larsen-keller.com

Telecommunications Engineering
Edited by Owen Thompson
ISBN: 978-1-63549-189-0 (Hardback)

☰ Larsen & Keller

Published by Larsen and Keller Education,
5 Penn Plaza,
19th Floor,
New York, NY 10001, USA

Cataloging-in-Publication Data

Telecommunications engineering / edited by Owen Thompson.
    p. cm.
Includes bibliographical references and index.
ISBN 978-1-63549-189-0
  1. Telecommunication. 2. Telecommunication systems--Technological innovations.
3. Wireless communication system. I. Thompson, Owen.
TK5101 .T45 2017
621.382--dc23

The publisher's policy is to use permanent paper from mills that operate a sustainable forestry policy. Furthermore, the publisher ensures that the text paper and cover boards used have met acceptable environmental accreditation standards.

Printed and bound in the United States of America.

For more information regarding Larsen and Keller Education and its products, please visit the publisher's website www.larsen-keller.com

Table of Contents

Preface   **VII**

Chapter 1 **Introduction to Telecommunications Engineering**   **1**

Chapter 2 **Key Concepts of Telecommunications Engineering**   **12**
- Automatic Link Establishment   12
- Link Budget   20
- Grade of Service   24
- Loading Coil   26
- Electrical Length   36
- Standing Wave Ratio   41
- Reflections of Signals on Conducting Lines   48
- Network Architecture   57
- Primary Line Constants   59

Chapter 3 **Wireless Technology and its Concepts**   **68**
- Wireless   68
- Wireless Network   75
- Wireless LAN   82
- Wireless Access Point   88
- Wireless WAN   91
- Wireless Mesh Network   92
- Hotspot (Wi-Fi)   96
- Li-Fi   102
- Bluetooth   105

Chapter 4 **Wireless Security and Concerns**   **127**
- Wireless Security   127
- Wired Equivalent Privacy   141
- Wi-Fi Protected Access   146
- Wireless Intrusion Prevention System   150
- Computer Security   153
- Mobile Security   175

Chapter 5 **Applications of Telecommunications Engineering**   **194**
- Photophone   194
- Radio   200
- Radio Resource Management   214
- Digital Radio   217
- Communications Satellite   226
- Fiber-optic Communication   237
- Push-button Telephone   249
- Dual-tone Multi-frequency Signalling   253
- Free-space Optical Communication   256

Chapter 6   **Allied Fields of Telecommunications Engineering**   **265**
- Electrical Engineering   265
- Computer Engineering   280
- Telegraphy   284
- Informatics   301

Permissions

Index

Preface

As the world is becoming a global village, the need for advanced telecommunications is rapidly increasing. Telecommunications engineering is a branch of engineering which works with other fields like computer engineering, mechanical engineering, software engineering, etc. to bring these advances in the field of wireless telecommunication. This book is compiled in such a manner, that it will provide in-depth knowledge about the theory and practice of mobile, wireless and telecommunication engineering. It presents this complex subject in the most comprehensible and easy to understand language. Some of the diverse topics covered in this textbook address the varied branches that fall under this field. It aims to serve as a resource guide for students in the field of networking systems, data scheduling and radar and radio engineering.

Given below is the chapter wise description of the book:

Chapter 1- Telecommunication engineering brings together all electrical engineering disciplines with system engineering to enhance telecommunication systems. Telecommunication engineers provide high-speed data and information transmission services. This chapter will provide an integrated understanding of the subject matter.

Chapter 2- The major key concepts of telecommunications engineering are discussed in this chapter. Automatic link establishment, link budget, grade of service, loading coil and primary line constants are some of the significant and important topics related to telecommunications engineering. The following content unfolds its crucial theories and principles of telecommunications engineering.

Chapter 3- Wireless technology can be explained as the transferring of information between two or more point not connected by any electrical conductors. This text elucidates the concepts of wireless technology, by briefly explaining, wireless mesh network, hotspot, Wi-Fi, Bluetooth and wireless LAN. Wireless communication is best understood in confluence with the major topics listed in the following chapter.

Chapter 4- The prevention of unauthorized access or damage to computers is wireless security. The most common types of wireless security are wireless security, wired equivalent privacy, Wi-Fi protected access, mobile and computer security. This chapter explains to the reader the importance of wireless security and its various aspects.

Chapter 5- Digital radio is the technology used to communicate across the radio spectrum and push buttons is a simple mechanism for controlling some aspect of a machine. The applications of telecommunication engineering such as photo phone, radio and free space optical communication are dealt within this chapter. This chapter discusses the methods of telecommunication in a critical manner providing key analysis to this subject matter.

Chapter 6- Telecommunications engineering is an interdisciplinary subject. This section will provide a glimpse of the allied fields of telecommunications engineering. Some of these allied fields are electrical engineering, computer engineering, telegraphy and informatics. The topics discussed in the chapter are of great importance to broaden the existing knowledge on this field.

Indeed, my job was extremely crucial and challenging as I had to ensure that every chapter is informative and structured in a student-friendly manner. I am thankful for the support provided by my family and colleagues during the completion of this book.

Editor

Introduction to Telecommunications Engineering

Telecommunication engineering brings together all electrical engineering disciplines with system engineering to enhance telecommunication systems. Telecommunication engineers provide high-speed data and information transmission services. This chapter will provide an integrated understanding of the subject matter.

Telecommunications engineering, or telecoms engineering', is an engineering discipline that brings together all electrical engineering disciplines including computer engineering with systems engineering to enhance telecommunication systems. The work ranges from basic circuit design to strategic mass developments. A telecommunication engineer is responsible for designing and overseeing the installation of telecommunications equipment and facilities, such as complex electronic switching systems, copper wire telephone facilities,fiber optics cabling, IP data systems, Terrestrial radio link systems for conventional communications and process information. Telecommunication engineering also overlaps heavily with broadcast engineering.

Telecommunications engineer working to maintain London's phone service during World War 2, January 1942 while all the power was off so he needed to find a power supply

Telecommunication is a diverse field of engineering which is connected to electronics, civil, structural, and electrical engineering. Ultimately, telecom engineers are responsible for providing the method for customers to have telephone and high-speed data services. It helps people who are closely working in political and social fields, as well accounting and project management.

Telecom engineers use a variety of equipment and transport media available from a multitude of manufacturers to design the telecom network infrastructure. The most common media used by wired telecommunications companies today are copper wires, coaxial cable, and fiber optics. Telecommunications engineers use their technical expertise to also provide a range of services and engineering solutions revolving around wireless mode of communication and other information transfer, such as wireless telephony services, radio and satellite communications, internet and broadband technologies.

Most of a telecom engineer's work is carried out on a project basis with tight deadlines and well-defined milestones for the delivery of project objectives. Telecommunication engineers are involved across all aspects of service delivery, from carrying out feasibility exercises and determining connectivity to preparing detailed, technical and operational documentation. This often leads to creative solutions to problems that often would have been designed differently without the budget constraints dictated by modern society. In the earlier days of the telecom industry, massive amounts of cable were placed that were never used or have been replaced by modern technology such as fiber optic cable and digital multiplexing techniques.

Telecom engineers are also responsible for overseeing the companies' records of equipment and facility assets. Their work directly impacts assigning appropriate accounting codes for taxes and maintenance purposes, budgeting and overseeing projects.

History

Telecommunication systems are generally designed by telecommunication engineers which sprang from technological improvements in the telegraph industry in the late 19th century and the radio and the telephone industries in the early 20th century. Today, telecommunication is widespread and devices that assist the process, such as the television, radio and telephone, are common in many parts of the world. There are also many networks that connect these devices, including computer networks, public switched telephone network (PSTN), radio networks, and television networks. Computer communication across the Internet is one of many examples of telecommunication. Telecommunication plays a vital role in the part of world economy and the telecommunication industry's revenue has been placed at just under 3% of the gross world product.

Telegraph and Telephone

Alexander Graham Bell's big box telephone, 1876, one of the first commercially available telephones - National Museum of American History

Samuel Morse independently developed a version of the electrical telegraph that he unsuccessfully demonstrated on 2 September 1837. Soon after he was joined by Alfred Vail who developed the register — a telegraph terminal that integrated a logging device for recording messages to paper tape. This was demonstrated successfully over three miles (five kilometres) on 6 January 1838 and eventually over forty miles (sixty-four kilometres) between Washington, D.C. and Baltimore on 24 May 1844. The patented invention proved lucrative and by 1851 telegraph lines in the United States spanned over 20,000 miles (32,000 kilometres).

The first successful transatlantic telegraph cable was completed on 27 July 1866, allowing transatlantic telecommunication for the first time. Earlier transatlantic cables installed in 1857 and 1858 only operated for a few days or weeks before they failed. The international use of the telegraph has sometimes been dubbed the "Victorian Internet".

The first commercial telephone services were set up in 1878 and 1879 on both sides of the Atlantic in the cities of New Haven and London. Alexander Graham Bell held the master patent for the telephone that was needed for such services in both countries. The technology grew quickly from this point, with inter-city lines being built and telephone exchanges in every major city of the United States by the mid-1880s. Despite this, transatlantic voice communication remained impossible for customers until January 7, 1927 when a connection was established using radio. However no cable connection existed until TAT-1 was inaugurated on September 25, 1956 providing 36 telephone circuits.

In 1880, Bell and co-inventor Charles Sumner Tainter conducted the world's first wireless telephone call via modulated lightbeams projected by photophones. The scientific principles of their invention would not be utilized for several decades, when they were first deployed in military and fiber-optic communications.

Radio and Television

Marconi crystal radio receiver

Over several years starting in 1894 the Italian inventor Guglielmo Marconi built the first complete, commercially successful wireless telegraphy system based on airborne electromagnetic waves (radio transmission). In December 1901, he would go on to established wireless communication between Britain and Newfoundland, earning him the Nobel Prize in physics in 1909 (which he shared with Karl Braun). In 1900 Reginald Fessenden was able to wirelessly transmit a human voice. On March 25, 1925, Scottish inventor John Logie Baird publicly demonstrated the transmission of moving silhouette pictures at the London department store Selfridges. In October 1925, Baird was successful in obtaining moving pictures with halftone shades, which were by most accounts the first true television pictures. This led to a public demonstration of the improved device on 26 January 1926 again at Selfridges. Baird's first devices relied upon the Nipkow disk and thus became known as the mechanical television. It formed the basis of semi-experimental broadcasts done by the British Broadcasting Corporation beginning September 30, 1929.

Satellite

The first U.S. satellite to relay communications was Project SCORE in 1958, which used a tape recorder to store and forward voice messages. It was used to send a Christmas greeting to the world from U.S. President Dwight D. Eisenhower. In 1960 NASA launched an Echo satellite; the 100-foot (30 m) aluminized PET film balloon served as a passive reflector for radio communications. Courier 1B, built by Philco, also launched in 1960, was the world's first active repeater satellite. Satellites these days are used for many applications such as uses in GPS, television, internet and telephone uses.

Telstar was the first active, direct relay commercial communications satellite. Belonging to AT&T as part of a multi-national agreement between AT&T, Bell Telephone Laboratories, NASA, the British General Post Office, and the French National PTT (Post Office) to develop satellite communications, it was launched by NASA from Cape Canaveral on July 10, 1962, the first privately sponsored space launch. Relay 1 was launched on December 13, 1962, and became the first satellite to broadcast across the Pacific on November 22, 1963.

The first and historically most important application for communication satellites was in intercontinental long distance telephony. The fixed Public Switched Telephone Network relays telephone calls from land line telephones to an earth station, where they are then transmitted a receiving satellite dish via a geostationary satellite in Earth orbit. Improvements in submarine communications cables, through the use of fiber-optics, caused some decline in the use of satellites for fixed telephony in the late 20th century, but they still exclusively service remote islands such as Ascension Island, Saint Helena, Diego Garcia, and Easter Island, where no submarine cables are in service. There are also some continents and some regions of countries where landline telecommunications are rare to nonexistent, for example Antarctica, plus large regions of Australia, South America, Africa, Northern Canada, China, Russia and Greenland.

After commercial long distance telephone service was established via communication satellites, a host of other commercial telecommunications were also adapted to similar satellites starting in 1979, including mobile satellite phones, satellite radio, satellite television and satellite Internet access. The earliest adaption for most such services occurred in the 1990s as the pricing for commercial satellite transponder channels continued to drop significantly.

Computer Networks and the Internet

Symbolic representation of the Arpanet as of September 1974

On 11 September 1940, George Stibitz was able to transmit problems using teleprinter to his Complex Number Calculator in New York and receive the computed results back at Dartmouth College

in New Hampshire. This configuration of a centralized computer or mainframe computer with remote "dumb terminals" remained popular throughout the 1950s and into the 1960s. However, it was not until the 1960s that researchers started to investigate packet switching — a technology that allows chunks of data to be sent between different computers without first passing through a centralized mainframe. A four-node network emerged on 5 December 1969. This network soon became the ARPANET, which by 1981 would consist of 213 nodes.

ARPANET's development centered around the Request for Comment process and on 7 April 1969, RFC 1 was published. This process is important because ARPANET would eventually merge with other networks to form the Internet, and many of the communication protocols that the Internet relies upon today were specified through the Request for Comment process. In September 1981, RFC 791 introduced the Internet Protocol version 4 (IPv4) and RFC 793 introduced the Transmission Control Protocol (TCP) — thus creating the TCP/IP protocol that much of the Internet relies upon today.

Optical Fiber

Optical fiber can be used as a medium for telecommunication and computer networking because it is flexible and can be bundled as cables. It is especially advantageous for long-distance communications, because light propagates through the fiber with little attenuation compared to electrical cables. This allows long distances to be spanned with few repeaters.

In 1966 Charles K. Kao and George Hockham proposed optical fibers at STC Laboratories (STL) at Harlow, England, when they showed that the losses of 1000 dB/km in existing glass (compared to 5-10 dB/km in coaxial cable) was due to contaminants, which could potentially be removed.

Optical fiber was successfully developed in 1970 by Corning Glass Works, with attenuation low enough for communication purposes (about 20dB/km), and at the same time GaAs (Gallium arsenide) semiconductor lasers were developed that were compact and therefore suitable for transmitting light through fiber optic cables for long distances.

After a period of research starting from 1975, the first commercial fiber-optic communications system was developed, which operated at a wavelength around 0.8 μm and used GaAs semiconductor lasers. This first-generation system operated at a bit rate of 45 Mbps with repeater spacing of up to 10 km. Soon on 22 April 1977, General Telephone and Electronics sent the first live telephone traffic through fiber optics at a 6 Mbit/s throughput in Long Beach, California.

The first wide area network fibre optic cable system in the world seems to have been installed by Rediffusion in Hastings, East Sussex, UK in 1978. The cables were placed in ducting throughout the town, and had over 1000 subscribers. They were used at that time for the transmission of television channels,not available because of local reception problems.

The first transatlantic telephone cable to use optical fiber was TAT-8, based on Desurvire optimized laser amplification technology. It went into operation in 1988.

In the late 1990s through 2000, industry promoters, and research companies such as KMI, and RHK predicted massive increases in demand for communications bandwidth due to increased use of the Internet, and commercialization of various bandwidth-intensive consumer services, such as

video on demand. Internet protocol data traffic was increasing exponentially, at a faster rate than integrated circuit complexity had increased under Moore's Law.

Concepts

Radio Transmitter room

Basic Elements of a Telecommunication System

Transmitter

Transmitter (information source) that takes information and converts it to a signal for transmission. In electronics and telecommunications a transmitter or radio transmitter is an electronic device which, with the aid of an antenna, produces radio waves. In addition to their use in broadcasting, transmitters are necessary component parts of many electronic devices that communicate by radio, such as cell phones,

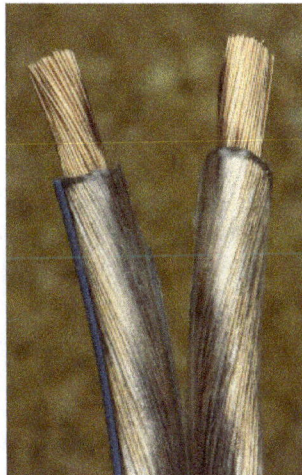

Copper wires

Transmission Medium

Transmission medium over which the signal is transmitted. For example, the transmission medium for sounds is usually air, but solids and liquids may also act as transmission media for sound. Many transmission media are used as communications channel. One of the most common physical medias used in networking is copper wire. Copper wire to carry signals to long distances using

relatively low amounts of power.Another example of a physical medium is optical fiber, which has emerged as the most commonly used transmission medium for long-distance communications. Optical fiber is a thin strand of glass that guides light along its length.

The absence of a material medium in vacuum may also constitute a transmission medium for electromagnetic waves such as light and radio waves.

Receiver

Receiver (information sink) that receives and converts the signal back into required information. In radio communications, a radio receiver is an electronic device that receives radio waves and converts the information carried by them to a usable form. It is used with an antenna. The information produced by the receiver may be in the form of sound (an audio signal), images (a video signal) or data (a digital signal).

Wireless communication tower, cell site

Wired Communication

Wired communications make use of underground communications cables (less often, overhead lines), electronic signal amplifiers (repeaters) inserted into connecting cables at specified points, and terminal apparatus of various types, depending on the type of wired communications used.

Wireless Communication

Wireless communication involves the transmission of information over a distance without help of wires, cables or any other forms of electrical conductors. Wireless operations permit services, such as long-range communications, that are impossible or impractical to implement with the use of wires. The term is commonly used in the telecommunications industry to refer to telecommunications systems (e.g. radio transmitters and receivers, remote controls etc.) which use some form of energy (e.g. radio waves, acoustic energy, etc.) to transfer information without the use of wires. Information is transferred in this manner over both short and long distances.

Roles

Telecom Equipment Engineer

A telecom equipment engineer is an electronics engineer that designs equipment such as routers, switches, multiplexers, and other specialized computer/electronics equipment designed to be used in the telecommunication network infrastructure.

Network Engineer

A network engineer is a computer engineer who is in charge of designing, deploying and maintaining computer networks. In addition, they oversee network operations from a network operations center, designs backbone infrastructure, or supervises interconnections in a data center.

Central-office Engineer

Typical Northern Telecom DMS100 Telephone Central Office Installation

A central-office engineer is responsible for designing and overseeing the implementation of telecommunications equipment in a central office (CO for short), also referred to as a wire center or telephone exchange A CO engineer is responsible for integrating new technology into the existing network, assigning the equipment's location in the wire center, and providing power, clocking (for digital equipment), and alarm monitoring facilities for the new equipment. The CO engineer is also responsible for providing more power, clocking, and alarm monitoring facilities if there are currently not enough available to support the new equipment being installed. Finally, the CO engineer is responsible for designing how the massive amounts of cable will be distributed to various equipment and wiring frames throughout the wire center and overseeing the installation and turn up of all new equipment.

Subroles

As structural engineers, CO engineers are responsible for the structural design and placement of racking and bays for the equipment to be installed in as well as for the plant to be placed on.

As electrical engineers, CO engineers are responsible for the resistance, capacitance, and inductance (RCL) design of all new plant to ensure telephone service is clear and crisp and data service is clean as well as reliable. Attenuation or gradual loss in intensity and loop loss calculations are

required to determine cable length and size required to provide the service called for. In addition, power requirements have to be calculated and provided to power any electronic equipment being placed in the wire center.

Overall, CO engineers have seen new challenges emerging in the CO environment. With the advent of Data Centers, Internet Protocol (IP) facilities, cellular radio sites, and other emerging-technology equipment environments within telecommunication networks, it is important that a consistent set of established practices or requirements be implemented.

Installation suppliers or their sub-contractors are expected to provide requirements with their products, features, or services. These services might be associated with the installation of new or expanded equipment, as well as the removal of existing equipment.

Several other factors must be considered such as:

- Regulations and safety in installation
- Removal of hazardous material
- Commonly used tools to perform installation and removal of equipment

Outside-plant Engineer

Engineers working on a cross-connect box, also known as a serving area interface.

Outside plant (OSP) engineers are also often called field engineers because they frequently spend much time in the field taking notes about the civil environment, aerial, above ground, and below ground. OSP engineers are responsible for taking plant (copper, fiber, etc.) from a wire center to a distribution point or destination point directly. If a distribution point design is used, then a cross-connect box is placed in a strategic location to feed a determined distribution area.

The cross-connect box, also known as a serving area interface, is then installed to allow connections to be made more easily from the wire center to the destination point and ties up fewer facilities by not having dedication facilities from the wire center to every destination point. The plant is then taken directly to its destination point or to another small closure called a terminal, where access can also be gained to the plant if necessary. These access points are preferred as they allow faster repair times for customers and save telephone operating companies large amounts of money.

The plant facilities can be delivered via underground facilities, either direct buried or through conduit or in some cases laid under water, via aerial facilities such as telephone or power poles, or via microwave radio signals for long distances where either of the other two methods is too costly.

Subroles

Engineer (OSP) climbing the telephone pole.

As structural engineers, OSP engineers are responsible for the structural design and placement of cellular towers and telephone poles as well as calculating pole capabilities of existing telephone or power poles onto which new plant is being added. Structural calculations are required when boring under heavy traffic areas such as highways or when attaching to other structures such as bridges. Shoring also has to be taken into consideration for larger trenches or pits. Conduit structures often include encasements of slurry that needs to be designed to support the structure and withstand the environment around it (soil type, high traffic areas, etc.).

As electrical engineers, OSP engineers are responsible for the resistance, capacitance, and inductance (RCL) design of all new plant to ensure telephone service is clear and crisp and data service is clean as well as reliable. Attenuation or gradual loss in intensity and loop loss calculations are required to determine cable length and size required to provide the service called for. In addition power requirements have to be calculated and provided to power any electronic equipment being placed in the field. Ground potential has to be taken into consideration when placing equipment, facilities, and plant in the field to account for lightning strikes, high voltage intercept from improperly grounded or broken power company facilities, and from various sources of electromagnetic interference.

As civil engineers, OSP engineers are responsible for drafting plans, either by hand or using Computer-aided design (CAD) software, for how telecom plant facilities will be placed. Often when working with municipalities trenching or boring permits are required and drawings must be made for these. Often these drawings include about 70% or so of the detailed information required to pave a road or add a turn lane to an existing street. Structural calculations are required when boring under heavy traffic areas such as highways or when attaching to other structures such as bridges. As civil engineers, telecom engineers provide the modern communications backbone for all technological communications distributed throughout civilizations today.

Unique to telecom engineering is the use of air-core cable which requires an extensive network of air handling equipment such as compressors, manifolds, regulators and hundreds of miles of air pipe per system that connects to pressurized splice cases all designed to pressurize this special form of copper cable to keep moisture out and provide a clean signal to the customer.

As political and social ambassador, the OSP engineer is a telephone operating company's face and voice to the local authorities and other utilities. OSP engineers often meet with municipalities, construction companies and other utility companies to address their concerns and educate them about how the telephone utility works and operates. Additionally, the OSP engineer has to secure real estate to place outside facilities on, such as an easement to place a cross-connect box on.

References

- Hafner, Katie (1998). Where Wizards Stay Up Late: The Origins Of The Internet. Simon & Schuster. ISBN 0-684-83267-4.

- "Program criteria for telecommunications engineering technology or similarly named programs" (PDF). Criteria for accrediting engineering technology programs 2012-2013. ABET. October 2011. p. 23. Retrieved September 22, 2012.

- Burnham, Gerald O.; et al. (October 2001). "The First Telecommunications Engineering Program in the United States" (PDF). Journal of Engineering Education (American Society for Engineering Education) 90 (4): 653–657. Retrieved September 22, 2012.

Key Concepts of Telecommunications Engineering

The major key concepts of telecommunications engineering are discussed in this chapter. Automatic link establishment, link budget, grade of service, loading coil and primary line constants are some of the significant and important topics related to telecommunications engineering. The following content unfolds its crucial theories and principles of telecommunications engineering.

Automatic Link Establishment

Automatic Link Establishment, commonly known as ALE, is the worldwide de facto standard for digitally initiating and sustaining HF radio communications. ALE is a feature in an HF communications radio transceiver system, that enables the radio station to make contact, or initiate a circuit, between itself and another HF radio station or network of stations. The purpose is to provide a reliable rapid method of calling and connecting during constantly changing HF ionospheric propagation, reception interference, and shared spectrum use of busy or congested HF channels.

Mechanism

A standalone ALE radio combines an HF SSB radio transceiver with an internal microprocessor and MFSK modem. It is programmed with a unique ALE Address, similar to a phone number (or on newer generations, a username). When not actively in contact with another station, the HF SSB transceiver constantly scans through a list of HF frequencies called channels, listening for any ALE signals transmitted by other radio stations. It decodes calls and soundings sent by other stations and uses the Bit error rate to store a quality score for that frequency and sender-address.

To reach a specific station, the caller enters the ALE Address. On many ALE radios this is similar to dialing a phone number. The ALE controller selects the best available idle channel for that destination address. After confirming the channel is indeed idle, it then sends a brief selective calling signal identifying the intended recipient. When the distant scanning station detects ALE activity, it stops scanning and stays on that channel until it can confirm whether or not the call is for it. The two stations' ALE controllers automatically handshake to confirm that a link of sufficient quality has been established, then notify the operators that the link is up. If the callee fails to respond or the handshaking fails, the originating ALE node usually selects another frequency either at random or by making a guess of varying sophistication.

Upon successful linking, the receiving station generally emits an audible alarm and shows a visual alert to the operator, thus indicating the incoming call. It also indicates the callsign or other identifying information of the linked station, similar to Caller ID. The operator then un-mutes the radio and answers the call then can talk in a regular conversation or negotiates a data link using voice or the ALE built-in short text message format. Alternatively, digital data can be exchanged via a built-in or external modem (such as a STANAG 5066 or MIL-STD-188-110B serial tone modem) depending on needs and availability. The ALE built-in text messaging facility can be used to transfer short text messages as an "orderwire" to allow operators to coordinate external equipment such as phone patches or non-embedded digital links, or for short tactical messages.

Operator Skill

Due to the vagaries of ionospheric communications, HF radio as used by large governmental organizations in the mid-20th century was traditionally the domain of highly skilled and trained radio operators. One of the new characteristics that embedded microprocessors and computers brought to HF radio via ALE, was alleviation of the need for the radio operator to constantly monitor and change the radio frequency manually to compensate for ionospheric conditions or interference. For the average user of ALE, after learning how to work the basic functions of the HF transceiver, it became similar to operating a cellular mobile phone. For more advanced functions and programming of ALE controllers and networks, it became similar to the use of menu-enabled consumer equipment or the optional features typically encountered in software. In a professional or military organization, this does not eliminate the need for skilled and trained communicators to coordinate the per-unit authorized frequency lists and node addresses - it merely allows the deployment of relatively unskilled technicians as "field communicators" and end-users of the existing coordinated architecture.

Common Applications

An ALE radio system enables connection for voice conversation, alerting, data exchange, texting, instant messaging, email, file transfer, image, geo-position tracking, or telemetry. With a radio operator initiating a call, the process normally takes a few minutes for the ALE to pick an HF frequency that is optimum for both sides of the communication link. It signals the operators audibly and visually on both ends, so they can begin communicating with each other immediately. In this respect, the longstanding need in HF radio for repetitive calling on pre-determined time schedules or tedious monitoring static is eliminated. It is useful as a tool for finding optimum channels to communicate between stations in real-time. In modern HF communications, ALE has largely replaced HF prediction charts, propagation beacons, chirp sounders, propagation prediction software, and traditional radio operator educated guesswork. ALE is most commonly used for hooking up operators for voice contacts on SSB (single sideband modulation), HF internet connectivity for email, SMS phone texting or text messaging, real-time chat via HF text, Geo Position Reporting, and file transfer. High Frequency Internet Protocol or HFIP may be used with ALE for internet access via HF.

Techniques

The essence of ALE techniques is the use of automatic channel selection, scanning receivers, selective calling, handshaking, and robust burst modems. An ALE node decodes all received ALE signals heard on the channel(s) it monitors. It utilizes the fact that all ALE messages utilize Forward error correction (FEC) redundancy. By noting how much error-correction occurred in each received and decoded message, an ALE node can detect the "quality" of the path between the sending station and itself. This information is coupled with the ALE address of the sending node and the channel the message was received on, and stored in the node's Link Quality Analysis (LQA) memory. When a call is initiated, the LQA lookup table is searched for matches involving the target ALE address and the best historic channel is used to call the target station. This reduces the likelihood that the call has to be repeated on alternate frequencies. Once the target station has heard the call and responded, a bell or other signalling device will notify both operators that a link has been established. At this point, the operators may coordinate further communication via orderwire text messages, voice, or other means. If further digital communication is desired, it may take place via external data modems or via optional modems built into the ALE terminal.

This unusual usage of FEC redundancy is the primary innovation that differentiates ALE from previous selective calling systems which either decoded a call or failed to decode due to noise or interference. A binary outcome of "Good enough" or not gave no way of automatically choosing between two channels, both of which are currently good enough for minimum communications. The redundancy-based scoring inherent in ALE thus allows for selecting the "best" available channel and (in more advanced ALE nodes) using all decoded traffic over some time window to sort channels into a list of decreasing probability-to-contact, significantly reducing co-channel interference to other users as well as dramatically decreasing the time needed to successfully link with the target node.

Techniques used in the ALE standard include automatic signaling, automatic station identification (sounding), polling, message store-and-forward, linking protection and anti-spoofing to prevent hostile denial of service by ending the channel scanning process. Optional ALE functions include polling and the exchange of orderwire commands and messages. The orderwire message, known as AMD (Automatic Message Display), is the most commonly used text transfer method of ALE, and the only universal method that all ALE controllers have in common for displaying text. It is common for vendors to offer extensions to AMD for various non-standard features, although dependency on these extensions undermines interoperability. As in all interoperability scenarios, care should be taken to determine if this is acceptable before using such extensions.

History and Precedents

ALE evolved from older HF radio selective calling technology. It combined existing channel-scanning selective calling concepts with microprocessors (enabling FEC decoding and quality scoring decisions), burst transmissions (minimizing co-channel interference), and transponding (allowing unattended operation and incoming-call signalling). Early ALE systems were developed in the late 1970s and early 1980s by several radio manufacturers. The first ALE-family controller units were external rack mounted controllers connected to control military radios, and were rarely interoperable across vendors.

Various methods and proprietary digital signaling protocols were used by different manufacturers in first generation ALE, leading to incompatibility. Later, a cooperative effort among manufacturers and the US government resulted in a second generation of ALE that included the features of first generation systems, while improving performance. The second generation 2G ALE system standard in 1986, MIL-STD-188-141A, was adopted in FED-STD-1045 for US federal entities. In the 1980s, military and other entities of the US government began installing early ALE units, using ALE controller products built primarily by US companies. The primary application during the first

10 years of ALE use was government and military radio systems, and the limited customer base combined with the necessity to adhere to MILSPEC standards kept prices extremely high. Over time, demand for ALE capabilities spread and by the late 1990s, most new government HF radios purchased were designed to meet at least the minimum ALE interoperability standard, making them eligible for use with standard ALE node gear. Radios implementing at least minimum ALE node functionality as an option internal to the radio became more common and significantly more affordable. As the standards were adopted by other governments worldwide, more manufacturers produced competitively priced HF radios to meet this demand. The need to interoperate with government organizations prompted many non-government organizations (NGOs) to at least partially adopt ALE standards for communication. As non-military experience spread and prices came down, other civilian entities started using 2G ALE. By the year 2000, there were enough civilian and government organizations worldwide using ALE that it became a de facto HF interoperability standard for situations where a priori channel and address coordination is possible.

In the late 1990s, a third generation 3G ALE with significantly improved capability and performance was included in MIL-STD-188-141B, retaining backward compatibility with 2G ALE, and was adopted in NATO STANAG 4538. Civilian and non-government adoption rates are much lower than 2G ALE due to the extreme cost as compared to surplus or entry-level 2G gear as well as the significantly increased system and planning complexity necessary to realize the benefits inherent in the 3G specification. For many militaries, whose needs for maximized intra-organizational capability and capacity always strain existing systems, the additional cost and complexity of 3G is far more compelling.

Reliability

ALE enables rapid unscheduled communication and message passing without requiring complex message centers, multiple radios and antennas, or highly trained operators. With the removal of these potential sources of failure, the tactical communication process becomes much more robust and reliable. The effects extend beyond mere Force multiplication of existing communications methods; units such as helicopters, when outfitted with ALE radios, can now reliably communicate in situations where the crew are too busy to operate a traditional non-line of sight radio. This ability to enable tactical communication in conditions where dedicated trained operators and hardware are inappropriate is often considered to be the true improvement offered by ALE.

ALE is a critical path toward increased interoperability between organizations. By enabling a station to participate nearly simultaneously in many different HF networks, ALE allows for convenient cross-organization message passing and monitoring without requiring dedicated separate equipment and operators for each partner organization. This dramatically reduces staffing and equipment considerations, while enabling small mobile or portable stations to participate in multiple networks and subnetworks. The result is increased resilience, decreased fragility, increased ability to communicate information effectively, and the ability to rapidly add to or replace communication points as the situation demands.

When combined with Near Vertical Incidence Skywave (NVIS) techniques and sufficient channels spread across the spectrum, an ALE node can provide greater than 95% success linking on the first call, nearly on par with SATCOM systems. This is significantly more reliable than cellphone infrastructure during disasters or wars yet is mostly immune to such considerations itself.

Standards and Protocols

Global standards for ALE are based on the original US MIL-STD 188-141A and FED-1045, known as 2nd Generation (2G) ALE. 2G ALE uses non-synchronised scanning of channels, and it takes several seconds to half a minute to repeatedly scan through an entire list of channels looking for calls. Thus it requires sufficient duration of transmission time for calls to connect or link with another station that is unsynchronised with its calling signal. The vast majority of ALE systems in use in the world at the present time are 2G ALE.

2G Technical Characteristics

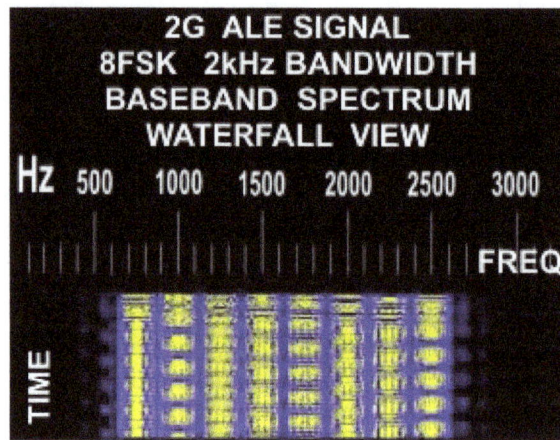

2G ALE Signal

The more common 2G ALE signal waveform is designed to be compatible with standard 3 kHz SSB narrowband voice channel transceivers. The modulation method is 8ary Frequency Shift Keying or 8FSK, also sometimes called Multi Frequency Shift Keying MFSK, with eight orthogonal tones between 750 and 2500 Hz. Each tone is 8 ms long, resulting in a transmitted over-the-air symbol rate of 125 baud or 125 symbols per second, with a raw data rate of 375 bits per second. The ALE data is formatted in 24-bit frames, which consist of a 3 bit preamble followed by three ASCII characters, each seven bits long. The received signal is usually decoded using digital signal processing techniques that are capable of recovering the 8FSK signal at a negative decibel signal to noise ratio (i.e., the signal may be recovered even when it is below the noise level). The over-the-air layers of the protocol involve the use of forward error correction, redundancy, and handshaking transponding similar to those used in ARQ techniques.

3G Technical Characteristics

Newer standards of ALE called 3rd Generation or 3G ALE, use accurate time synchronization (via a defined time-synch protocol as well as the option of GPS-locked clocks) to achieve faster and more dependable linking. Through synchronization, the calling time to achieve a link may be reduced to less than 10 seconds. The 3G ALE modem signal also provides better robustness and can work in channel conditions that are less favorable than 2G ALE. Dwell groups, limited callsigns, and shorter burst transmissions enable more rapid intervals of scanning. All stations in the same group scan and receive each channel at precisely the same time window. Although 3G ALE is more reliable and has significantly enhanced channel-time efficiency, the existence of a large installed base of 2G ALE radio systems and the wide availability of moderately priced (often military surplus) equipment, has made 2G the baseline standard for global interoperability.

Basis for HF Interoperability Communications

Interoperability is a critical issue for the disparate entities which use radiocommunications to fulfill the needs of organizations. Largely due to the ubiquity of 2G ALE, it became the primary method for providing interoperability on HF between governmental and non-governmental disaster relief and emergency communications entities, and amateur radio volunteers. With digital techniques increasingly employed in communications equipment, a universal digital calling standard was needed, and ALE filled the gap. Nearly every major HF radio manufacturer in the world builds ALE radios to the 2G standard to meet the high demand that new installations of HF radio systems conform to this standard protocol. Disparate entities that historically used incompatible radio methods were then able to call and converse with each other using the common 2G ALE platform. Some manufacturers and organizations have utilized the AMD feature of ALE to expand the performance and connectivity. In some cases, this has been successful, and in other cases, the use of proprietary preamble or embedded commands has led to interoperability problems.

Tactical Communication and Resource Management

ALE serves as a convenient method of beyond line of sight communication. Originally developed to support military requirements, ALE is useful to many organizations who find themselves managing widely located units. United States Immigration and Customs Enforcement and United States Coast Guard are two members of the Customs Over the Horizon Enforcement Network (COTHEN), a MIL-STD 188-141A ALE network. All U.S. armed forces operate multiple similar networks. Similarly, shortwave utility listeners have documented frequency and callsign lists for many nations' military and guard units, as well as networks operated by oil exploration and production companies and public utilities in many countries.

Emergency / Disaster Relief or Extraordinary Situation Response Communications

ALE radio communication systems for both HF regional area networks and HF interoperability communications are in service among emergency and disaster relief agencies as well as military and guard forces. Extraordinary response agencies and organizations use ALE to respond to situations in the world where conventional communications may have been temporarily overloaded or damaged. In many cases, it is in place as alternative back-channel for organizations that may have to respond to situations or scenarios involving the loss of conventional communications. Earthquakes, storms, volcanic eruptions, and power or communication infrastructure failures are typical situations in which organizations may deem ALE necessary to operations. ALE networks are common among organizations engaged in extraordinary situation response such as: natural and man-made disasters, transportation, power, or telecommunication network failures, war, peacekeeping, or stability operations. Organizations known to use ALE for Emergency management, disaster relief, ordinary communication or extraordinary situation response include: Red Cross, FEMA, Disaster Medical Assistance Teams, NATO, Federal Bureau of Investigation, United Nations, AT&T, Civil Air Patrol, SHARES, State of California Emergency Management Agency

(CalEMA), other US States' Offices of Emergency Services or Emergency Management Agencies, and Amateur Radio Emergency Service (ARES).

International HF Telecommunications for Disaster Relief

The International Telecommunications Union (ITU), in response to the need for interoperation in international disaster response spurred largely by humanitarian relief, included ALE in its Telecommunications for Disaster Relief recommendations. The increasing need for instant connectivity for logistical and tactical disaster relief response communications, such as the 2004 Indian Ocean earthquake tsunami led to ITU actions of encouragement to countries around the world toward loosening restrictions on such communications and equipment border transit during catastrophic disasters. The IARU Global Amateur Radio Emergency Communications Conferences (GAREC) and IARU Global Simulated Emergency Tests have included ALE.

Use in Amateur Radio

Amateur radio operators began sporadic ALE operation on a limited basis in the early to mid-1990s, with commercial ALE radios and ALE controllers. In 2000, the first widely available software ALE controller for the Personal Computer, *PCALE*, became available, and hams started to set up stations based on it. In 2001, the first organized and coordinated global ALE nets for International Amateur Radio began. In August 2005, ham radio operators supporting communications for emergency Red Cross shelters used ALE for Disaster Relief operations during the Hurricane Katrina disaster. After the event, hams developed more permanent ALE emergency/disaster relief networks, including internet connectivity, with a focus on interoperation between organizations. The amateur radio HFLink Automatic Link Establishment system uses an open net protocol to enable all amateur radio operators and amateur radio nets worldwide to participate in ALE and share the same ALE channels legally and interoperably. Amateur radio operators may use it to call each other for voice or data communications.

Amateur Radio Interoperability Adaptations

Amateur radio operators commonly provide local, regional, national, and international emergency / disaster relief communications. The need for interoperability on HF led to the adoption of Automatic Link Establishment ALE open networks by hams. Amateur radio adapted 2G ALE techniques, by utilizing the common denominators of the 2G ALE protocol, with a limited subset of features found in the majority of all ALE radios and controllers. Each amateur radio ALE station uses the operator's call sign as the address, also known as the ALE Address, in the ALE radio controller. The lowest common denominator technique enables any manufacturer's ALE radios or software to be utilized for HF interoperability communications and networking. Known as Ham-Friendly ALE, the amateur radio ALE standard is used to establish radio communications, through a combination of active ALE on internationally recognized automatic data frequencies, and passive ALE scanning on voice channels. In this technique, active ALE frequencies include pseudorandom periodic polite station identification, while passive ALE frequencies are silently scanned for selective calling. ALE systems include Listen Before Transmit as a standard function, and in most cases this feature provides better busy channel detection of voice and data signals than the human ear. Ham-Friendly ALE technique is also known as 2.5G ALE, because it maintains 2G

ALE compatibility while employing some of the adaptive channel management features of 3G ALE, but without the accurate GPS time synchronization of 3G ALE.

Disaster Relief HF Network

Hot standby nets are in constant operation 24/7/365 for International Emergency and Disaster Relief communications. The Ham Radio Global ALE High Frequency Network, which began service in June 2007, is the world's largest intentionally open ALE network for internet connectivity. It is a free open network staffed by volunteers, and utilized by amateur radio operators supporting disaster relief organizations.

International Coordination

International amateur radio ALE High Frequency channels are frequency coordinated with all Regions of the International Amateur Radio Union(IARU entity of ITU), for international, regional, national, and local use in the Amateur Radio Service. All Amateur Radio ALE channels use "USB" Upper Sideband standard. Different rules, regulations, and bandplans of the region and local country of operation apply to use of various channels. Some channels may not be available in every country. Primary or global channels are in common with most countries and regions.

International Channels

Channel	Freq (kHz)	SSB	Common use	NET	Description
01	3596.0	USB	PRIMARY DATA	HFN	Global ALE High Frequency Network, HF Relay, Traffic, Internet connectivity, sounding
02	3791.0	USB	VOICE	HFL	International Emergency/Disaster Relief
03	3996.0	USB	VOICE	HFL	North America
04	5357.0	USB	VOICE or TEXT	HFL	Regional Interoperability North America
05	5371.5	USB	VOICE	HFL	International Emergency/Disaster Relief
06	7102.0	USB	PRIMARY DATA	HFN	Global ALE High Frequency Network, HF Relay, Traffic, Internet connectivity, sounding
07	7185.5	USB	VOICE	HFL	International Emergency/Disaster Relief
08	7296.0	USB	VOICE	HFL	North America
09	10145.5	USB	PRIMARY DATA	HFN	International Emergency/Relief, Internet
10	14109.0	USB	PRIMARY DATA	HFN	Global ALE High Frequency Network, HF Relay, Traffic, Internet connectivity, sounding
11	14346.0	USB	VOICE	HFL	International Emergency/Disaster Relief
12	18106.0	USB	PRIMARY DATA	HFN	Global ALE High Frequency Network, HF Relay, Traffic, Internet connectivity, sounding
13	18117.5	USB	VOICE/ DATA	HFL	International Emergency/Disaster Relief
14	21096.0	USB	PRIMARY DATA	HFN	Global ALE High Frequency Network, HF Relay, Traffic, Internet connectivity, sounding
15	21432.5	USB	VOICE	HFL	International Emergency/Disaster Relief

16	24926.0	USB	PRIMARY DATA	HFN	Global ALE High Frequency Network, HF Relay, Traffic, Internet connectivity, sounding
17	24932.0	USB	VOICE	HFL	International Emergency/Disaster Relief
18	28146.0	USB	PRIMARY DATA	HFN	Global ALE High Frequency Network, HF Relay, Traffic, Internet connectivity, sounding
19	28312.5	USB	VOICE / DATA	HFL	International Emergency/Disaster Relief

Standard Configurations

Note	Configuration	Standard
1	ALE System	MIL-STD 188-141A ; FED-1045 (8FSK, 2kHzBW)
2	Transmission duration	Calling optimum 22 seconds; Maximum 30 seconds.
3	Scan rate	1 or 2 channels per second.
4	Sounding Interval	60 Minutes or more (for same channel)
5	Audio Centre Frequency	1625 Hz for digital mode text and data
6	Messaging standard	AMD (Automatic Message Display) Universal short text
7	Sounding Type	TWS Sounding (This Was Sound)

International Nets

Net	Member Slots	Entity or Purpose
HFL	10	All ALE voice SSB stations, open selective calling
HFN	10	Ham Radio Global ALE High Frequency Network
QRZ	3	Open calling on all channels
GPR	3	Geo Position Reporting
RPT	3	Station Status Reporting

Link Budget

A link budget is accounting of all of the gains and losses from the transmitter, through the medium (free space, cable, waveguide, fiber, etc.) to the receiver in a telecommunication system. It accounts for the attenuation of the transmitted signal due to propagation, as well as the antenna gains, feedline and miscellaneous losses. Randomly varying channel gains such as fading are taken into account by adding some margin depending on the anticipated severity of its effects. The amount of margin required can be reduced by the use of mitigating techniques such as antenna diversity or frequency hopping.

A simple link budget equation looks like this:

Received Power (dB) = Transmitted Power (dB) + Gains (dB) − Losses (dB)

Note that decibels are logarithmic measurements, so adding decibels is equivalent to multiplying the actual numeric ratios.

In Radio Systems

For a line-of-sight radio system, the primary source of loss is the decrease of the signal power due to uniform propagation, proportional to the inverse square of the distance (geometric spreading).

- Transmitting antennas are for the most part not isotropic aka omnidirectional.

- Completely omnidirectional antennas are rare in telecommunication systems, so almost every link budget equation must consider antenna gain.

- Transmitting antennas typically concentrate the signal power in a favoured direction, normally that in which the receiving antenna is placed.

- Transmitter power is effectively increased (in the direction of highest antenna gain). This systemic gain is expressed by including the antenna gain in the link budget.

- The receiving antenna is also typically directional, and when properly oriented collects more power than an isotropic antenna would; as a consequence, the receiving antenna gain (in decibels from isotropic, dBi) adds to the received power.

- The antenna gains (transmitting or receiving) are scaled by the wavelength of the radiation in question. This step may not be required if adequate systemic link budgets are achieved.

Simplifications Needed

Often link budget equations can become messy and complex, so there have evolved some standard practices to simplify the link budget equation

- The wavelength term is often considered part of the free space loss equation. This complexity reduction is acceptable for terrestrial communication systems, where only line of sight is considered.

- Considering all carrier wave propagation to be wavelength-independent. This is justified by the conservation of energy law that requires that the electric field decrease in power as the square of the distance regardless of frequency (in free space propagation conditions).

Transmission Line and Polarization Loss

In practical situations (Deep Space Telecommunications, Weak signal DXing etc. ...) other sources of signal loss must also be accounted for

- The transmitting and receiving antennas may be partially cross-polarized.

- The cabling between the radios and antennas may introduce significant additional loss.

- Doppler shift induced signal power losses in the receiver.

Endgame

If the estimated received power is sufficiently large (typically relative to the receiver sensitivity), which may be dependent on the communications protocol in use, the link will be useful for sending data. The amount by which the received power exceeds receiver sensitivity is called the link margin.

Equation

A link budget equation including all these effects, expressed logarithmically, might look like this:

$$P_{RX} = P_{TX} + G_{TX} - L_{TX} - L_{FS} - L_{M} + G_{RX} - L_{RX}$$

where:

P_{RX} = received power (dBm)

P_{TX} = transmitter output power (dBm)

G_{TX} = transmitter antenna gain (dBi)

L_{TX} = transmitter losses (coax, connectors...) (dB)

L_{FS} = path loss, usually free space loss (dB)

L_{M} = miscellaneous losses (fading margin, body loss, polarization mismatch, other losses...) (dB)

G_{RX} = receiver antenna gain (dBi)

L_{RX} = receiver losses (coax, connectors...) (dB)

The loss due to propagation between the transmitting and receiving antennas, often called the path loss, can be written in dimensionless form by normalizing the distance to the wavelength:

L_{FS} (dB) = 20×log[4×π×distance/wavelength] (where distance and wavelength are in the same units)

When substituted into the link budget equation above, the result is the logarithmic form of the Friis transmission equation.

In some cases it is convenient to consider the loss due to distance and wavelength separately, but in that case it is important to keep track of which units are being used, as each choice involves a differing constant offset. Some examples are provided below.

L_{FS} (dB) = 32.45 dB + 20×log[frequency(MHz)] + 20×log[distance(km)]

L_{FS} (dB) = - 27.55 dB + 20×log[frequency(MHz)] + 20×log[distance(m)]

L_{FS} (dB) = 36.6 dB + 20×log[frequency(MHz)] + 20×log[distance(miles)]

These alternative forms can be derived by substituting wavelength with the ratio of propagation velocity (c, approximately 3×10^8 m/s) divided by frequency, and by inserting the proper conversion factors between km or miles and meters, and between MHz and (1/sec).

Non-line-of-sight Radio

Because of building obstructions such as walls and ceilings, propagation losses indoors can be significantly higher. This occurs because of a combination of attenuation by walls and ceilings, and blockage due to equipment, furniture, and even people.

- For example, a "2 x 4" wood stud wall with drywall on both sides results in about 6 dB loss per wall.

- Older buildings may have even greater internal losses than new buildings due to materials and line of sight issues.

Experience has shown that line-of-sight propagation holds only for about the first 3 meters. Beyond 3 meters propagation losses indoors can increase at up to 30 dB per 30 meters in dense office environments.

This is a good "rule-of-thumb", in that it is conservative (it overstates path loss in most cases). Actual propagation losses may vary significantly depending on building construction and layout.

The attenuation of the signal is highly dependent on the frequency of the signal.

In Waveguides and Cables

Guided media such as coaxial and twisted pair electrical cable, radio frequency waveguide and optical fiber have losses that are exponential with distance.

The path loss will be in terms of dB per unit distance.

This means that there is always a crossover distance beyond which the loss in a guided medium will exceed that of a line-of-sight path of the same length.

Long distance fiber-optic communication became practical only with the development of ultra-transparent glass fibers. A typical path loss for single mode fiber is 0.2 dB/km, far lower than any other guided medium.

Examples

Earth–Moon–Earth Communications

Link budgets are important in Earth–Moon–Earth communications. As the albedo of the Moon is very low (maximally 12% but usually closer to 7%), and the path loss over the 770,000 kilometre return distance is extreme (around 250 to 310 dB depending on VHF-UHF band used, modulation format and Doppler shift effects), high power (more than 100 watts) and high-gain antennas (more than 20 dB) must be used.

- In practice, this limits the use of this technique to the spectrum at VHF and above.

- The Moon must be above the horizon in order for EME communications to be possible.

Voyager Program

The Voyager Program spacecraft have the highest known path loss and lowest link budgets of any telecommunications circuit. Although the Deep Space Network has been able to maintain the necessary technological advances to maintain the link, the received field strength is still many billions of times weaker than a battery-powered wristwatch.

Grade of Service

In telecommunication engineering, and in particular teletraffic engineering, the quality of voice service is specified by two measures: the grade of service (GoS) and the quality of service (QoS).

Grade of service is the probability of a call in a circuit *group* being blocked or delayed for more than a specified interval, expressed as a vulgar fraction or decimal fraction. This is always with reference to the busy hour when the traffic intensity is the greatest. Grade of service may be viewed independently from the perspective of incoming versus outgoing calls, and is not necessarily equal in each direction or between different source-destination pairs.

On the other hand, the quality of service which a *single* circuit is designed or conditioned to provide, e.g. voice grade or program grade is called the quality of service. Quality criteria for such circuits may include equalization for amplitude over a specified band of frequencies, or in the case of digital data transported via analogue circuits, may include equalization for phase. Criteria for mobile quality of service in cellular telephone circuits include the probability of abnormal termination of the call.

What is Grade of Service and how is it Measured?

When a user attempts to make a telephone call, the routing equipment handling the call has to determine whether to accept the call, reroute the call to alternative equipment, or reject the call entirely. Rejected calls occur as a result of heavy traffic loads (congestion) on the system and can result in the call either being delayed or lost. If a call is delayed, the user simply has to wait for the traffic to decrease, however if a call is lost then it is removed from the system.

The Grade of Service is one aspect of the quality a customer can expect to experience when making a telephone call. In a Loss System, the Grade of Service is described as that proportion of calls that are lost due to congestion in the busy hour. For a Lost Call system, the Grade of Service can be measured using *Equation 1*.

$$\text{Grade of Service} = \frac{\text{number of blocked calls}}{\text{number of offered calls}} \qquad (1)$$

For a delayed call system, the Grade of Service is measured using three separate terms:

- The mean delay t_d – Describes the average time a user spends waiting for a connection if their call is delayed.

- The mean delay t_o – Describes the average time a user spends waiting for a connection whether or not their call is delayed.

- The probability that a user may be delayed longer than time t while waiting for a connection. Time t is chosen by the telecommunications service provider so that they can measure whether their services conform to a set Grade of Service.

Where and when is Grade of Service Measured?

The Grade of Service can be measured using different sections of a network. When a call is routed from one end to another, it will pass through several exchanges. If the Grade of Service is calculated based on the number of calls rejected by the final circuit group, then the Grade of Service is determined by the final circuit group blocking criteria. If the Grade of Service is calculated based on the number of rejected calls between exchanges, then the Grade of Service is determined by the exchange-to-exchange blocking criteria.

The Grade of Service should be calculated using both the access networks and the core networks as it is these networks that allow a user to complete an end-to-end connection. Furthermore, the Grade of Service should be calculated from the average of the busy hour traffic intensities of the 30 busiest traffic days of the year. This will cater for most scenarios as the traffic intensity will seldom exceed the reference level.

The grade of service is a measure of the ability of a user to access a trunk system during the busiest hour. The busy is based upon customer demand at the busiest hour during a week month or year.

Class of Service

Different telecommunications applications require different Qualities of Service. For example, if a telecommunications service provider decides to offer different qualities of voice connection, then a premium voice connection will require a better connection quality compared to an ordinary voice connection. Thus different Qualities of Service are appropriate, depending on the intended use. To help telecommunications service providers to market their different services, each service is placed into a specific class. Each Class of Service determines the level of service required.

To identify the Class of Service for a specific service, the network's switches and routers examine the call based on several factors. Such factors can include:

- The type of service and priority due to precedence
- The identity of the initiating party
- The identity of the recipient party

Quality of Service in Broadband Networks

In broadband networks, the Quality of Service is measured using two criteria. The first criterion is the probability of packet losses or delays in already accepted calls. The second criterion refers to the probability that a new incoming call will be rejected or blocked. To avoid the former, broadband networks limit the number of active calls so that packets from established calls will not be lost due to new calls arriving. As in circuit-switched networks, the Grade of Service can be calculated for individual switches or for the whole network.

Maintaining a Grade of Service

The telecommunications provider is usually aware of the required Grade of Service for a particular product. To achieve and maintain a given Grade of Service, the operator must ensure that sufficient telecommunications circuits or routes are available to meet a specific level of demand. It should also be kept in mind that too many circuits will create a situation where the operator is providing excess capacity which may never be used, or at the very least may be severely underutilized. This adds costs which must be borne by other parts of the network. To determine the correct number of circuits that are required, telecommunications service providers make use of Traffic Tables. An example of a Traffic Table can be viewed in *Figure 1*. It follows that in order for a telecommunications network to continue to offer a given Grade of Service, the number of circuits provided in a circuit group must increase (non-linearly) if the traffic intensity increases.

Erlang's Lost Call Assumptions

To calculate the Grade of Service of a specified group of circuits or routes, Agner Krarup Erlang used a set of assumptions that relied on the network losing calls when all circuits in a group were busy. These assumptions are:

- All traffic through the network is pure-chance traffic, i.e. all call arrivals and terminations are independent random events

- There is statistical equilibrium, i.e., the average number of calls does not change

- Full availability of the network, i.e., every outlet from a switch is accessible from every inlet

- Any call that encounters congestion is immediately lost.

From these assumptions Erlang developed the Erlang-B formula which describes the probability of congestion in a circuit group. The probability of congestion gives the Grade of Service experienced.

Calculating the Grade of Service

To determine the Grade of Service of a network when the traffic load and number of circuits are known, telecommunications network operators make use of *Equation 2*, which is the Erlang-B equation.

$$\text{Grade of Service} = \frac{\left(\dfrac{A^N}{N!}\right)}{\left(\displaystyle\sum_{k=0}^{N} \dfrac{A^k}{k!}\right)} \quad (2)$$

A = Expected traffic intensity in Erlangs, N = Number of circuits in group.

This equation allows operators to determine whether each of their circuit groups meet the required Grade of Service, simply by monitoring the reference traffic intensity.

(For delay networks, the Erlang-C formula allows network operators to determine the probability of delay depending on peak traffic and the number of circuits.)

Loading Coil

A loading coil or load coil is an inductor that is inserted into an electronic circuit to increase its inductance. A loading coil is not a transformer as it does not provide coupling to another circuit. The term originated in the 19th century for inductors used to prevent signal distortion in long-distance telegraph transmission cables. The term is also used for inductors in radio antennas, or between the antenna and its feedline, to make an electrically short antenna resonant at its operating frequency.

A loading coil in a small cell phone antenna on a car roof.

Loading coils are historically also known as Pupin coils after Mihajlo Pupin, especially when used for the Heaviside condition and the process of inserting them is sometimes called *pupinization*.

The concept of loading coils was discovered by Oliver Heaviside in studying the problem of slow signalling speed of the first transatlantic telegraph cable in the 1860s. He concluded additional inductance was required to prevent amplitude and time delay distortion of the transmitted signal. The mathematical condition for distortion-free transmission is known as the Heaviside condition. Previous telegraph lines were overland or shorter and hence had less delay, and the need for extra inductance was not as great. Submarine communications cables are particularly subject to the problem, but early 20th century installations using balanced pairs were often continuously loaded with iron wire or tape rather than discretely with loading coils, which avoided the sealing problem.

Applications

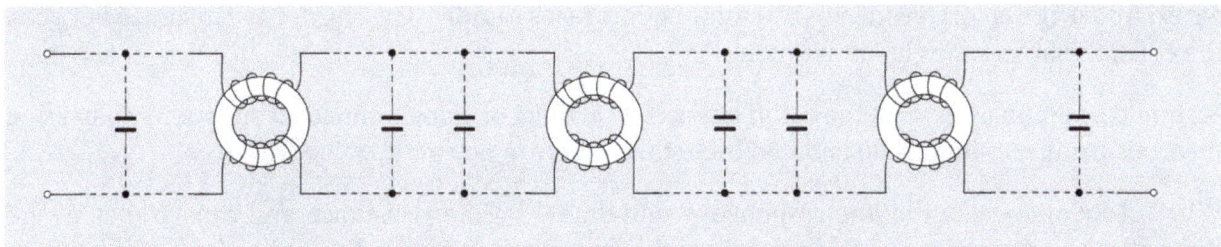

Schematic of a balanced loaded telephone line. The capacitors are not discrete components but represent the distributed capacitance between the closely spaced wire conductors of the line, this is indicated by the dotted lines. The loading coils prevent the audio (voice) signal from being distorted by the line capacitance. The windings of the loading coil are wound such that the magnetic flux induced in the core is in the same direction for both windings.

Telephone Lines

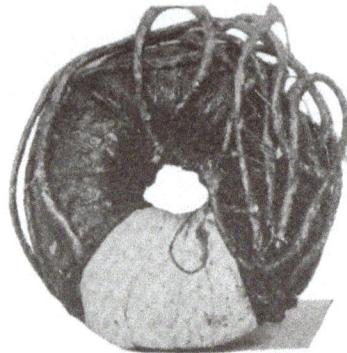

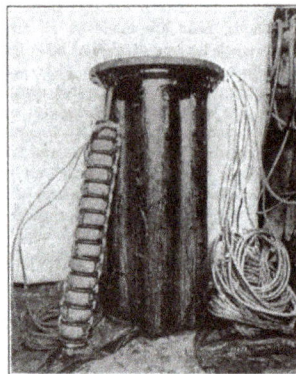

(left) Toroidal 0.175 H loading coil for an AT&T long distance telephone trunkline from New York to Chicago 1922. Each of the 108 twisted pairs in the cable required a coil. The coils were enclosed in an oil-filled steel tank *(right)* on the telephone pole. The cable required loading coils every 6000 ft (1.83 km).

A common application of loading coils is to improve the voice-frequency amplitude response characteristics of the twisted balanced pairs in a telephone cable. Because twisted pair is a balanced format, half the loading coil must be inserted in each leg of the pair to maintain the balance. It is common for both these windings to be formed on the same core. This increases the flux linkages, without which the number of turns on the coil would need to be increased.

Loading coils inserted periodically in series with a pair of wires reduce the attenuation at the higher voice frequencies up to the cutoff frequency of the low-pass filter formed by the inductance of the coils (plus the distributed inductance of the wires) and the distributed capacitance between the wires. Above the cutoff frequency, attenuation increases rapidly. The shorter the distance between the coils, the higher the cut-off frequency.

It should be emphasised that the cutoff effect is an artifact of using lumped inductors. With loading methods using continuous distributed inductance there is no cutoff.

Without loading coils, the line response is dominated by the resistance and capacitance of the line with the attenuation gently increasing with frequency. With loading coils of exactly the right inductance, neither capacitance nor inductance dominate: the response is flat, waveforms are undistorted and the characteristic impedance is resistive up to the cutoff frequency. The coincidental formation of an audio frequency filter is also beneficial in that noise is reduced.

DSL

When loading coils are in place, signal attenuation remains low for signals within the passband of the transmission line but increases rapidly for frequencies above the audio cutoff frequency. Thus, if the pair is subsequently reused to support applications that require higher frequencies (such as analog or digital carrier systems or DSL), any loading coils that were present on the line must be removed or replaced with ones which are transparent to DSL. Using coils with parallel capacitors will form a filter with the topology of an m-derived filter and a band of frequencies above the cutoff will also be passed.

If the coils are not removed, and the subscriber is an extended distance (e.g. over 4 miles or 6.4 km) from the Central Office, DSL can not be supported. This sometimes happens in dense, growing areas such as Southern California in the late 1990s and early 21st century.

Enormous antenna loading coil used in a powerful longwave radiotelegraphy transmitter in New Jersey in 1912.

Carrier Systems

American early and middle 20th century telephone cables had load coils at intervals of a mile (1.61 km), usually in coil cases holding many. The coils had to be removed to pass higher frequencies, but the coil cases provided convenient places for repeaters of digital T-carrier systems, which could then transmit a 1.5 Mbit/s signal that distance. Due to narrower streets and higher cost of copper, European cables had thinner wires and used closer spacing. Intervals of a kilometer allowed European systems to carry 2 Mbit/s.

Radio Antenna

Another type of loading coil is used in radio antennas. Monopole and dipole radio antennas are designed to act as resonators for radio waves; the power from the transmitter, applied to the antenna through the antenna's transmission line, excites standing waves of voltage and current in the antenna element. To be resonant, the antenna must have a physical length of one quarter of the wavelength of the radio waves used (or a multiple of that length). At resonance the antenna acts electrically as a pure resistance, absorbing all the power applied to it from the transmitter.

In many cases for practical reasons it is necessary to make the antenna shorter than the resonant length. An antenna shorter than a quarter wavelength presents capacitive reactance to the transmission line. Some of the applied power is reflected back into the transmission line and travels back toward the transmitter. This causes standing waves on the transmission line (a standing wave ratio (SWR) greater than one) which waste energy, and can even overheat the transmitter.

So to make an electrically short antenna resonant, an inductor called a loading coil is inserted in series with the antenna. The inductive reactance of the coil is equal and opposite to, and cancels, the capacitive reactance of the antenna, so the loaded antenna presents a pure resistance to the transmission line, preventing energy from being reflected.

The loading coil is usually inserted at the base of the antenna, between it and the transmission line (*base loading*), but sometimes it is inserted in the center of the antenna element itself (*center loading*).

Campbell Equation

The Campbell equation is a relationship due to George Ashley Campbell for predicting the propagation constant of a loaded line. It is stated as;

$$\cosh(\gamma'd) = \cosh(\gamma d) + \frac{Z}{2Z_0}\sinh(\gamma d)$$

where,

γ is the propagation constant of the unloaded line

γ' is the propagation constant of the loaded line

d is the interval between coils on the loaded line

Z is the impedance of a loading coil and

Z_0 is the characteristic impedance of the unloaded line.

A more engineer friendly rule of thumb is that the approximate requirement for spacing loading coils is ten coils per wavelength of the maximum frequency being transmitted. This approximation can be arrived at by treating the loaded line as a constant k filter and applying image filter theory to it. From basic image filter theory the angular cutoff frequency and the characteristic impedance of a low-pass constant k filter are given by;

$$\omega_c = \frac{1}{\sqrt{L_{\frac{1}{2}}C_{\frac{1}{2}}}} \quad \text{and,} \quad Z_0 = \sqrt{\frac{L_{\frac{1}{2}}}{C_{\frac{1}{2}}}}$$

where $L_{\frac{1}{2}}$ and $C_{\frac{1}{2}}$ are the half section element values.

From these basic equations the necessary loading coil inductance and coil spacing can be found;

$$L = \frac{Z_0}{\omega_c} \quad \text{and,} \quad d = \frac{2}{\omega_c Z_0 C}$$

where C is the capacitance per unit length of the line.

Expressing this in terms of number of coils per cutoff wavelength yields;

$$\frac{\lambda_c}{d} = \pi v Z_0 C$$

where v is the velocity of propagation of the cable in question.

The phenomenon of cutoff whereby frequencies above the cutoff frequency are not transmitted is an undesirable side effect of loading coils (although it proved highly useful in the development of filters). Cutoff is avoided by the use of continuous loading since it arises from the lumped nature of the loading coils.

History

Oliver Heaviside

Oliver Heaviside

The origin of the loading coil can be found in the work of Oliver Heaviside on the theory of transmission lines. Heaviside (1881) represented the line as a network of infinitesimally small circuit elements. By applying his operational calculus to the analysis of this network he discovered (1887) what has become known as the Heaviside condition. This is the condition that must be fulfilled in order for a transmission line to be free from distortion. The Heaviside condition is that the series impedance, Z, must be proportional to the shunt admittance, Y, at all frequencies. In terms of the primary line coefficients the condition is:

$$\frac{R}{G} = \frac{L}{C}$$

where;

R is the series resistance of the line per unit length

L is the series self-inductance of the line per unit length

G is the shunt leakage conductance of the line insulator per unit length

C is the shunt capacitance between the line conductors per unit length

Heaviside was aware that this condition was not met in the practical telegraph cables in use in his day. In general, a real cable would have,

$$\frac{R}{G} \gg \frac{L}{C}$$

This is mainly due to the low value of leakage through the cable insulator, which is even more pronounced in modern cables which have better insulators than in Heaviside's day. In order to meet the condition, the choices are therefore to try to increase G or L or to decrease R or C. Decreasing R requires larger conductors. Copper was already in use in telegraph cables and this is the very best conductor available short of using silver. Decreasing R means using more copper and a more expensive cable. Decreasing C would also mean a larger cable (although not necessarily more copper). Increasing G is highly undesirable; while it would reduce distortion, it would at the same time increase the signal loss. Heaviside considered, but rejected, this possibility which left him with the strategy of increasing L as the way to reduce distortion.

Heaviside immediately (1887) proposed several methods of increasing the inductance, including spacing the conductors further apart and loading the insulator with iron dust. Finally, Heaviside made the proposal (1893) to use discrete inductors at intervals along the line. However, he never succeeded in persuading the British GPO to take up the idea. Brittain attributes this to Heaviside's failure to provide engineering details on the size and spacing of the coils for particular cable parameters. Heaviside's eccentric character and setting himself apart from the establishment may also have played a part in their ignoring of him.

John Stone

John S. Stone worked for the American Telephone & Telegraph Company (AT&T) and was the first to attempt to apply Heaviside's ideas to real telecommunications. Stone's idea (1896) was to use a bimetallic iron-copper cable which he had patented. This cable of Stone's would increase the line inductance due to the iron content and had the potential to meet the Heaviside condition. However, Stone left the company in 1899 and the idea was never implemented. Stone's cable was an example of continuous loading, a principle that was eventually put into practice is other forms.

George Campbell

George Campbell was another AT&T engineer working in their Boston facility. Campbell was tasked with continuing the investigation into Stone's bimetallic cable, but soon abandoned it in favour of the loading coil. His was an independent discovery, Campbell was aware of Heaviside's work in discovering the Heaviside condition, but unaware of Heaviside's suggestion of using loading coils to enable a line to meet it. The motivation for the change of direction was Campbell's limited budget.

Campbell was struggling to set up a practical demonstration over a real telephone route with the budget he had been allocated. After considering that his artificial line simulators used lumped components rather than the distributed quantities found in a real line, he wondered if he could not insert the inductance with lumped components instead of using Stone's distributed line. When his calculations showed that the manholes on telephone routes were sufficiently close together to be able to insert the loading coils without the expense of either having to dig up the route or lay in new cables he changed to this new plan. The very first demonstration of loading coils on a telephone cable was on a 46-mile length of the so-called Pittsburgh cable (the test was actually in Boston, the cable had previously been used for testing in Pittsburgh) on 6 September 1899 carried out by Campbell himself and his assistant. The first telephone cable using loaded lines put into public service was between Jamaica Plain and West Newton in Boston on 18 May 1900.

Campbell's work on loading coils provided the theoretical basis for his subsequent work on filters which proved to be so important for frequency-division multiplexing. The cut-off phenomena of loading coils, an undesirable side-effect, can be exploited to produce a desirable filter frequency response.

Michael Pupin

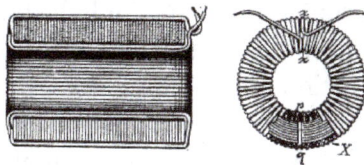

Pupin's design of loading coil

Michael Pupin, inventor and Serbian immigrant to the USA, also played a part in the story of loading coils. Pupin filed a rival patent to the one of Campbell's. This patent of Pupin's dates from 1899. There is an earlier patent (1894, filed December 1893) which is sometimes cited as Pupin's loading coil patent but is, in fact, something different. The confusion is easy to understand, Pupin himself claims that he first thought of the idea of loading coils while climbing a mountain in 1894, although there is nothing from him published at that time.

Pupin's 1894 patent "loads" the line with capacitors rather than inductors, a scheme that has been criticised as being theoretically flawed and never put into practice. To add to the confusion, one variant of the capacitor scheme proposed by Pupin does indeed have coils. However, these are not intended to compensate the line in any way. They are there merely to restore DC continuity to the line so that it may be tested with regular equipment. Pupin states that the inductance is to be so large that it will block all AC signals above 50 Hz. Consequently, only the capacitor is adding any significant impedance to the line and "the coils will not exercise any material influence on the results before noted".

Legal Battle

Heaviside never patented his idea; indeed, he took no commercial advantage of any of his work. Despite the legal disputes surrounding this invention, it is unquestionable that Campbell was the first to actually construct a telephone circuit using loading coils. There also can be little doubt that Heaviside was the first to publish and many would dispute Pupin's priority.

AT&T fought a legal battle with Pupin over his claim. Pupin was first to patent but Campbell had already conducted practical demonstrations before Pupin had even filed his patent (December 1899). Campbell's delay in filing was due to the slow internal machinations of AT&T.

However, AT&T foolishly deleted from Campbell's proposed patent application all the tables and graphs detailing the exact value of inductance that would be required before the patent was submitted. Since Pupin's patent contained a (less accurate) formula, AT&T was open to claims of incomplete disclosure. Fearing that there was a risk that the battle would end with the invention being declared unpatentable due to Heaviside's prior publication, they decided to desist from the challenge and buy an option on Pupin's patent for a yearly fee so that AT&T would control both patents. By January 1901 Pupin had been paid $200,000 ($13 million in 2011) and by 1917, when the AT&T monopoly ended and payments ceased, he had received a total of $455,000 ($25 million in 2011).

Benefit to AT&T

The invention was of enormous value to AT&T. Telephone cables could now be used to twice the distance previously possible, or alternatively, a cable of half the previous quality (and cost) could be used over the same distance. When considering whether to allow Campbell to go ahead with the demonstration, their engineers had estimated that they stood to save $700,000 in new installation costs in New York and New Jersey alone. It has been estimated that AT&T saved $100 million in the first quarter of the 20th century. Heaviside, who began it all, came away with nothing. He was offered a token payment but would not accept, wanting the credit for his work. He remarked ironically that if his prior publication had been admitted it would "interfere . . . with the flow of dollars in the proper direction . . .".

Krarup Cable

Loading coils were not without their problems. In heavy submarine cables, loading coils were difficult to lay. Discontinuities where the coils were installed caused stresses in the cable during laying. Without great care, the cable might part and would be difficult to repair. A second problem was that the material science of the time had difficulties sealing the joint between coil and cable against ingress of seawater. When this occurred the cable was ruined.

A Danish engineer, Carl Emil Krarup, invented a form of continuously loaded cable which solved these problems and the cable is named for him. Krarup cable has iron wires continuously wound around the central copper conductor with adjacent turns in contact with each other. This cable was the first use of continuous loading on any telecommunication cable. In 1902, Krarup both wrote his paper on this subject and saw the installation of the first cable between Helsingør (Denmark) and Helsingborg (Sweden).

Permalloy Cable

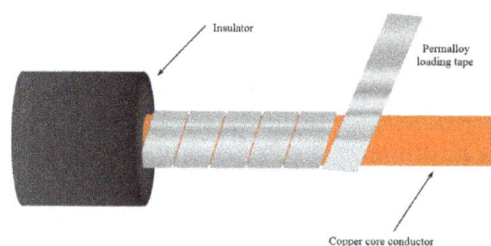

Permalloy cable construction

Even though the Krarup cable added inductance to the line, this was insufficient to meet the Heaviside condition. AT&T searched for a better material with higher magnetic permeability. In 1914, Gustav Elmen discovered permalloy, a magnetic nickel-iron annealed alloy. In c. 1915, Oliver E. Buckley, H. D. Arnold, and Elmen, all at Bell Labs, greatly improved transmission speeds by suggesting a method of constructing submarine communications cable using permalloy tape wrapped around the copper conductors.

The cable was tested in a trial in Bermuda in 1923. The first permalloy cable placed in service connected New York City and Horta (Azores) in September 1924. Permalloy cable enabled signalling speed on submarine telegraph cables to be increased to 400 words/min at a time when 40 words/min was considered good. The first transatlantic cable achieved only two words/min.

Mu-metal Cable

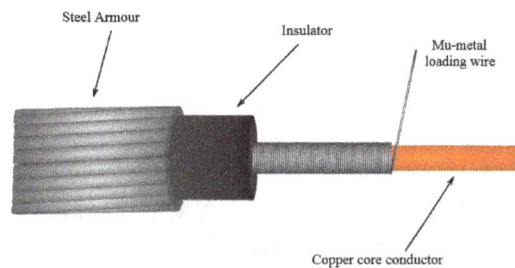

Mu-metal cable construction

Mu-metal has similar magnetic properties to permalloy but the addition of copper to the alloy increases the ductility and allows the metal to be drawn into wire. Mu-metal cable is easier to construct than permalloy cable, the mu-metal being wound around the core copper conductor in much the same way as the iron wire in Krarup cable. A further advantage with mu-metal cable is that the construction lends itself to a variable loading profile whereby the loading is tapered towards the ends.

Mu-metal was invented in 1923 by The Telegraph Construction and Maintenance Company Ltd., London, who made the cable, initially, for the Western Union Telegraph Co. Western Union were in competition with AT&T and the Western Electric Company who were using permalloy. The patent for permalloy was held by Western Electric which prevented Western Union from using it.

Patch Loading

Continuous loading of cables is expensive and hence is only done when absolutely necessary. Lumped loading with coils is cheaper but has the disadvantages of difficult seals and a definite cutoff frequency. A compromise scheme is patch loading whereby the cable is continuously loaded in repeated sections. The intervening sections are left unloaded.

Current Practice

Loaded cable is no longer a useful technology for submarine communication cables, having first been superseded by co-axial cable using electrically powered in-line repeaters and then by fibre-optic cable. Manufacture of loaded cable declined in the 1930s and was then superseded by other technologies post-war. Loading coils can still be found in some telephone landlines today but new installations would use more modern technology.

Electrical Length

In telecommunications and electrical engineering, electrical length (or phase length) refers to the length of an electrical conductor in terms of the phase shift introduced by transmission over that conductor at some frequency.

Usage of the Term

Depending on the specific usage, the term "electrical length" is used rather than simple physical length to incorporate one or more of the following three concepts:

- When one is concerned with the number of wavelengths, or phase, involved in a wave's transit over a segment of transmission line especially, one may simply specify that electrical length, while specification of a physical length, frequency, or velocity factor is omitted. The electrical length is then typically expressed as N wavelengths or as the phase φ expressed in degrees or radians. Thus in a microstrip design one might specify a shorted stub of 60° phase length, which will correspond to different physical lengths when applied to different frequencies. Or one might consider a 2-meter section of coax which has an electrical length of one quarter wavelength (90°) at 25 MHz and ask what its electrical length becomes when the circuit is operated at a different frequency.

- Due to the velocity factor of a particular transmission line, for instance, the transit time of a signal in a certain length of cable is equal to the transit time over a *longer* distance when travelling at the speed of light. So a pulse sent down a 2-meter section of coax (whose velocity factor is 2/3) would arrive at the end of the coax at the same time that the same pulse arrives at the end of a bare wire of length 3 meters (over which it propagates at the speed of light), and one might refer to the 2 meter section of coax as having an electrical length of 3 meters, or an electrical length of 1/2 wavelength at 50 MHz (since a 50 MHz radio wave has a wavelength of 6 meters).

- Since resonant antennas are usually specified in terms of the electrical length of their conductors (such as the *half wave* dipole), the attainment of such an electrical length is loosely equated with electrical resonance, that is, a purely resistive impedance at the antenna's input, as is usually desired. An antenna that has been made slightly too long, for instance, will present an inductive reactance, which can be corrected by physically shortening the antenna. Based on this understanding, a common jargon in the antenna trade refers to the achievement of resonance (cancellation of reactance) at the antenna terminals as *electrically shortening* that too-long antenna (or *electrically lengthening* a too-short antenna) when an electrical matching network (or antenna tuner) has performed that task without *physically* altering the antenna's length. Although a very inexact use of terminology, this usage is widespread, especially as applied to the use of a loading coil at the bottom of a short monopole (a vertical, or whip antenna) to "electrically lengthen" it and achieve electrical resonance as seen through the loading coil.

Phase Length

The first usage of the term "electrical length" assumes a sine wave of some frequency, or at least a narrowband waveform centered around some frequency f. The sine wave will repeat with a period of $T = 1/f$. The frequency f will correspond to a particular wavelength λ along a particular conductor. For conductors (such as bare wire or air-filled coax) which transmit signals at the speed of light c, the wavelength is given by $\lambda = c/f$. A distance L along that conductor corresponds to N wavelengths where $N = L / \lambda$.

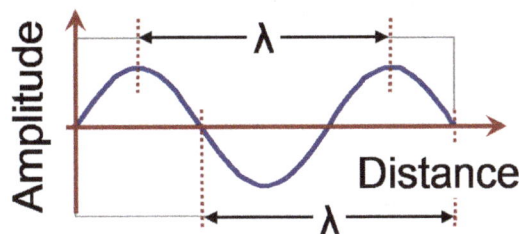

In the figure at the right, the wave shown is seen to be $N=1.5$ wavelengths long. A wave crest at the beginning of the graph, moving towards the right, will arrive at the end after a time $1.5T$. The *electrical length* of that segment is said to be "1.5 wavelengths" or, expressed as a phase angle, "540°" (or 3π radians) where N wavelengths corresponds to $\varphi = 360° \cdot N$ (or $\varphi = 2\pi \cdot N$ radians). In radio frequency applications, when a delay is introduced due to a transmission line, it is often the phase shift φ that is of importance, so specifying a design in terms of the phase or electrical length allows one to adapt that design to an arbitrary frequency by employing the wavelength λ applying to that frequency.

Velocity Factor

In a transmission line, a signal travels at a rate controlled by the effective capacitance and inductance per unit of length of the transmission line. Some transmission lines consist only of bare conductors, in which case their signals propagate at the speed of light, c. More often the signal travels at a reduced velocity κc, where κ is the *velocity factor*, a number less than 1 representing the ratio of that velocity to the speed of light.

Most transmission lines contain a dielectric material (insulator) filling some or all of the space in between the conductors. The relative permittivity or *dielectric constant* of that material increases the distributed capacitance in the cable, which reduces the velocity factor below unity. It is also possible for κ to be reduced due to a relative permeability of that material which increases the distributed inductance, but this is almost never the case. Now, if one fills a space with a dielectric of relative permittivity ϵ_r, then the velocity of an electromagnetic plane wave is reduced by the velocity factor:

$$\kappa = \frac{v_p}{c} = \frac{1}{\sqrt{\epsilon_r}}.$$

This reduced velocity factor would also apply to propagation of signals along wires immersed in a large space filled with that dielectric. However, with only part of the space around the conductors filled with that dielectric, there is less reduction of the wave velocity. Part of the electromagnetic wave surrounding each conductor "feels" the effect of the dielectric, and part is in free space. Then it is possible to define an *effective relative permittivity* ϵ_{eff} which then predicts the velocity factor according to

$$\kappa = \frac{1}{\sqrt{\grave{o}_{eff}}}$$

ϵ_{eff} is computed as a weighted average of the relative permittivity of free space (1) and that of the dielectric:

$$\epsilon_{eff} = (1 - F) + F\epsilon_{r}$$

where the *fill factor* F expresses the effective proportion of space so affected by the dielectric.

In the case of coaxial cable, where all of the volume in between the inner conductor and the shield is filled with a dielectric, the fill factor is unity, since the electromagnetic wave is confined to that region. In other types of cable, such as twin lead, the fill factor can be much smaller. Regardless, any cable intended for radio frequencies will have its velocity factor (as well as its characteristic impedance) specified by the manufacturer. In the case of coaxial cable, where F=1, the velocity factor is solely determined by the sort of dielectric used as specified here.

For example, a typical velocity factor for coaxial cable is .66, corresponding to a dielectric constant of 2.25. Suppose we wish to send a 30 MHz signal down a short section of such a cable, and delay it by a quarter wave (90°). In free space, this frequency corresponds to a wavelength of λ_{o}=10m, so a delay of .25λ would require an *electrical length* of 2.5 m. Applying the velocity factor of .66, this results in a *physical* length of cable 1.67 m long.

The velocity factor likewise applies to antennas in cases where the antenna conductors are (partly) surrounded by a dieletric. This particularly applies to microstrip antennas such as the patch antenna. Waves on microstrip are affected by the dielectric of the circuit board beneath them, but not the air above them. Their velocity factors thus depend not directly on the permittivity of the circuit board material but on the *effective* permittivity ϵ_{eff} which is often specified for a circuit board material (or can be calculated). Note that the fill factor and therefore ϵ_{eff} are somewhat dependent on the width of the trace compared to the thickness of the board.

Antennas

While there are certain wideband antenna designs, many antennas are classified as resonant and perform according to design around a particular frequency. This applies especially to broadcasting stations and communication systems which are confined to one frequency or narrow frequency band. This includes the dipole and monopole antennas and all of the designs based on them (Yagi, dipole or monopole arrays, folded dipole, etc.). In addition to the directive gain in beam antennas suffering away from the design frequency, the antenna feedpoint impedance is very sensitive to frequency offsets. Especially for transmitting, the antenna is often intended to operate at the resonant frequency. At the resonant frequency, by definition, that impedance is a pure resistance which matches the characteristic impedance of the transmission line and the output (or input) impedance of the transmitter (or receiver). At frequencies away from the resonant frequency, the impedance includes some reactance (capacitance or inductance). It is possible for an antenna tuner to be used to cancel that reactance (and to change the resistance to match the transmission line), however that is often avoided as an extra complication (and needs to be controlled at the antenna side of the transmission line).

The condition for resonance in a monopole antenna is for the element to be an odd multiple of a quarter-wavelength, $\lambda/4$. In a dipole antenna both driven conductors must be that long, for a total dipole length of $(2N+1)\lambda/2$.

The electrical length of an antenna element is, in general, different from its physical length For example, increasing the diameter of the conductor, or the presence of nearby metal objects, will decrease the velocity of the waves in the element, increasing the electrical length.

An antenna which is shorter than its resonant length is described as *"electrically short"*, and exhibits capacitive reactance. Similarly, an antenna which is longer than its resonant length is described as *"electrically long"* and exhibits inductive reactance.

Changing Electrical Length by Loading

An antenna's effective electrical length can be changed without changing its physical length by adding reactance, (inductance or capacitance) in series with it. This is called *lumped-impedance matching* or *loading*.

For example, a monopole antenna such as a metal rod fed at one end, will be resonant when its electrical length is equal to a quarter wavelength, $\lambda/4$, of the frequency used. If the antenna is shorter than a quarter wavelength, the feedpoint impedance will include capacitive reactance; this causes reflections on the feedline and a mismatch at the transmitter or receiver, even if the resistive component of the impedance is correct. To cancel the capacitive reactance, an inductance, called a loading coil, is inserted in between the feedline and the antenna terminal. Selecting an inductance with the same reactance as the (negative) capacitative reactance seen at the antenna terminal, cancels that capacitance, and the *antenna system* (antenna and coil) will again be resonant. The feedline sees a purely resistive impedance. Since an antenna which had been too short now appears as if it were resonant, the addition of the loading coil is sometimes referred to as "electrically lengthening" the antenna.

Similarly, the feedpoint impedance of a monopole antenna longer than $\lambda/4$ (or a dipole with arms longer than $\lambda/4$) will include inductive reactance. A capacitor in series with the antenna can cancel this reactance to make it resonant, which can be referred to as "electrically shortening" the antenna.

Inductive loading is widely used to reduce the length of whip antennas on portable radios such as walkie-talkies and short wave antennas on cars, to meet physical requirements.

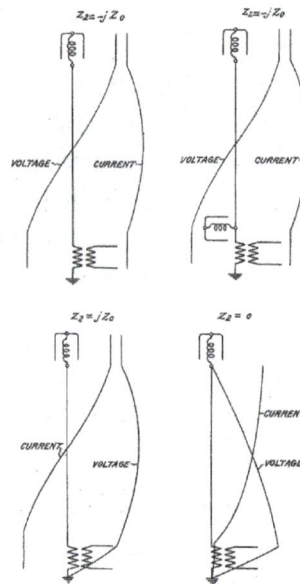

Vertical antenna which may be of any desired height : less than about one-half wavelength of the frequency at which the antenna operates. These antennas may operate either as transmitting or receiving antennas

Advantages

The electrical lengthening allows the construction of shorter aerials. It is applied in particular for aerials for VLF, longwave and medium-wave transmitters. Because those radio waves are several hundred meters to many kilometers long, mast radiators of the necessary height cannot be realised economically. It is also used widely for whip antennas on portable devices such as walkie-talkies to allow antennas much shorter than the standard quarter-wavelength to be used. The most widely used example is the rubber ducky antenna.

Disadvantages

The electrical lengthening reduces the bandwidth of the antenna if other phase control measures are not undertaken. An electrically extended aerial is less efficient than a non-extended antenna.

Technical Realization

There are two possibilities for the realisation of the electric lengthening.

1.   switching in inductive coils in series with the aerial

2.   switching in metal surfaces, known as roof capacitance, at the aerial ends which form capacitors to earth.

Often both measures are combined. The coils switched in series must sometimes be placed in the middle of the aerial construction. The cabin installed at a height of 150-metres on the Blosenbergturm in Beromünster is such a construction, in which a lengthening coil is installed for the supply of the upper tower part (the Blosenbergturm has in addition a ring-shaped roof capacitor on its top)

Application

Transmission aerials of transmitters working at frequencies below the longwave broadcasting band always apply electric lengthening. Broadcasting aerials of longwave broadcasting stations apply it often. However, for transmission aerials of NDBs electrical lengthening is extensively applied, because these use antennas which are considerably less tall than a quarter of the radiated wavelength.

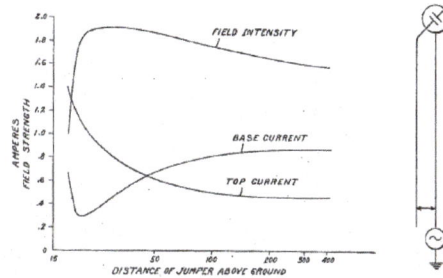

On the left, characteristics plotted from experimentally obtained data on coordinates with logarithmic abscissa. On the right, an antenna with increased effective inductance between the two points in accordance with the well known operation of shunt tuned circuits adjusted somewhat off resonance.

Standing Wave Ratio

In radio engineering and telecommunications, standing wave ratio (SWR) is a measure of impedance matching of loads to the characteristic impedance of a transmission line or waveguide. Impedance mismatches result in standing waves along the transmission line, and SWR is defined as the ratio of the partial standing wave's amplitude at an antinode (maximum) to the amplitude at a node (minimum) along the line.

The SWR is usually thought of in terms of the maximum and minimum AC voltages along the transmission line, thus called the voltage standing wave ratio or VSWR (sometimes pronounced "vizwar"). For example, the VSWR value 1.2:1 denotes an AC voltage due to standing waves along the transmission line reaching a peak value 1.2 times that of the minimum AC voltage along that line. The SWR can as well be defined as the ratio of the maximum amplitude to minimum amplitude of the transmission line's currents, electric field strength, or the magnetic field strength. Neglecting transmission line loss, these ratios are identical.

The power standing wave ratio (PSWR) is defined as the square of the VSWR, however this terminology has no physical relation to actual powers involved in transmission.

SWR is usually measured using a dedicated instrument called an SWR meter. Since SWR is a measure of the load impedance relative to the characteristic impedance of the transmission line in use (which together determine the reflection coefficient as described below), a given SWR meter can only interpret the impedance it sees in terms of SWR if it has been designed for that particular characteristic impedance. In practice most transmission lines used in these applications are coaxial cables with an impedance of either 50 or 75 ohms, so most SWR meters correspond to one of these.

Checking the SWR is a standard procedure in a radio station. Although the same information could be obtained by measuring the load's impedance with an impedance analyzer (or "impedance bridge"), the SWR meter is simpler and more robust for this purpose. By measuring the magnitude of the impedance mismatch at the transmitter output it reveals problems due to either the antenna or the transmission line.

Impedance Matching

SWR is used as a measure of impedance matching of a load to the characteristic impedance of a transmission line carrying radio frequency (RF) signals. This especially applies to transmission lines connecting radio transmitters and receivers with their antennas, as well as similar uses of RF cables such as cable television connections to TV receivers and distribution amplifiers. Impedance matching is achieved when the source impedance is the complex conjugate of the load impedance. The easiest way of achieving this, and the way that minimizes losses along the transmission line, is for both the source and load to be real, that is, pure resistances, equal to the characteristic impedance of the transmission line. When there is a mismatch between the load impedance and the transmission line, part of the forward wave sent toward the load is reflected back along the transmission line towards the source. The source then sees a different impedance than it expects which can lead to lesser (or in some cases, more) power being supplied by it, the result being very sensitive to the electrical length of the transmission line.

Such a mismatch is usually undesired and results in standing waves along the transmission line which magnifies transmission line losses (significant at higher frequencies and for longer cables). The SWR is a measure of the depth of those standing waves and is therefore a measure of the matching of the load to the transmission line. A matched load would result in an SWR of 1:1 implying no reflected wave. An infinite SWR represents complete reflection by a load unable to absorb electrical power, with all the incident power reflected back towards the source.

It should be understood that the match of a load to the transmission line is different from the match of a *source* to the transmission line or the match of a source to the load *seen through* the transmission line. For instance, if there is a perfect match between the load impedance Z_{load} and the source impedance $Z_{source}=Z^*_{load}$, that perfect match will remain if the source and load are connected through a transmission line with an electrical length of one half wavelength (or a multiple of one half wavelengths) using a transmission line of *any* characteristic impedance Z_o. However the SWR will generally not be 1:1, depending only on Z_{load} and Z_o. With a different length of transmission line, the source will see a different impedance than Z_{load} which may or may not be a good match to the source. Sometimes this is deliberate, as when a quarter-wave matching section is used to improve the match between an otherwise mismatched source and load.

However typical RF sources such as transmitters and signal generators are designed to look into a purely resistive load impedance such as 50Ω or 75Ω, corresponding to common transmission lines' characteristic impedances. In those cases, matching the load to the transmission line, $Z_{load}=Z_o$, *always* insures that the source will see the same load impedance as if the transmission line weren't there. This is identical to a 1:1 SWR. This condition ($Z_{load}=Z_o$) also means that the load seen by the source is independent of the transmission line's electrical length. Since the electrical length of a physical segment of transmission line depends on the signal frequency, violation of this condition means that the impedance seen by the source through the transmission line becomes a function of frequency (especially if the line is long), even if Z_{load} is frequency-independent. So in practice, a good SWR (near 1:1) implies a transmitter's output seeing the exact impedance it expects for optimum and safe operation.

Relationship to the Reflection Coefficient

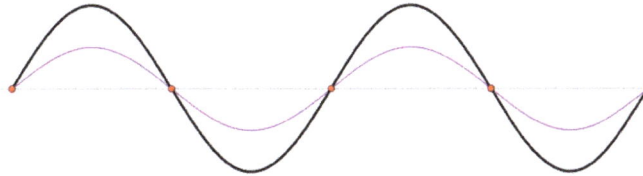

Incident wave (blue) is fully reflected (red wave) out of phase at short-circuited end of transmission line creating a net voltage (black) standing wave. Γ=-1, SWR=∞.

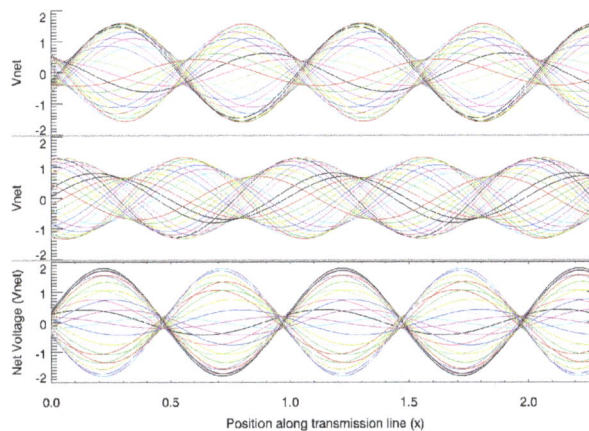

Standing waves on transmission line, net voltage shown in different colors during one period of oscillation. Incoming wave from left (amplitude = 1) is partially reflected with (top to bottom) Γ= .6, -.333, and .8∠60°. Resulting SWR = 4, 2, 9.

The voltage component of a standing wave in a uniform transmission line consists of the forward wave (with complex amplitude V_f) superimposed on the reflected wave (with complex amplitude V_r).

A wave is partly reflected when a transmission line is terminated with other than a pure resistance equal to its characteristic impedance. The reflection coefficient Γ is defined thus:

$$\tilde{A} = \frac{V_r}{V_f}.$$

Γ is a complex number that describes both the magnitude and the phase shift of the reflection. The simplest cases with Γ *measured at the load* are:

- $\Gamma = -1$: complete negative reflection, when the line is short-circuited,

- $\Gamma = 0$: no reflection, when the line is perfectly matched,

- $\Gamma = +1$: complete positive reflection, when the line is open-circuited.

The SWR directly corresponds to the magnitude of Γ.

At some points along the line the forward and reflected waves interfere constructively, exactly in phase, with the resulting amplitude V_{max} given by the sum of their those waves' amplitudes:

$$|V_{max}| = |V_f| + |V_r|$$

$$= |V_f| + |\Gamma V_f|$$

$$= (1 + |\Gamma|)|V_f|$$

At other points, the waves interfere 180° out of phase with the amplitudes partially cancelling:

$$|V_{min}| = |V_f| - |V_r|$$

$$= |V_f| - |\Gamma V_f|$$

$$= (1 - |\Gamma|)|V_f|$$

The voltage standing wave ratio is then equal to:

$$\mathrm{VSWR} = \frac{|V_{max}|}{|V_{min}|} = \frac{1 + |\Gamma|}{1 - |\Gamma|}.$$

Since the magnitude of Γ always falls in the range [0,1], the SWR is always greater than or equal to unity. Note that the *phase* of V_f and V_r vary along the transmission line in opposite directions to each other. Therefore, the complex valued reflection coefficient Γ varies as well, but only in phase. With the SWR dependent *only* on the complex magnitude of Γ, it can be seen that the SWR measured at *any* point along the transmission line (neglecting transmission line losses) obtains an identical reading.

Since the power of the forward and reflected waves are proportional to the square of the voltage components due to each wave, SWR can be expressed in terms of forward and reflected power as follows:

$$\mathrm{SWR} = \frac{1 + \sqrt{P_r/P_f}}{1 - \sqrt{P_r/P_f}}$$

By sampling the complex voltage and current at the point of insertion, an SWR meter is able to compute the effective forward and reflected voltages on the transmission line for the characteristic impedance for which the SWR meter has been designed. Since the forward and reflected power is related to the square of the forward and reflected voltages, some SWR meters also display the forward and reflected power.

In the special case of a load R_L which is purely resistive but unequal to the characteristic impedance of the transmission line Z_o, the SWR is given simply by their ratio:

$$\mathrm{SWR} = \left(\frac{R_L}{Z_0} \right)^{\pm 1}$$

with the ±1 chosen to obtain a value greater than unity.

The Standing Wave Pattern

Using complex notation for the voltage amplitudes, for a signal at frequency ν, the actual (real) voltages V_{actual} as a function of time t are understood to relate to the complex voltages according to:

$$V_{actual} = \Re(e^{i2\pi\nu t}V).$$

Thus taking the real part of the complex quantity inside the parenthesis, the actual voltage consists of a sine wave at frequency ν with a peak amplitude equal to the complex magnitude of V, and with a phase given by the phase of the complex V. Then with the position along a transmission line given by x, with the line ending in a load located at x_0, the complex amplitudes of the forward and reverse waves would be written as:

$$V_f(x) = e^{-ik(x-x_0)}A$$
$$V_r(x) = \Gamma e^{ik(x-x_0)}A$$

for some complex amplitude A (corresponding to the forward wave at x_0). Here k is the wave-number due to the guided wavelength along the transmission line. Note that some treatments use phasors where the time dependence is according to $e^{-i2\pi\nu t}$ and spatial dependence (for a wave in the +x direction) of $e^{+ik(x-x_0)}$. Either convention obtains the same result for V_{actual}.

According to the superposition principle the net voltage present at any point x on the transmission line is equal to the sum of the voltages due to the forward and reflected waves:

$$V_{net}(x) = V_f(x) + V_r(x)$$

$$= e^{-ik(x-x_0)}\left(1 + \Gamma e^{i2k(x-x_0)}\right)A$$

Since we are interested in the variations of the *magnitude* of V_{net} along the line (as a function of x), we shall solve instead for the squared magnitude of that quantity, which simplifies the mathematics. To obtain the squared magnitude we multiply the above quantity by its complex conjugate:

$$|V_{net}(x)|^2 = V_{net}(x)V_{net}^*(x)$$
$$= e^{-i(x-x_0)}\left(1 + \Gamma e^{i2k(x-x_0)}\right)Ae^{+i(x-x_0)}\left(1 + \Gamma^* e^{-i2k(x-x_0)}\right)A^*$$
$$= \left[1 + |\Gamma|^2 + 2\Re(\Gamma e^{i2k(x-x_0)})\right]|A|^2$$

Depending on the phase of the third term, the maximum and minimum values of V_{net} (the square root of the quantity in the equations) are $(1 + |\Gamma|)|A|$ and $(1 - |\Gamma|)|A|$ respectively, for a standing wave ratio of:

$$SWR = \frac{|V_{max}|}{|V_{min}|} = \frac{1 + |\Gamma|}{1 - |\Gamma|}$$

as earlier asserted. Along the line, the above expression for $|V_{net}(x)|^2$ is seen to oscillate sinusoidally between $|V_{min}|^2$ and $|V_{max}|^2$ with a period of $2\pi/2k$. This is *half* of the guided wavelength $\lambda = 2\pi/k$ for the frequency ν. That can be seen as due to interference between two waves of that frequency which are travelling in *opposite* directions.

For example, at a frequency v=20 MHz (free space wavelength of 15 m) in a transmission line whose velocity factor is 2/3, the guided wavelength (distance between voltage peaks of the forward wave alone) would be λ = 10 m. At instances when the forward wave at x = 0 is at zero phase (peak voltage) then at x = 10 m it would also be at zero phase, but at x = 5 m it would be at 180° phase (peak *negative* voltage). On the other hand, the magnitude of the voltage due to a standing wave produced by its addition to a reflected wave, would have a wavelength between peaks of only $\lambda/2$ = 5 m. Depending on the location of the load and phase of reflection, there might be a peak in the magnitude of V_{net} at x = 1.3 m. Then there would be another peak found where $|V_{net}|=V_{max}$ at x = 6.3 m, whereas it would find minima of the standing wave $|V_{net}| = V_{min}$ at x = 3.8 m, 8.8 m, etc.

Practical Implications of SWR

The most common case for measuring and examining SWR is when installing and tuning transmitting antennas. When a transmitter is connected to an antenna by a feed line, the driving point impedance of the antenna must be resistive and matching the characteristic impedance of the feed line in order for the transmitter to see the impedance it was designed for (the impedance of the feed line, usually 50 or 75 ohms).

The impedance of a particular antenna design can vary due to a number of factors that cannot always be clearly identified. This includes the transmitter frequency (as compared to the antenna's design or resonant frequency), the antenna's height above the ground and proximity to large metal structures, and variations in the exact size of the conductors used to construct the antenna.

When an antenna and feed line do not have matching impedances, the transmitter sees an unexpected impedance, where it might not be able to produce its full power, and can even damage the transmitter in some cases. The reflected power in the transmission line increases the average current and therefore losses in the transmission line compared to power actually delivered to the load. It is the interaction of these reflected waves with forward waves which causes standing wave patterns, with the negative repercussions we have noted.

Matching the impedance of the antenna to the impedance of the feed line can sometimes be accomplished through adjusting the antenna itself, but otherwise is possible using an antenna tuner, an impedance matching device. Installing the tuner between the feed line and the antenna allows for the feed line to see a load close to its characteristic impedance, while sending most of the transmitter's power (a small amount may be dissipated within the tuner) to be radiated by the antenna despite its otherwise unacceptable feed point impedance. Installing a tuner in between the transmitter and the feed line can also transform the impedance seen at the transmitter end of the feed line to one preferred by the transmitter. However, in the latter case, the feed line still has a high SWR present, with the resulting increased feed line losses unmitigated.

The magnitude of those losses are dependent on the type of transmission line, and its length. They always increase with frequency. For example, a certain antenna used well away from its resonant frequency may have an SWR of 6:1. For a frequency of 3.5 MHz, with that antenna fed through 75 meters of RG-8A coax, the loss due to standing waves would be 2.2 dB. However the same 6:1 mismatch through 75 meters of RG-8A coax would incur 10.8 dB of loss at 146 MHz. Thus, a better match of the antenna to the feed line, that is, a lower SWR, becomes increasingly important with increasing frequency, even if the transmitter is able to accommodate the impedance seen (or an antenna tuner is used between the transmitter and feed line).

Certain types of transmissions can suffer other negative effects from reflected waves on a transmission line. Analog TV can experience "ghosts" from delayed signals bouncing back and forth on a long line. FM stereo can also be affected and digital signals can experience delayed pulses leading to bit errors. Whenever the delay times for a signal going back down and then again up the line are comparable to the modulation time constants, effects occur. For this reason, these types of transmissions require a low SWR on the feedline, even if SWR induced loss might be acceptable and matching is done at the transmitter.

Methods of Measuring Standing Wave Ratio

Many different methods can be used to measure standing wave ratio. The most intuitive method uses a slotted line which is a section of transmission line with an open slot which allows a probe to detect the actual voltage at various points along the line.Thus the maximum and minimum values can be compared directly. This method is used at VHF and higher frequencies. At lower frequencies, such lines are impractically long. Directional couplers can be used at HF through microwave frequencies and respond only to waves travelling either in the forward or reverse direction. From these two values we can derive SWR. The computations can be done mathematically or using graphical methods built into the meter as an additional scale or by reading from the crossing point between two needles on the same meter. Various types of circuits can be used to measure the complex voltage and current at the measuring point and to use those values to derive SWR. These methods can provide more information than just SWR or forward and reflected power. Some modern transmitters include a graphical display of complex load vs frequency. Stand alone antenna analyzers also are available that measure much more than simple SWR using various measuring methods.

Power Standing Wave Ratio

The term *power standing wave ratio* (PSWR) is sometimes referred to, and defined as the square of the voltage standing wave ratio. The term is widely cited as "misleading." In the words of Gridley:

The expression "power standing-wave ratio", which may sometimes be encountered is even more misleading, for the power distribution along a loss-free line is constant.....

—J. H. Gridley

In other words, there are no actual powers being compared. Patently a misnomer, the term *power standing wave ratio* is not the ratio of any two physical quantities.

However it does correspond to one type of measurement of SWR using what was formerly a standard measuring instrument at microwave frequencies. A slotted line involves a waveguide (or air-filled coaxial line) in which a small sensing antenna measures the electric field along the transmission line *directly*. The electric field strength is commonly measured using a crystal detector or Schottky barrier diode. These detectors have a square law output for low levels of input. Readings therefore corresponded to the square of the electric field along the slot, $E^2(x)$, with maximum and minimum readings of E^2_{max} and E^2_{min} found as the probe is moved along the slot. The ratio of these yields the *square* of the SWR, the so-called PSWR.

Implications of SWR on Medical Applications

SWR can also have a detrimental impact upon the performance of microwave based medical applications. In microwave electrosurgery an antenna that is placed directly into tissue may not always have an optimal match with the feedline resulting in an SWR. The presence of SWR can affect monitoring components used to measure power levels impacting the reliability of such measurements.

Reflections of Signals on Conducting Lines

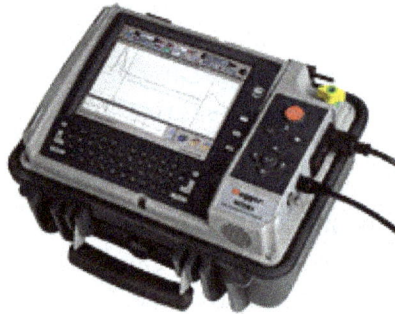

A time-domain reflectometer; an instrument used to locate the position of faults on lines from the time taken for a reflected wave to return from the discontinuity.

A signal travelling along an electrical transmission line will be partly, or wholly, reflected back in the opposite direction when the travelling signal encounters a discontinuity in the characteristic impedance of the line, or if the far end of the line is not terminated in its characteristic impedance. This can happen, for instance, if two lengths of dissimilar transmission lines are joined together.

This article is about signal reflections on electrically conducting lines. Such lines are loosely referred to as copper lines, and indeed, in telecommunications are generally made from copper, but other metals are used, notably aluminium in power lines. Although this article is limited to describing reflections on conducting lines, this is essentially the same phenomenon as optical reflections in fibre-optic lines and microwave reflections in waveguides.

Reflections cause several undesirable effects, including modifying frequency responses, causing overload power in transmitters and overvoltages on power lines. However, the reflection phenomenon can also be made use of in such devices as stubs and impedance transformers. The special cases of open circuit and short circuit lines are of particular relevance to stubs.

Reflections cause standing waves to be set up on the line. Conversely, standing waves are an indication that reflections are present. There is a relationship between the measures of reflection coefficient and standing wave ratio.

Specific Cases

There are several approaches to understanding reflections, but the relationship of reflections to the conservation laws is particularly enlightening. A simple example is a step voltage, $Vu(t)$ (where V is the height of the step and $u(t)$ is the unit step function with time t), applied to one end of a lossless line, and consider what happens when the line is terminated in various ways. The step will be propagated down the line according to the telegrapher's equation at some velocity κ and the incident voltage, v_i, at some point x on the line is given by

$$v_i = Vu(\kappa t - x)$$

The incident current, i_i can be found by dividing the characteristic impedance, Z_0

$$i_i = \frac{v_i}{Z_0} = Iu(\kappa t - x)$$

Open Circuit Line

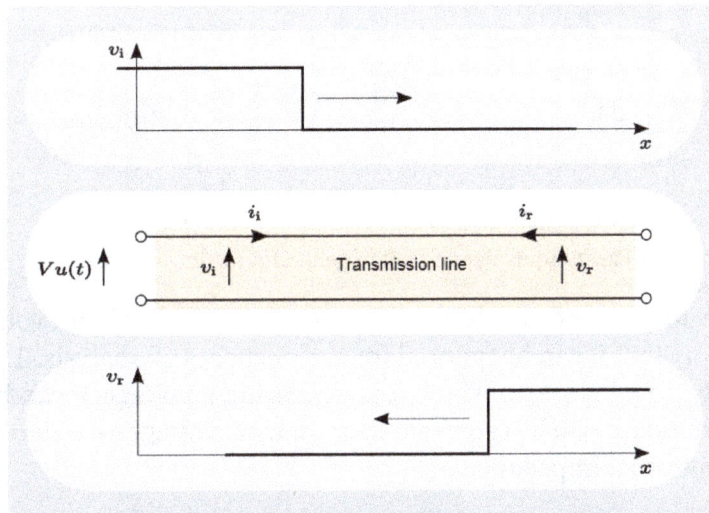

Fig. 1. Step voltage disturbance $Vu(t)$ is injected into the input of the line, v_i travels down the line and is reflected back at the far end as v_r.

The incident wave travelling down the line is not affected in any way by the open circuit at the end of the line. It cannot have any effect until the step actually reaches that point. The signal cannot have any foreknowledge of what is at the end of the line and is only affected by the local characteristics of the line. However, if the line is of length l the step will arrive at the open circuit at time $t = l/\kappa$, , at which point the current in the line is zero (by the definition of an open circuit). Since charge continues to arrive at the end of the line through the incident current, but no current is leaving the line, then conservation of electric charge requires that there must be an equal and opposite current into the end of the line. Essentially, this is Kirchhoff's current law in operation. This equal and opposite current is the reflected current, i_r, and since

$$i_r = \frac{v_r}{Z_0}$$

there must also be a reflected voltage, v_r to drive the reflected current down the line. This reflected voltage must exist by reason of conservation of energy. The source is supplying energy to the line at a rate of $v_i i_i$. None of this energy is dissipated in the line or its termination and it must go somewhere. The only available direction is back up the line. Since the reflected current is equal in magnitude to the incident current, it must also be so that

$$v_r = v_i$$

These two voltages will *add* to each other so that after the step has been reflected, twice the incident voltage appears across the output terminals of the line. As the reflection proceeds back up the line the reflected voltage continues to add to the incident voltage and the reflected current continues to subtract from the incident current. After a further interval of $t = l / \kappa$ the reflected step arrives at the generator end and the condition of double voltage and zero current will pertain there also as well as all along the length of the line. If the generator is matched to the line with an impedance of z_0 the step transient will be absorbed in the generator internal impedance and there will be no further reflections.

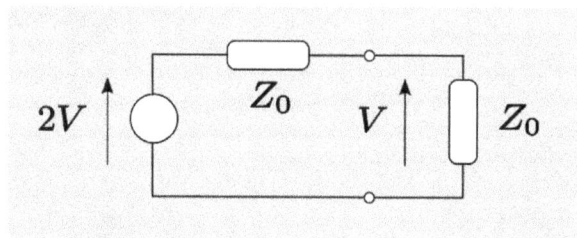

Fig. 2. Equivalent circuit of generator feeding a line.

This counter-intuitive doubling of voltage may become clearer if the circuit voltages are considered when the line is so short that it can be ignored for the purposes of analysis. The equivalent circuit of a generator matched to a load z_0 to which it is delivering a voltage V can be represented as in figure 2. That is, the generator can be represented as an ideal voltage generator of twice the voltage it is to deliver and an internal impedance of z_0.

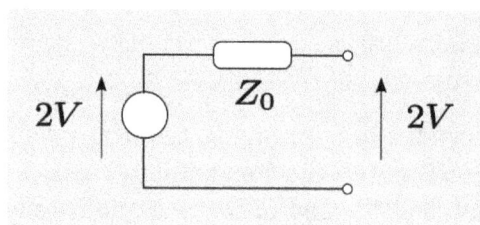

Fig. 3. Open circuit generator

However, if the generator is left open circuit, a voltage of $2V$ appears at the generator output terminals as in figure 3. The same situation pertains if a very short transmission line is inserted between the generator and the open circuit. If, however, a longer line with a characteristic impedance of z_0 and noticeable end-to-end delay is inserted, the generator – being initially matched to the impedance of the line – will have V at the output. But after an interval, a reflected transient will return from the end of the line with the "information" on what the line is actually terminated with, and the voltage will become $2V$ as before.

Short Circuit Line

The reflection from a short-circuited line can be described in similar terms to that from an open-circuited line. Just as in the open circuit case the current must be zero at the end of the line, in the short circuit case the voltage must be zero since there can be no volts across a short circuit. Again, all of the energy must be reflected back up the line and the reflected voltage must be equal and opposite to the incident voltage by Kirchhoff's voltage law:

$v_r = -v_i$ and,

$i_r = -i_i$

As the reflection travels back up the line, the two voltages subtract and cancel, while the currents will add (the reflection is double negative - a negative current travelling in the reverse direction), the dual situation to the open circuit case.

Arbitrary Impedance

For the general case of a line terminated in some arbitrary impedance it is usual to describe the signal as a wave travelling down the line and analyse it in the frequency domain. The impedance is consequently represented as a frequency dependant complex function.

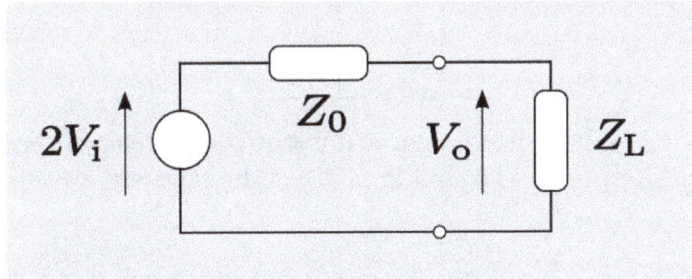

Fig. 4. Equivalent circuit of an incident wave on a transmission line arriving at an arbitrary load impedance.

For a line terminated in its own characteristic impedance there is no reflection. By definition, terminating in the characteristic impedance has the same effect as an infinitely long line. Any other impedance will result in a reflection. The magnitude of the reflection will be smaller than the magnitude of the incident wave if the terminating impedance is wholly or partly resistive since some of the energy of the incident wave will be absorbed in the resistance. The voltage, V_o, across the terminating impedance, Z_L, may be calculated by replacing the output of the line with an equivalent generator (figure 4) and is given by

$$V_o = 2V_i \frac{Z_L}{Z_0 + Z_L}$$

The reflection, V_r must be the exact amount required to make $V_i + V_r = V_o$,

$$V_r = V_o - V_i = 2V_i \frac{Z_L}{Z_0 + Z_L} - V_i = V_i \frac{Z_L - Z_0}{Z_L + Z_0}$$

The reflection coefficient, Γ, is defined as

$$\Gamma := \frac{V_r}{V_i}$$

and substituting in the expression for V_r,

$$\Gamma = \frac{V_r}{V_i} = \frac{I_r}{I_i} = \frac{Z_L - Z_0}{Z_L + Z_0}$$

In general Γ is a complex function but the above expression shows that the magnitude is limited to

$$|\Gamma| \leq 1 \text{ when } \Re(Z_L), \Re(Z_0) > 0$$

The physical interpretation of this is that the reflection cannot be greater than the incident wave when only passive elements are involved. For the special cases described above,

Termination	Γ
Open circuit	$\Gamma = 1$
Short circuit	$\Gamma = -1$
$Z_L = R_L$ $Z_0 = R_0$	$\Re(\Gamma) < 1,$ $\Im(\Gamma) = 0$

When both Z_0 and Z_L are purely resistive then Γ must be purely real. In the general case when Γ is complex, this is to be interpreted as a shift in phase of the reflected wave relative to the incident wave.

Reactive Termination

Another special case occurs when Z_0 is purely real (R_0) and Z_L is purely imaginary (jX_L), that is, it is a reactance. In this case,

$$\Gamma = \frac{jX_L - R_0}{jX_L + R_0}$$

Since

$$|jX_L - R_0| = |jX_L + R_0|$$

then

$$|\Gamma| = 1$$

showing that all the incident wave is reflected, and none of it is absorbed in the termination, as is to be expected from a pure reactance. There is, however, a change of phase, θ, in the reflection given by

$$\theta = \begin{cases} \pi - 2\arctan\dfrac{X_L}{R_0} & \text{if } X_L > 0 \\[2mm] -\pi - 2\arctan\dfrac{X_L}{R_0} & \text{if } X_L < 0 \end{cases}$$

Discontinuity Along Line

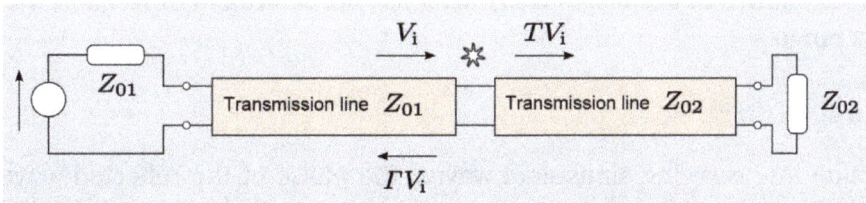

Fig. 5. Mismatch of transmission line characteristic impedances causes a discontinuity (marked with a star) in the line parameters and results in a reflected wave.

A discontinuity, or mismatch, somewhere along the length of the line results in part of the incident wave being reflected and part being transmitted onward in the second section of line as shown in figure 5. The reflection coefficient in this case is given by

$$\Gamma = \frac{Z_{02} - Z_{01}}{Z_{02} + Z_{01}}$$

In a similar manner, a transmission coefficient, T, can be defined to describe the portion of the wave, V_t, that it is transmitted in the forward direction:

$$T = \frac{V_t}{V_i} = \frac{2Z_{02}}{Z_{02} + Z_{01}}$$

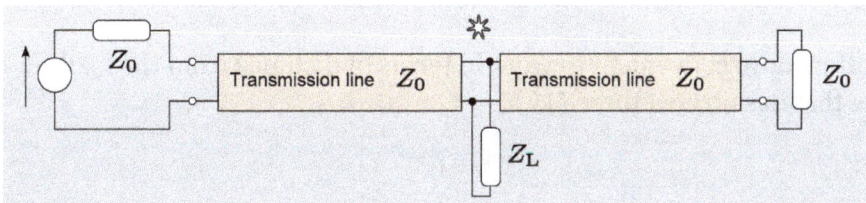

Fig. 6. Lumped components or networks connected to the line also cause a discontinuity (marked with a star).

Another kind of discontinuity is caused when both sections of line have an identical characteristic impedance but there is a lumped element, Z_L, at the discontinuity. For the example shown (figure 6) of a shunt lumped element,

$$\Gamma = \frac{-Z_0}{Z_0 + 2Z_L}$$

$$T = \frac{2Z_L}{Z_0 + 2Z_L}$$

Similar expressions can be developed for a series element, or any electrical network for that matter.

Networks

Reflections in more complex scenarios, such as found on a network of cables, can result in very complicated and long lasting waveforms on the cable. Even a simple overvoltage pulse entering a cable system as uncomplicated as the power wiring found in a typical private home can result in an oscillatory disturbance as the pulse is reflected to and fro from multiple circuit ends. These *ring waves* as they are known persist for far longer than the original pulse and their waveforms bears little obvious resemblance to the original disturbance, containing high frequency components in the tens of MHz range.

Standing Waves

For a transmission line carrying sinusoidal waves, the phase of the reflected wave is continually changing with distance, with respect to the incident wave, as it proceeds back down the line. Because of this continuous change there are certain points on the line that the reflection will be in phase with the incident wave and the amplitude of the two waves will add. There will be other points where the two waves are in anti-phase and will consequently subtract. At these latter points the amplitude is at a minimum and they are known as nodes. If the incident wave has been totally reflected and the line is lossless, there will be complete cancellation at the nodes with zero signal present there despite the ongoing transmission of waves in both directions. The points where the waves are in phase are anti-nodes and represent a peak in amplitude. Nodes and anti-nodes alternate along the line and the combined wave amplitude varies continuously between them. The combined (incident plus reflected) wave appears to be standing still on the line and is called a standing wave.

The incident wave can be characterised in terms of the line's propagation constant, γ, source voltage, V and distance from the source, x', by

$$V_i = Ve^{-\gamma x'}$$

However, it is often more convenient to work in terms of distance from the load ($x = l - x'$) and the incident voltage that has arrived there (V_{iL}).

$$V_i = V_{iL}e^{\gamma x}$$

The exponent is positive because x is measured in the reverse direction back up the line and the voltage is increasing closer to the source. Likewise the reflected voltage is given by

$$V_r = \Gamma V_{iL}e^{-\gamma x}$$

The total voltage on the line is given by

$$V_T = V_i + V_r = V_{iL}(e^{\gamma x} + \Gamma e^{-\gamma x})$$

It is often convenient to express this in terms of hyperbolic functions

$$V_T = V_{iL}[(1+\Gamma)\cosh(\gamma x) + (1-\Gamma)\sinh(\gamma x)]$$

Similarly, the total current on the line is

$$I_{\tau} = I_{iL}[(1-\Gamma)\cosh(\gamma x)+(1+\Gamma)\sinh(\gamma x)]$$

The voltage nodes (current nodes are not at the same locations) and anti-nodes occur when

$$\frac{\partial |V_{\tau}|}{\partial x} = 0$$

This does not have an easy analytical solution in the general case, but in the case of lossless lines (or lines that are short enough to be considered so) γ can be replaced by $i\beta$ where β is the phase change constant. The voltage equation then reduces to trigonometric functions

$$V_{\tau} = V_{iL}[(1+\Gamma)\cos(\beta x)+i(1-\Gamma)\sin(\beta x)]$$

and the partial differential of the magnitude of this yields the condition,

$$-2\Im(\Gamma) = \tan(2\beta x)$$

Expressing β in terms of wavelength, λ, allows x to be solved in terms of λ:

$$-2\Im(\Gamma) = \tan\left(\frac{4\pi}{\lambda}x\right)$$

Γ is purely real when the termination is short circuit or open circuit, or when both Z_0 and Z_L are purely resistive. In those cases the nodes and anti-nodes are given by

$$\tan\left(\frac{4\pi}{\lambda}x\right) = 0$$

which solves for x at

$$x = 0, \frac{\lambda}{4}, \frac{\lambda}{2}, \frac{3\lambda}{4},\ldots$$

For $R_L < R_0$ the first point is a node, for $R_L < R_0$ the first point is an anti-node and thenceforth they will alternate. For terminations that are not purely resistive the spacing and alternation remain the same but the whole pattern is shifted along the line by a constant amount related to the phase of Γ.

Voltage Standing Wave Ratio

The ratio of $|V_{\tau}|$ at anti-nodes and nodes is called the voltage standing wave ratio (VSWR) and is related to the reflection coefficient by

$$\text{VSWR} = \frac{1+|\Gamma|}{1-|\Gamma|}$$

for a lossless line. For a lossy line the expression is only valid adjacent to the termination; VSWR asymptotically approaches unity with distance from the termination or discontinuity.

VSWR and the positions of the nodes are parameters that can be directly measured with an instrument called a slotted line. This instrument makes use of the reflection phenomenon to make many different measurements at microwave frequencies. One use is that VSWR and node position can be used to calculate the impedance of a test component terminating the slotted line. This is a useful method because measuring impedances by directly measuring voltages and currents is difficult at these frequencies.

VSWR is the conventional means of expressing the match of a radio transmitter to its antenna. It is an important parameter because power reflected back in to a high power transmitter can damage its output circuitry.

Input Impedance

The input impedance looking into a transmission line which is not terminated with its characteristic impedance at the far end will be something other than Z_0 and will be a function of the length of the line. The value of this impedance can be found by dividing the expression for total voltage by the expression for total current given above:

$$Z_{in} = \frac{V_T}{I_T} = Z_0 \frac{(1+\Gamma)\cosh(\gamma x)+(1-\Gamma)\sinh(\gamma x)}{(1-\Gamma)\cosh(\gamma x)+(1+\Gamma)\sinh(\gamma x)}$$

Substituting $x = l$, the length of the line and dividing through by $(1+\Gamma)\cosh(\gamma x)$ reduces this to

$$Z_{in} = Z_0 \frac{Z_L + Z_0 \tanh(\gamma l)}{Z_0 + Z_L \tanh(\gamma l)}$$

As before, when considering just short pieces of transmission line, γ can be replaced by $j\beta$ and the expression reduces to trigonometric functions

$$Z_{in} = Z_0 \frac{Z_L + jZ_0 \tan(\beta l)}{Z_0 + jZ_L \tan(\beta l)}$$

Applications

There are two structures that are of particular importance which use reflected waves to modify impedance. One is the stub which is a short length of line terminated in a short-circuit (or it can be an open-circuit). This produces a purely imaginary impedance at its input, that is, a reactance

$$X_{in} = Z_0 \tan(\beta l)$$

By suitable choice of length, the stub can be used in place of a capacitor, an inductor or a resonant circuit.

The other structure is the quarter wave impedance transformer. As its name suggests, this is a line exactly $\lambda/4$ in length. Since $\beta l = \pi/2$ this will produce the inverse of its terminating impedance

$$Z_{in} = \frac{Z_0^2}{Z_L}$$

Both of these structures are widely used in distributed element filters and impedance matching networks.

Network Architecture

Network architecture is the design of a communication network. It is a framework for the specification of a network's physical components and their functional organization and configuration, its operational principles and procedures, as well as data formats used. In telecommunication, the specification of a network architecture may also include a detailed description of products and services delivered via a communications network, as well as detailed rate and billing structures under which services are compensated.

The network architecture of the Internet is predominantly expressed by its use of the Internet Protocol Suite, rather than a specific model for interconnecting networks or nodes in the network, or the usage of specific types of hardware links.

OSI Network Model

The Open Systems Interconnection model (OSI model) is a product of the Open Systems Interconnection effort at the International Organization for Standardisation (ISO) . It is a way of sub-dividing a communications system into smaller parts called layers. A layer is a collection of similar functions that provide services to the layer above it and receives services from the layer below it. On each layer, an instance provides services to the instances at the layer above and requests service from the layer below.

Physical Layer

The physical layer defines the electrical and physical specifications for devices. In particular, it defines the relationship between a device and a transmission medium, such as a copper or optical cable. This includes the layout of pins, voltages, cable specifications, hubs, repeaters, network adapters, host bus adapters (HBA used in storage area networks) and more. Its main task is the transmission of a stream of bits over a communication channel.

Data-linking Layer

The data link layer provides the functional and procedural means to transfer data between network entities and to detect and possibly correct errors that may occur in the physical layer. Originally, this layer was intended for point-to-point and point-to-multipoint media, characteristic of wide area media in the telephone system. Local area network architecture, which included broadcast-capable multi-access media, was developed independently of the ISO work in IEEE Project 802. IEEE work assumed sublayering and management functions not required for WAN use. In modern practice, only error detection, not flow control using sliding window, is present in data link protocols such as Point-to-Point Protocol (PPP), and, on local area networks, the IEEE 802.2 LLC layer is not used for most protocols on the Ethernet, and on other local area networks, its flow control and acknowledgment mechanisms are rarely used. Sliding-window flow control and acknowledgment is used at the transport layer by protocols such as TCP, but is still used in niches where X.25 offers performance advantages. Simply, its main job is to create and recognize the frame boundary. This can be done by attaching special bit patterns to the beginning and the end of the frame. The input data is broken up into frames.

Network Layer

The network layer provides the functional and procedural means of transferring variable length data sequences from a source host on one network to a destination host on a different network, while maintaining the quality of service requested by the transport layer (in contrast to the data link layer which connects hosts within the same network). The network layer performs network routing functions, and might also perform fragmentation and reassembly, and report delivery errors. Routers operate at this layer—sending data throughout the extended network and making the Internet possible. This is a logical addressing scheme; values are chosen by the network engineer. The addressing scheme is not hierarchical. It controls the operation of the subnet and determine the routing strategies between IMP and ensures that all the packs are correctly received at the destination in the proper order.

Transport Layer

The transport layer provides transparent transfer of data between end users, providing reliable data transfer services to the upper layers. The transport layer controls the reliability of a given link through flow control, segmentation/desegmentation, and error control. Some protocols are state and connection oriented. This means that the transport layer can keep track of the segments and retransmit those that fail. The transport layer also provides the acknowledgement of the successful data transmission and sends the next data if no errors occurred. Some transport layer protocols (such as TCP, but not UDP) support virtual circuits that provide connection-oriented communication over an underlying packet-oriented datagram network, where it assures the delivery of packets in the order in which they were sent and that they are free of errors. The datagram transportation deliver the packets randomly and broadcast it to multiple nodes.

The transport layer multiplexes several streams on to one physical channel. The transport headers indicate which message belongs to which connection.

Session Layer

This layer provides a user interface to the network where the user negotiates to establish a connection. The user must provide the remote address to be contacted. The operation of setting up a session between two processes is known as "binding". In some protocols, it is merged with the transport layer. Its main work is to transfer data from the other application to this application so this application is mainly used for transferred layer.

Presentation Layer

The presentation layer establishes context between entities on the application layer, in which the higher-layer entities may use different syntax and semantics if the presentation service provides a mapping between them. If a mapping is available, presentation service data units are encapsulated into session protocol data units, and passed down the stack. This layer provides independence from data representation (e.g. encryption) by translating between application and network formats. The presentation layer transforms data into the form that the application accepts. This layer formats and encrypts data to be sent across a network. It is sometimes called the syntax layer. The original presentation structure used the basic encoding rules of Abstract Syntax Notation One (ASSN), with capabilities such as converting an BODICE-coded text file to an ASCII-coded file, or serialization of objects and other data structures from and to XML.

Application Layer

The application layer is the OSI layer closest to the end user, which means that both the OSI application layer and the user interact directly with the software application. This layer interacts with software applications that implement a communicating component. Such application programs fall outside the scope of the OSI model. Application layer functions typically include identifying communication partners, determining resource availability, and synchronizing communication. When identifying communication partners, the application layer determines the identity and availability of communication partners for an application with data to transmit.

Distributed Computing

In distinct usage in distributed computing, the term "network architecture" often describes the structure and classification of a distributed application architecture, as the participating nodes in a distributed application are often referred to as a "network". For example, the applications architecture of the public switched telephone network (PSTN) has been termed the Advanced Intelligent Network. There are any number of specific classifications but all lie on a continuum between the dumb network (e.g. Internet) and the intelligent computer network (e.g. the telephone network). Other networks contain various elements of these two classical types to make them suitable for various types of applications. Recently the context aware network, which is a synthesis of two, has gained much interest with its ability to combine the best elements of both.

A popular example of such usage of the term in distributed applications, as well as PVCs (permanent virtual circuits), is the organization of nodes in peer-to-peer (P2P) services and networks. P2P networks usually implement overlay networks running over an underlying physical or logical network. These overlay network may implement certain organizational structures of the nodes according to several distinct models, the network architecture of the system.

Network architecture is a broad plan that specifies everything necessary for two application programs on different networks on an Internet to be able to work together effectively.

Primary Line Constants

The primary line constants are parameters that describe the characteristics of conductive transmission lines, such as pairs of copper wires, in terms of the physical electrical properties of the line. The primary line constants are only relevant to such lines and are to be contrasted with the secondary line constants, which can be derived from them, and are more generally applicable. The secondary line constants can be used, for instance, to compare the characteristics of a waveguide to a copper line, whereas the primary constants have no meaning for a waveguide.

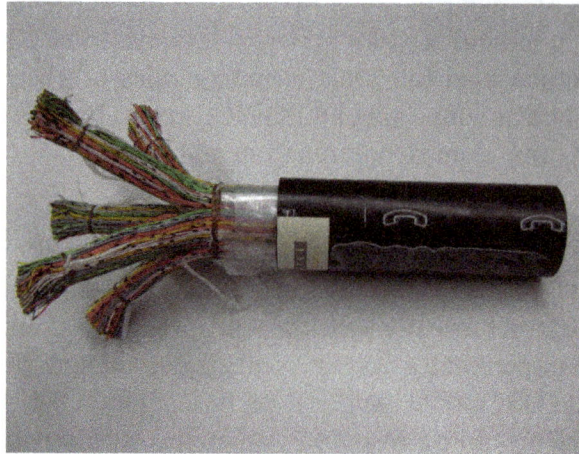

Telephone cable containing multiple twisted-pair lines

The constants are conductor resistance and inductance, and insulator capacitance and conductance, which are by convention given the symbols R, L, C, and G respectively. The constants are enumerated in terms of per unit length. The circuit representation of these elements requires a distributed element model and consequently calculus must be used to analyse the circuit. The secondary constants of characteristic impedance and propagation constant can be derived in this way.

A number of special cases have particularly simple solutions and important practical applications. Low loss cable requires only L and C to be included in the analysis, useful for short lengths of cable. Low frequency applications, such as twisted pair telephone lines, are dominated by R and C only. High frequency applications, such as RF co-axial cable, are dominated by L and C. Lines loaded to prevent distortion need all four elements in the analysis, but have a simple, elegant solution.

The Constants

There are four primary line constants, but in some circumstances some of them are small enough to be ignored and the analysis can be simplified. These four, and their symbols and units are as follows:

Name	Symbol	Units	Unit symbol
loop resistance	R	ohms per metre	Ω/m
loop inductance	L	henries per metre	H/m
insulator capacitance	C	farads per metre	F/m
insulator conductance	G	siemens per metre	S/m

R and L are elements in series with the line (because they are properties of the conductor) and C and G are elements shunting the line (because they are properties of the dielectric material between the conductors). G represents leakage current through the dielectric and in most cables is very small. The word loop is used to emphasise that the resistance and inductance of both conductors must be taken into account. For instance, if a line consists of two identical wires that have a resistance of 25 mΩ/m each, the *loop* resistance is double that, 50 mΩ/m. Because the values of the constants are quite small, it is common for manufacturers to quote them per kilometre rather than per metre; in the English-speaking world "per mile" can also be used.

The word "constant" can be misleading since there is some variation with frequency. In particular, R is heavily influenced by the skin effect. Furthermore, while G has virtually no effect at audio frequency, it can cause noticeable losses at high frequency with many of the dielectric materials used in cables due to a high loss tangent. Avoiding the losses caused by G is the reason many cables designed for use at UHF are air-insulated or foam-insulated (which makes them virtually air-insulated). The actual meaning of constant in this context is that the parameter is constant with *distance*. That is the line is assumed to be homogenous lengthwise. This condition is true for the vast majority of transmission lines in use today.

Typical Values for some Common Cables

Designation	Cable form	Application	R	L[†]	G	C	Z_0
			Ω/km	µH/km	nS/km	nF/km	Ω
CAT5	Twisted pair	Data transmission	176	490	<2	49	100
CAT5e	Twisted pair	Data transmission	176		<2		100
CW1308	Twisted pair	Telephony	98		<20		
RG59	Coaxial	Video	36	430		69	75
RG59	Coaxial (foam dielectric)	Video	17	303		54	75
RG58	Coaxial	Radio frequency	48	253	<0.01	101	50
Low loss	Coaxial (Foam dielectric)	Radio frequency transmitter feed	2.86	188		75	50
DIN VDE 0816	Star quad	Telephony (trunk lines)	31.8		<0.1	35	

† Manufacturers commonly omit a value for inductance in their data sheets. Some of these values are estimated from the figures for capacitance and characteristic impedance by $Z_0^2 = L / C$.

Circuit Representation

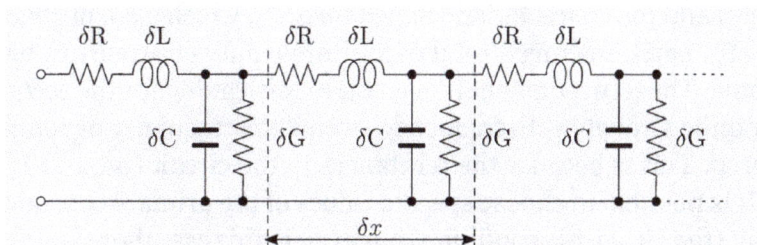

Fig. 1. Equivalent circuit representation of a transmission line using distributed elements. δL, δR, δC and δG are to be read as, *Lδx, Rδx, Cδx and Gδx respectively*

The line constants cannot be simply represented as lumped elements in a circuit; they must be described as distributed elements. For instance "pieces" of the capacitance are in between "pieces" of the resistance. However many pieces the R and C are broken into, it can always be argued they should be broken apart further to properly represent the circuit, and after each division the number of meshes in the circuit is increased. This is shown diagramtically in figure 1. To give a true representation of the circuit, the elements must be made infinitesimally small so that each element is distributed along the line. The infinitesimal elements in an infinitesimal distance dx are given by;

$$dL = \lim_{\delta x \to 0}(L\delta x) = Ldx$$

$$dR = \lim_{\delta x \to 0}(R\delta x) = Rdx$$

$$dC = \lim_{\delta x \to 0}(C\delta x) = Cdx$$

$$dG = \lim_{\delta x \to 0}(G\delta x) = Gdx$$

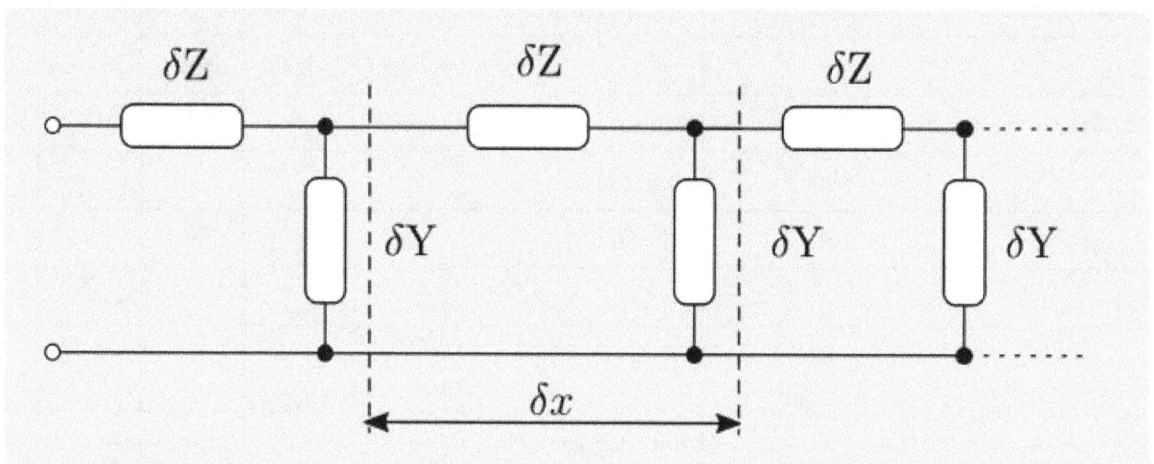

Fig. 2. Representation of a transmission line using generalised distributed impedance and admittance elements.

It is convenient for the purposes of analysis to roll up these elements into general series impedance, Z, and shunt admittance, Y elements such that;

$$dZ = (R + i\omega L)dx = Zdx, \quad \text{and,}$$

$$dY = (G + i\omega C)dx = Ydx.$$

Analysis of this network (figure 2) will yield the secondary line constants: the propagation constant, γ, (whose real and imaginary parts are the attenuation constant, α, and phase change constant, β, respectively) and the characteristic impedance, Z_0, which also, in general, will have real, R_0, and imaginary, X_0, parts, making a total of four secondary constants to be derived from the four primary constants. The term constant is even more misleading for the secondary constants as they all usually vary quite strongly with frequency, even if the frequency dependence of the primary constants is ignored. This is because the reactances in the circuit (ωL and $1/(\omega C)$) introduce a dependence on ω. It is possible to choose specific values of the primary constants that result in α and Z_0 being constant (the Heaviside condition) but even in this case there is still β which is directly proportional to ω. As with the primary constants, the meaning is that the secondary constants do not vary with distance along the line, not that they are independent of frequency.

Characteristic Impedance

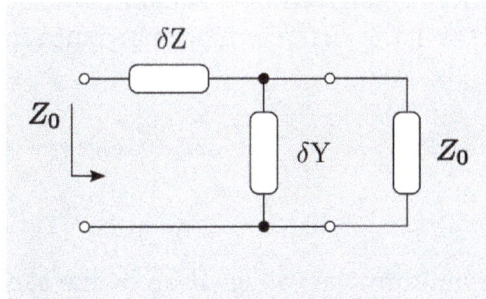

Fig. 3. Equivalent circuit of a transmission line for the calculation of Z_0 from the primary line constants

The characteristic impedance of a transmission line, Z_0, is defined as the impedance looking into an infinitely long line. Such a line will never return a reflection since the incident wave will never reach the end to be reflected. When considering a finite initial length of the line, the remainder of the line can be replaced by Z_0 as its equivalent circuit. This is so because the remainder of the line is still infinitely long. Considering just the first section of the equivalent circuit of the line (this is an L-network consisting of one element each of dZ and dY) the remainder can be replaced by Z_0. This results in the network shown in figure 3, which can be analysed for Z_0 using the usual network analysis theorems,

$$Z_0 = \delta Z + \frac{Z_0}{1 + Z_0 \delta Y}$$

which re-arranges to,

$$Z_0^2 - Z_0 \delta Z = \frac{\delta Z}{\delta Y}$$

Taking limits of both sides

$$\lim_{\delta x \to 0} (Z_0^2 - Z_0 \delta Z) = Z_0^2 = \frac{dZ}{dY}$$

and since the line was assumed to be homogenous lengthwise,

$$Z_0^2 = \frac{Z}{Y}$$

Propagation Constant

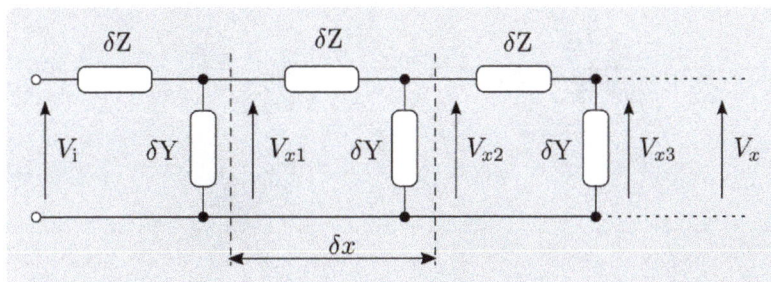

Fig. 4. Each infinitesimal section of the transmission line causes an infinitesimal drop in the line voltage as it is propagated along the line. Integrating these drops enables the propagation constant to be found.

The ratio of the line input voltage to the voltage a distance δx further down the line (that is, after one section of the equivalent circuit) is given by a standard voltage divider calculation. The remainder of the line to the right, as in the characteristic impedance calculation, is replaced with Z_0,

$$\frac{V_i}{V_{x1}} = \frac{\delta Z + \dfrac{Z_0 / \delta Y}{Z_0 + 1/\delta Y}}{\dfrac{Z_0 / \delta Y}{Z_0 + 1/\delta Y}} = 1 + \frac{\delta Z}{Z_0} + \delta Z \delta Y$$

Each infinitesimal section will multiply the voltage drop by the same factor. After n sections the voltage ratio will be,

$$\frac{V_i}{V_{xn}} = \left(1 + \frac{\delta Z}{Z_0} + \delta Z \delta Y\right)^n$$

At a distance x along the line, the number of sections is $x/\delta x$ so that,

$$\frac{V}{V_{xn}} = \left(1 + \frac{\delta Z}{Z_0} + \delta Z \delta Y\right)^{\frac{x}{\delta x}}$$

In the limit as $\delta x \to 0$,

$$\frac{V_i}{V_x} = \lim_{\delta x \to 0} \frac{V_i}{V_{xn}} = \lim_{\delta x \to 0}\left(1 + \frac{\delta Z}{Z_0} + \delta Z \delta Y\right)^{\frac{x}{\delta x}}$$

The second order term $\delta Z \delta Y$ will disappear in the limit, so we can write without loss of accuracy,

$$\frac{V_i}{V_x} = \lim_{\delta x \to 0}\left(1 + \frac{\delta Z}{Z_0}\right)^{\frac{x}{\delta x}}$$

and comparing with the mathematical identity,

$$e^x \equiv \lim_{p \to \infty}(1 + 1/p)^{px}$$

yields,

$$V_i = V_x e^{\frac{Z}{Z_0}x}$$

From the definition of propagation constant,

$$V_i = V_x e^{\gamma x}$$

Hence,

$$\gamma = \frac{Z}{Z_0} = \sqrt{ZY}$$

Special Cases

An ideal transmission line will have no loss, which implies that the resistive elements are zero. It also results in a purely real (resistive) characteristic impedance. The ideal line cannot be realised in practice, but it is a useful approximation in many circumstances. This is especially true, for instance, when short pieces of line are being used as circuit components such as stubs. A short line has very little loss and this can then be ignored and treated as an ideal line. The secondary constants in these circumstances are;

$$\gamma = i\omega\sqrt{LC}$$

$$\alpha = 0$$

$$\beta = \omega\sqrt{LC}$$

$$Z_0 = \sqrt{\frac{L}{C}}$$

Twisted Pair

Typically, twisted pair cable used for audio frequencies or low data rates has line constants dominated by R and C. The dielectric loss is usually negligible at these frequencies and G is close to zero. It is also the case that, at a low enough frequency, $R \gg \omega L$ which means that L can also be ignored. In those circumstances the secondary constants become,

$$\gamma \approx \sqrt{i\omega CR}$$

$$\alpha \approx \sqrt{\frac{\omega CR}{2}}$$

$$\beta \approx \sqrt{\frac{\omega CR}{2}}$$

$$Z_0 \approx \sqrt{\frac{R}{i\omega C}} = \sqrt{\frac{R}{2\omega C}} - i\sqrt{\frac{R}{2\omega C}}$$

The attenuation of this cable type increases with frequency, causing distortion of waveforms. Not so obviously, the variation of β with frequency also causes a distortion of a type called dispersion. To avoid dispersion the requirement is that β is directly proportional to ω. However, it is actually proportional to $\sqrt{\omega}$ and dispersion results. Z_0 also varies with frequency and is also partly reactive; both these features will be the cause of reflections from a resistive line termination. This is another undesirable effect. The nominal impedance quoted for this type of cable is, in this case, very nominal, being valid at only one spot frequency, usually quoted at 800 Hz or 1 kHz.

Co-axial Cable

Cable operated at a high enough frequency (VHF radio frequency or high data rates) will meet the conditions $R \ll \omega L$ and $G \ll \omega C$. This must eventually be the case as the frequency is increased for any cable. Under those conditions R and G can both be ignored (except for the purpose of calculating the cable loss) and the secondary constants become;

$$\gamma \approx i\omega\sqrt{LC}$$

$$\alpha \approx \frac{LG+RC}{2\sqrt{LC}} = \tfrac{1}{2}\left(Z_0 G + \frac{R}{Z_0}\right) \approx \frac{R}{2Z_0}$$

$$\beta \approx \omega\sqrt{LC}$$

$$Z_0 \approx \sqrt{\frac{L}{C}}$$

Loaded Line

Loaded lines are lines designed with deliberately increased inductance. This is done by adding iron or some other magnetic metal to the cable or adding coils. The purpose is to ensure that the line meets the Heaviside condition, which eliminates distortion caused by frequency-dependent attenuation and dispersion, and ensures that Z_0 is constant and resistive. The secondary constants are here related to the primary constants by;

$$\gamma = \sqrt{RG} + i\omega\sqrt{LC}$$

$$\alpha = \sqrt{RG}$$

$$\beta = \omega\sqrt{LC}$$

$$Z_0 = \sqrt{\frac{L}{C}} = \sqrt{\frac{R}{G}}$$

Velocity

The velocity of propagation is given by,

$$v = \lambda f.$$

Since,

$$\omega = 2\pi f \text{ and } \beta = \frac{2\pi}{\lambda}$$

then,

$$v = \frac{\omega}{\beta}.$$

In cases where β can be taken as,

$$\beta = \omega\sqrt{LC}$$

the velocity of propagation is given by,

$$v = \frac{1}{\sqrt{LC}}.$$

The lower the capacitance the higher the velocity. With an air dielectric cable, which is approximated to with low-loss cable, the velocity of propagation is very close to c, the speed of light *in vacuo*.

References

- Klingenfuss, J. (2003). Radio Data Code Manual (17th Ed.). Klingenfuss Publications. pp. 72–78. ISBN 3-924509-56-5.

- Bakshi, V. A.; Bakshi, A. V., Transmission Lines And Waveguide, Technical Publications, 2009 ISBN 8184316348.

- Weik, Martin (1997). Fiber Optics Standard Dictionary. Springer Science & Business Media. p. 270. ISBN 0412122413.

- Helfrick, Albert D. (2012). Electrical Spectrum & Network Analyzers: A Practical Approach. Academic Press. p. 192. ISBN 0080918670.

- Knott, Eugene F.; Shaeffer, John F.; Tuley, Michael T. (2004). Radar cross section. SciTech Radar and Defense Series (2nd ed.). SciTech Publishing. p. 374. ISBN 978-1-891121-25-8.

- Schaub, Keith B.; Kelly, Joe (2004). Production testing of RF and system-on-a-chip devices for wireless communications. Artech House microwave library. Artech House. p. 93. ISBN 978-1-58053-692-9.

- Samuel Silver, Microwave Antenna Theory and Design, p. 28, IEE, 1984 (originally published 1949) ISBN 0863410170.

- Hutchinson, Chuck, ed. (2000). The ARRL Handbook for Radio Amateurs 2001. Newington, CT: ARRL—the national association for Amateur Radio. p. 20.2. ISBN 0-87259-186-7.

- J. H. Gridley, Principles of Electrical Transmission Lines in Power and Communication, p. 265, Elsevier, 2014 ISBN 1483186032.

- "Electrical length". Federal Standard 1037C, Glossary of Telecommunication Terms. National Telecommunications & Information Admin., Dept. of Commerce, US Government. 1996. Retrieved November 23, 2014.

- Ford, Steve (April 1997). "The SWR Obsession" (PDF). QST. Newington, CT: ARRL—The national association for amateur radio. 78 (4): 70–72. Retrieved 2014-11-04.

- Amlaner, Charles J. Jr. (March 1979). "The design of antennas for use in radio telemetry". A Handbook on Biotelemetry and Radio Tracking: Proceedings of an International Conference on Telemetry and Radio Tracking in Biology and Medicine, Oxford, 20–22 March 1979. Elsevier. p. 260. Retrieved November 23, 2013.

Wireless Technology and its Concepts

Wireless technology can be explained as the transferring of information between two or more point not connected by any electrical conductors. This text elucidates the concepts of wireless technology, by briefly explaining, wireless mesh network, hotspot, Wi-Fi, Bluetooth and wireless LAN. Wireless communication is best understood in confluence with the major topics listed in the following chapter.

Wireless

Wireless communication is the transfer of information or power between two or more points that are not connected by an electrical conductor.

A handheld On-board communication station of the maritime mobile service

The most common wireless technologies use radio. With radio waves distances can be short, such as a few meters for television or as far as thousands or even millions of kilometers for deep-space radio communications. It encompasses various types of fixed, mobile, and portable applications, including two-way radios, cellular telephones, personal digital assistants (PDAs), and wireless networking. Other examples of applications of radio *wireless technology* include GPS units, garage door openers, wireless computer mice, keyboards and headsets, headphones, radio receivers, satellite television, broadcast television and cordless telephones.

Somewhat less common methods of achieving wireless communications include the use of other electromagnetic wireless technologies, such as light, magnetic, or electric fields or the use of sound.

The term *wireless* has been used twice in communications history, with slightly different meaning. It was initially used from about 1890 for the first radio transmitting and receiving technology, as in *wireless telegraphy*, until the new word *radio* replaced it around 1920. The term was revived in the 1980s and 1990s mainly to distinguish digital devices that communicate without wires, such as the examples listed in the previous paragraph, from those that require wires. This is its primary usage today.

LTE, LTE-Advanced, Wi-Fi, Bluetooth are some of the most common modern wireless technologies.

Introduction

Wireless operations permit services, such as long-range communications, that are impossible or impractical to implement with the use of wires. The term is commonly used in the telecommunications industry to refer to telecommunications systems (e.g. radio transmitters and receivers, remote controls, etc.) which use some form of energy (e.g. radio waves, acoustic energy, etc.) to transfer information without the use of wires. Information is transferred in this manner over both short and long distances.

History

Photophone

Bell and Tainter's photophone, of 1880.

The world's first wireless telephone conversation occurred in 1880, when Alexander Graham Bell and Charles Sumner Tainter invented and patented the photophone, a telephone that conducted audio conversations wirelessly over modulated light beams (which are narrow projections of electromagnetic waves). In that distant era, when utilities did not yet exist to provide electricity and lasers had not even been imagined in science fiction, there were no practical applications for their invention, which was highly limited by the availability of both sunlight and good weather. Similar to free-space optical communication, the photophone also required a clear line of sight between its transmitter and its receiver. It would be several decades before the photophone's principles found their first practical applications in military communications and later in fiber-optic communications.

Early Wireless Work

David E. Hughes transmitted radio signals over a few hundred yards using a clockwork keyed transmitter in 1878. As this was before Maxwell's work was understood, Hughes' contemporaries dismissed his achievement as mere "Induction." In 1885, Thomas Edison used a vibrator magnet for induction transmission. In 1888, Edison deployed a system of signaling on the Lehigh Valley Railroad. In 1891, Edison obtained the wireless patent for this method using inductance (U.S. Patent 465,971).

In 1888, Heinrich Hertz demonstrated the existence of electromagnetic waves, the underlying basis of most wireless technology. The theory of electromagnetic waves was predicted from the research of James Clerk Maxwell and Michael Faraday. Hertz demonstrated that electromagnetic waves traveled through space in straight lines, could be transmitted, and could be received by an experimental apparatus. Hertz did not follow up on the experiments. Jagadish Chandra Bose around this time developed an early wireless detection device and helped increase the knowledge of millimeter-length electromagnetic waves. Later inventors implemented practical applications of wireless radio communication and radio remote control technology.

Radio

Marconi transmitting the first radio signal across the Atlantic.

The term "wireless" came into public use to refer to a radio receiver or transceiver (a dual purpose receiver and transmitter device), establishing its use in the field of wireless telegraphy early on; now the term is used to describe modern wireless connections such as in cellular networks and wireless broadband Internet. It is also used in a general sense to refer to any operation that is implemented without the use of wires, such as "wireless remote control" or "wireless energy transfer", regardless of the specific technology (e.g. radio, infrared, ultrasonic) used. Guglielmo Marconi and Karl Ferdinand Braun were awarded the 1909 Nobel Prize for Physics for their contribution to wireless telegraphy.

Modes

Wireless communications can be via:

Radio

radio communication, microwave communication, for example long-range line-of-sight via highly directional antennas, or short-range communication,

Free-space Optical

An 8-beam free space optics laser link, rated for 1 Gbit/s at a distance of approximately 2 km. The receptor is the large disc in the middle, the transmitters the smaller ones. To the top and right corner a monocular for assisting the alignment of the two heads.

Free-space optical communication (FSO) is an optical communication technology that uses light propagating in free space to transmit wirelessly data for telecommunications or computer networking. "Free space" means the light beams travel through the open air or outer space. This contrasts with other communication technologies that use light beams traveling through transmission lines such as optical fiber or dielectric "light pipes".

The technology is useful where physical connections are impractical due to high costs or other considerations. For example, free space optical links are used in cities between office buildings which are not wired for networking, where the cost of running cable through the building and under the street would be prohibitive.

Another widely used example is consumer IR devices such as remote controls and IrDA (Infrared Data Association) networking, which is used as an alternative to WiFi networking to allow laptops, PDAs, printers, and digital cameras to exchange data.

Sonic

Sonic, especially ultrasonic short range communication involves the transmission and reception of sound.

Electromagnetic Induction

Electromagnetic induction short range communication and power. This has been used in biomedical situations such as pacemakers, as well as for short-range Rfid tags.

Wireless Services

Common examples of wireless equipment include:

- Infrared and ultrasonic remote control devices
- Professional LMR (Land Mobile Radio) and SMR (Specialized Mobile Radio) typically used

by business, industrial and Public Safety entities.

- Consumer Two-way radio including FRS Family Radio Service, GMRS (General Mobile Radio Service) and Citizens band ("CB") radios.

- The Amateur Radio Service (Ham radio).

- Consumer and professional Marine VHF radios.

- Airband and radio navigation equipment used by aviators and air traffic control

- Cellular telephones and pagers: provide connectivity for portable and mobile applications, both personal and business.

- Global Positioning System (GPS): allows drivers of cars and trucks, captains of boats and ships, and pilots of aircraft to ascertain their location anywhere on earth.

- Cordless computer peripherals: the cordless mouse is a common example; wireless headphones, keyboards, and printers can also be linked to a computer via wireless using technology such as Wireless USB or Bluetooth

- Cordless telephone sets: these are limited-range devices.

- Satellite television: Is broadcast from satellites in geostationary orbit. Typical services use direct broadcast satellite to provide multiple television channels to viewers.

Computers

- WiFi

- Cordless computer peripherals:

    - mouse

    - headphones,

    - keyboards,

    - printers,

    - USB and,

    - Bluetooth

- Wireless networking

    - To span a distance beyond the capabilities of typical cabling,

    - To provide a backup communications link in case of normal network failure,

    - To link portable or temporary workstations,

    - To overcome situations where normal cabling is difficult or financially impractical, or

    - To remotely connect mobile users or networks.

Developers need to consider some parameters involving Wireless RF technology for better developing wireless networks:

- Sub-GHz versus 2.4 GHz frequency trends

- Operating range and battery life

- Sensitivity and data rate

- Network topology and node intelligence

Applications may involve point-to-point communication, point-to-multipoint communication, broadcasting, cellular networks and other wireless networks, Wi-Fi technology.

Cordless

The term "wireless" should not be confused with the term "cordless", which is generally used to refer to powered electrical or electronic devices that are able to operate from a portable power source (e.g., a battery pack) without any cable or cord to limit the mobility of the cordless device through a connection to the mains power supply.

Some cordless devices, such as cordless telephones, are also wireless in the sense that information is transferred from the cordless telephone to the phone's base unit via some wireless communications link. This has caused some disparity in the usage of the term "cordless", for example in Digital Enhanced Cordless Telecommunications.

Electromagnetic Spectrum

Light, colors, AM and FM radio, and electronic devices make use of the electromagnetic spectrum. The frequencies of the radio spectrum that are available for use for communication are treated as a public resource and are regulated by national organizations such as the Federal Communications Commission in the USA, or Ofcom in the United Kingdom. This determines which frequency ranges can be used for what purpose and by whom. In the absence of such control or alternative arrangements such as a privatized electromagnetic spectrum, chaos might result if, for example, airlines did not have specific frequencies to work under and an amateur radio operator were interfering with the pilot's ability to land an aircraft. Wireless communication spans the spectrum from 9 kHz to 300 GHz.

Applications of Wireless Technology

Mobile Telephones

One of the best-known examples of wireless technology is the mobile phone, also known as a cellular phone, with more than 4.6 billion mobile cellular subscriptions worldwide as of the end of 2010. These wireless phones use radio waves from signal-transmission towers to enable their users to make phone calls from many locations worldwide. They can be used within range of the mobile telephone site used to house the equipment required to transmit and receive the radio signals from these instruments.

Wireless Data Communications

Wireless data communications are an essential component of mobile computing. The various available technologies differ in local availability, coverage range and performance, and in some circumstances, users must be able to employ multiple connection types and switch between them. To simplify the experience for the user, connection manager software can be used, or a mobile VPN deployed to handle the multiple connections as a secure, single virtual network. Supporting technologies include:

Wi-Fi is a wireless local area network that enables portable computing devices to connect easily to the Internet. Standardized as IEEE 802.11 a,b,g,n, Wi-Fi approaches speeds of some types of wired Ethernet. Wi-Fi has become the de facto standard for access in private homes, within offices, and at public hotspots. Some businesses charge customers a monthly fee for service, while others have begun offering it for free in an effort to increase the sales of their goods.

Cellular data service offers coverage within a range of 10-15 miles from the nearest cell site. Speeds have increased as technologies have evolved, from earlier technologies such as GSM, CDMA and GPRS, to 3G networks such as W-CDMA, EDGE or CDMA2000.

Mobile Satellite Communications may be used where other wireless connections are unavailable, such as in largely rural areas or remote locations. Satellite communications are especially important for transportation, aviation, maritime and military use.

Wireless Sensor Networks are responsible for sensing noise, interference, and activity in data collection networks. This allows us to detect relevant quantities, monitor and collect data, formulate clear user displays, and to perform decision-making functions

Wireless Energy Transfer

Wireless energy transfer is a process whereby electrical energy is transmitted from a power source to an electrical load (Computer Load) that does not have a built-in power source, without the use of interconnecting wires. There are two different fundamental methods for wireless energy transfer. They can be transferred using either far-field methods that involve beaming power/lasers, radio or microwave transmissions or near-field using induction. Both methods utilize electromagnetism and magnetic fields.

Wireless Medical Technologies

New wireless technologies, such as mobile body area networks (MBAN), have the capability to monitor blood pressure, heart rate, oxygen level and body temperature. The MBAN works by sending low powered wireless signals to receivers that feed into nursing stations or monitoring sites. This technology helps with the intentional and unintentional risk of infection or disconnection that arise from wired connections.

Computer Interface Devices

Answering the call of customers frustrated with cord clutter, many manufacturers of computer peripherals turned to wireless technology to satisfy their consumer base. Originally these units used bulky, highly local transceivers to mediate between a computer and a keyboard and mouse; however, more recent generations have used small, high-quality devices, some even incorporating Bluetooth. These systems have become so ubiquitous that some users have begun complaining about a lack of wired peripherals. Wireless devices tend to have a slightly slower response time than their wired counterparts; however, the gap is decreasing.

A battery powers computer interface devices such as a keyboard or mouse and send signals to a receiver through a USB port by the way of a radio frequency (RF) receiver. The RF design makes it possible for signals to be transmitted wirelessly and expands the range of efficient use, usually up to 10 feet. Distance, physical obstacles, competing signals, and even human bodies can all degrade the signal quality.

Concerns about the security of wireless keyboards arose at the end of 2007, when it was revealed that Microsoft's implementation of encryption in some of its 27 MHz models was highly insecure.

Categories of Wireless Implementations, Devices and Standards

- Radio station in accordance with ITU RR

- Radiocommunication service in accordance with ITU RR

- Radio communication system

- Land Mobile Radio or Professional Mobile Radio: TETRA, P25, OpenSky, EDACS, DMR, dPMR

- Cordless telephony:DECT (Digital Enhanced Cordless Telecommunications)

- Cellular networks: 0G, 1G, 2G, 3G, Beyond 3G (4G), Future wireless

- List of emerging technologies

- Short-range point-to-point communication : Wireless microphones, Remote controls, IrDA, RFID (Radio Frequency Identification), TransferJet, Wireless USB, DSRC (Dedicated Short Range Communications), EnOcean, Near Field Communication

- Wireless sensor networks: ZigBee, EnOcean; Personal area networks, Bluetooth, TransferJet, Ultra-wideband (UWB from WiMedia Alliance).

- Wireless networks: Wireless LAN (WLAN), (IEEE 802.11 branded as Wi-Fi and Hiper-LAN), Wireless Metropolitan Area Networks (WMAN) and (LMDS, WiMAX, and Hiper-MAN)

Wireless Network

A wireless network is any type of computer network that uses wireless data connections for connecting network nodes.

Wireless networking is a method by which homes, telecommunications networks and enterprise (business) installations avoid the costly process of introducing cables into a building, or as a connection between various equipment locations. Wireless telecommunications networks are generally implemented and administered using radio communication. This implementation takes place at the physical level (layer) of the OSI model network structure.

Wireless icon

Examples of wireless networks include cell phone networks, Wireless local networks, wireless sensor networks, satellite communication networks, and terrestrial microwave networks.

History

Wireless Links

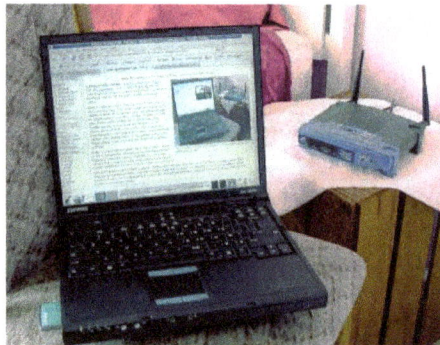

Computers are very often connected to networks using wireless links

- *Terrestrial microwave* – Terrestrial microwave communication uses Earth-based transmitters and receivers resembling satellite dishes. Terrestrial microwaves are in the low gigahertz range, which limits all communications to line-of-sight. Relay stations are spaced approximately 48 km (30 mi) apart.

- *Communications satellites* – Satellites communicate via microwave radio waves, which are not deflected by the Earth's atmosphere. The satellites are stationed in space, typically in geosynchronous orbit 35,400 km (22,000 mi) above the equator. These Earth-orbiting systems are capable of receiving and relaying voice, data, and TV signals.

- *Cellular and PCS systems* use several radio communications technologies. The systems divide the region covered into multiple geographic areas. Each area has a low-power transmitter or radio relay antenna device to relay calls from one area to the next area.

- *Radio and spread spectrum technologies* – Wireless local area networks use a high-frequency radio technology similar to digital cellular and a low-frequency radio technology. Wireless LANs use spread spectrum technology to enable communication between multiple devices in a limited area. IEEE 802.11 defines a common flavor of open-standards wireless radio-wave technology known as Wifi.

- *Free-space optical communication* uses visible or invisible light for communications. In most cases, line-of-sight propagation is used, which limits the physical positioning of communicating devices.

Types of Wireless Networks

Wireless PAN

Wireless personal area networks (WPANs) interconnect devices within a relatively small area, that is generally within a person's reach. For example, both Bluetooth radio and invisible infrared light provides a WPAN for interconnecting a headset to a laptop. ZigBee also supports WPAN applications. Wi-Fi PANs are becoming commonplace (2010) as equipment designers start to integrate Wi-Fi into a variety of consumer electronic devices. Intel "My WiFi" and Windows 7 "virtual Wi-Fi" capabilities have made Wi-Fi PANs simpler and easier to set up and configure.

Wireless LAN

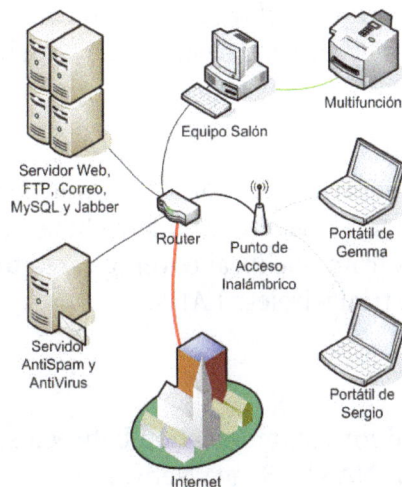

Wireless LANs are often used for connecting to local resources and to the Internet

A wireless local area network (WLAN) links two or more devices over a short distance using a wireless distribution method, usually providing a connection through an access point for internet access. The use of spread-spectrum or OFDM technologies may allow users to move around within a local coverage area, and still remain connected to the network.

Products using the IEEE 802.11 WLAN standards are marketed under the Wi-Fi brand name. Fixed wireless technology implements point-to-point links between computers or networks at two distant locations, often using dedicated microwave or modulated laser light beams over line of sight paths. It is often used in cities to connect networks in two or more buildings without installing a wired link.

Wireless Mesh Network

A wireless mesh network is a wireless network made up of radio nodes organized in a mesh topology. Each node forwards messages on behalf of the other nodes. Mesh networks can "self-heal", automatically re-routing around a node that has lost power.

Wireless MAN

Wireless metropolitan area networks are a type of wireless network that connects several wireless LANs.

  • WiMAX is a type of Wireless MAN and is described by the IEEE 802.16 standard.

Wireless WAN

Wireless wide area networks are wireless networks that typically cover large areas, such as between neighbouring towns and cities, or city and suburb. These networks can be used to connect branch offices of business or as a public Internet access system. The wireless connections between access points are usually point to point microwave links using parabolic dishes on the 2.4 GHz band, rather than omnidirectional antennas used with smaller networks. A typical system contains base station gateways, access points and wireless bridging relays. Other configurations are mesh systems where each access point acts as a relay also. When combined with renewable energy systems such as photovoltaic solar panels or wind systems they can be stand alone systems.

Global Area Network

A global area network (GAN) is a network used for supporting mobile across an arbitrary number of wireless LANs, satellite coverage areas, etc. The key challenge in mobile communications is handing off user communications from one local coverage area to the next. In IEEE Project 802, this involves a succession of terrestrial wireless LANs.

Space Network

Space networks are networks used for communication between spacecraft, usually in the vicinity of the Earth. The example of this is NASA's Space Network.

Different Uses

Some examples of usage include cellular phones which are part of everyday wireless networks, allowing easy personal communications. Another example, Intercontinental network systems, use radio satellites to communicate across the world. Emergency services such as the police utilize wireless networks to communicate effectively as well. Individuals and businesses use wireless networks to send and share data rapidly, whether it be in a small office building or across the world.

Properties

General

In a general sense, wireless networks offer a vast variety of uses by both business and home users.

"Now, the industry accepts a handful of different wireless technologies. Each wireless technology is defined by a standard that describes unique functions at both the Physical and the Data Link layers of the OSI model. These standards differ in their specified signaling methods, geographic ranges, and frequency usages, among other things. Such differences can make certain technologies better suited to home networks and others better suited to network larger organizations."

Performance

Each standard varies in geographical range, thus making one standard more ideal than the next depending on what it is one is trying to accomplish with a wireless network. The performance of wireless networks satisfies a variety of applications such as voice and video. The use of this technology also gives room for expansions, such as from 2G to 3G and, most recently, 4G technology, which stands for the fourth generation of cell phone mobile communications standards. As wireless networking has become commonplace, sophistication increases through configuration of network hardware and software, and greater capacity to send and receive larger amounts of data, faster, is achieved.

Space

Space is another characteristic of wireless networking. Wireless networks offer many advantages when it comes to difficult-to-wire areas trying to communicate such as across a street or river, a warehouse on the other side of the premises or buildings that are physically separated but operate as one. Wireless networks allow for users to designate a certain space which the network will be able to communicate with other devices through that network. Space is also created in homes as a result of eliminating clutters of wiring. This technology allows for an alternative to installing physical network mediums such as TPs, coaxes, or fiber-optics, which can also be expensive.

Home

For homeowners, wireless technology is an effective option compared to Ethernet for sharing printers, scanners, and high-speed Internet connections. WLANs help save the cost of installation of cable mediums, save time from physical installation, and also creates mobility for devices connected to the network. Wireless networks are simple and require as few as one single wireless access point connected directly to the Internet via a router.

Wireless Network Elements

The telecommunications network at the physical layer also consists of many interconnected wireline network elements (NEs). These NEs can be stand-alone systems or products that are either supplied by a single manufacturer or are assembled by the service provider (user) or system integrator with parts from several different manufacturers.

Wireless NEs are the products and devices used by a wireless carrier to provide support for the backhaul network as well as a mobile switching center (MSC).

Reliable wireless service depends on the network elements at the physical layer to be protected against all operational environments and applications.

What are especially important are the NEs that are located on the cell tower to the base station (BS) cabinet. The attachment hardware and the positioning of the antenna and associated closures and cables are required to have adequate strength, robustness, corrosion resistance, and resistance against wind, storms, icing, and other weather conditions. Requirements for individual components, such as hardware, cables, connectors, and closures, shall take into consideration the structure to which they are attached.

Difficulties

Interferences

Compared to wired systems, wireless networks are frequently subject to electromagnetic interference. This can be caused by other networks or other types of equipment that generate radio waves that are within, or close, to the radio bands used for communication. Interference can degrade the signal or cause the system to fail.

Absorption and Reflection

Some materials cause absorption of electromagnetic waves, preventing it from reaching the receiver, in other cases, particularly with metallic or conductive materials reflection occurs. This can cause dead zones where no reception is available. Aluminium foiled thermal isolation in modern homes can easily reduce indoor mobile signals by 10 dB frequently leading to complaints about the bad reception of long-distance rural cell signals.

Multipath Fading

In multipath fading two or more different routes taken by the signal, due to reflections, can cause the signal to cancel out at certain locations, and to be stronger in other places (upfade).

Hidden Node Problem

The hidden node problem occurs in some types of network when a node is visible from a wireless access point (AP), but not from other nodes communicating with that AP. This leads to difficulties in media access control.

Shared Resource Problem

The wireless spectrum is a limited resource and shared by all nodes in the range of its transmitters. Bandwidth allocation becomes complex with multiple participating users. Often users are not aware that advertised numbers (e.g., for IEEE 802.11 equipment or LTE networks) are not their capacity, but shared with all other users and thus the individual user rate is far lower. With increasing demand, the capacity crunch is more and more likely to happen. User-in-the-loop (UIL) may be an alternative solution to ever upgrading to newer technologies for over-provisioning.

Capacity

Channel

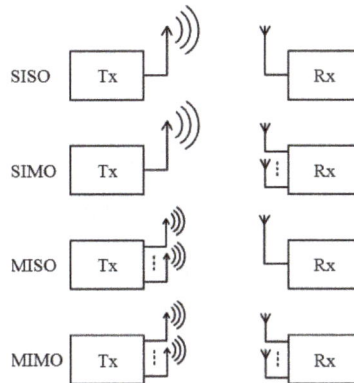

Understanding of SISO, SIMO, MISO and MIMO. Using multiple antennas and transmitting in different frequency channels can reduce fading, and can greatly increase the system capacity.

Shannon's theorem can describe the maximum data rate of any single wireless link, which relates to the bandwidth in hertz and to the noise on the channel.

One can greatly increase channel capacity by using MIMO techniques, where multiple aerials or multiple frequencies can exploit multiple paths to the receiver to achieve much higher throughput – by a factor of the product of the frequency and aerial diversity at each end.

Under Linux, the Central Regulatory Domain Agent (CRDA) controls the setting of channels.

Network

The total network bandwidth depends on how dispersive the medium is (more dispersive medium generally has better total bandwidth because it minimises interference), how many frequencies are available, how noisy those frequencies are, how many aerials are used and whether a directional antenna is in use, whether nodes employ power control and so on.

Cellular wireless networks generally have good capacity, due to their use of directional aerials, and their ability to reuse radio channels in non-adjacent cells. Additionally, cells can be made very small using low power transmitters this is used in cities to give network capacity that scales linearly with population density.

Safety

Wireless access points are also often close to humans, but the drop off in power over distance is fast, following the inverse-square law. The position of the United Kingdom's Health Protection Agency (HPA) is that "...radio frequency (RF) exposures from WiFi are likely to be lower than those from mobile phones." It also saw "...no reason why schools and others should not use WiFi equipment." In October 2007, the HPA launched a new "systematic" study into the effects of WiFi networks on behalf of the UK government, in order to calm fears that had appeared in the media in a recent period up to that time". Dr Michael Clark, of the HPA, says published research on mobile phones and masts does not add up to an indictment of WiFi.

Wireless LAN

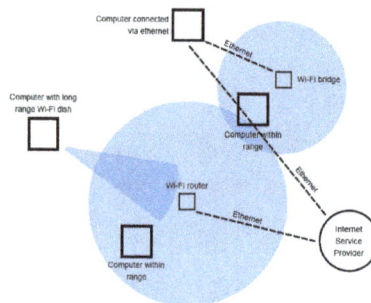

An example of a Wi-Fi network.

A wireless local area network (WLAN) is a wireless computer network that links two or more devices using a wireless distribution method (often spread-spectrum or OFDM radio) within a limited area such as a home, school, computer laboratory, or office building. This gives users the ability to move around within a local coverage area and still be connected to the network, and can provide a connection to the wider Internet. Most modern WLANs are based on IEEE 802.11 standards, marketed under the Wi-Fi brand name.

Wireless LANs have become popular in the home due to ease of installation and use, and in commercial complexes offering wireless access to their customers; often for free. New York City, for instance, has begun a pilot program to provide city workers in all five boroughs of the city with wireless Internet access.

An embedded RouterBoard 112 with U.FL-RSMA pigtail and R52 mini PCI Wi-Fi card widely used by wireless Internet service providers (WISPs)

History

Norman Abramson, a professor at the University of Hawaii, developed the world's first wireless computer communication network, ALOHAnet (operational in 1971), using low-cost ham-like radios. The system included seven computers deployed over four islands to communicate with the central computer on the Oahu Island without using phone lines.

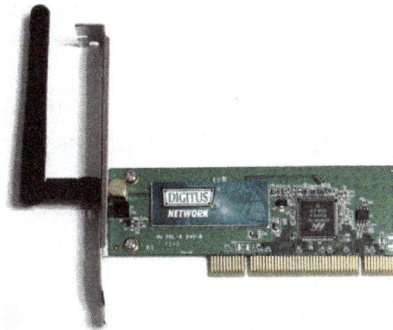

54 Mbit/s WLAN PCI Card (802.11g)

WLAN (Wireless Local Area Network) hardware initially cost so much that it was only used as an alternative to cabled LAN in places where cabling was difficult or impossible. Early development included industry-specific solutions and proprietary protocols, but at the end of the 1990s these were replaced by standards, primarily the various versions of IEEE 802.11 (in products using the Wi-Fi brand name). Beginning in 1991, a European alternative known as HiperLAN/1 was pursued by the European Telecommunications Standards Institute (ETSI) with a first version approved in 1996. This was followed by a HiperLAN/2 functional specification with ATM influences accomplished February 2000. Neither European standard achieved the commercial success of 802.11, although much of the work on HiperLAN/2 has survived in the PHY specification for IEEE 802.11a, which is nearly identical to the PHY of HiperLAN/2.

In 2009 802.11n was added to 802.11. It operates in both the 2.4 GHz and 5 GHz bands at a maximum data transfer rate of 600 Mbit/s. Most newer routers are able to utilise both wireless bands, known as dualband. This allows data communications to avoid the crowded 2.4 GHz band, which is also shared with Bluetooth devices and microwave ovens. The 5 GHz band is also wider than the 2.4 GHz band, with more channels, which permits a greater number of devices to share the space. Not all channels are available in all regions.

A HomeRF group formed in 1997 to promote a technology aimed for residential use, but it disbanded at the end of 2002.

Architecture

Stations

All components that can connect into a wireless medium in a network are referred to as stations (STA). All stations are equipped with wireless network interface controllers (WNICs). Wireless stations fall into one of two categories: wireless access points, and clients. Access points (APs), normally wireless routers, are base stations for the wireless network. They transmit and receive radio frequencies for wireless enabled devices to communicate with. Wireless clients can be mobile devices such as laptops, personal digital assistants, IP phones and other smartphones, or fixed devices such as desktops and workstations that are equipped with a wireless network interface.

Basic Service Set

The basic service set (BSS) is a set of all stations that can communicate with each other at PHY layer. Every BSS has an identification (ID) called the BSSID, which is the MAC address of the access point servicing the BSS.

There are two types of BSS: Independent BSS (also referred to as IBSS), and infrastructure BSS. An independent BSS (IBSS) is an ad hoc network that contains no access points, which means they cannot connect to any other basic service set.

Extended Service Set

An extended service set (ESS) is a set of connected BSSs. Access points in an ESS are connected by a distribution system. Each ESS has an ID called the SSID which is a 32-byte (maximum) character string.

Distribution System

A distribution system (DS) connects access points in an extended service set. The concept of a DS can be used to increase network coverage through roaming between cells.

DS can be wired or wireless. Current wireless distribution systems are mostly based on WDS or MESH protocols, though other systems are in use.

Types of Wireless Lans

The IEEE 802.11 has two basic modes of operation: infrastructure and **ad hoc** mode. In *ad hoc* mode, mobile units transmit directly peer-to-peer. In infrastructure mode, mobile units communicate through an access point that serves as a bridge to other networks (such as Internet or LAN).

Since wireless communication uses a more open medium for communication in comparison to wired LANs, the 802.11 designers also included encryption mechanisms: Wired Equivalent Privacy (WEP, now insecure), Wi-Fi Protected Access (WPA, WPA2), to secure wireless computer networks. Many access points will also offer Wi-Fi Protected Setup, a quick (but now insecure) method of joining a new device to an encrypted network.

Infrastructure

Most Wi-Fi networks are deployed in infrastructure mode.

In infrastructure mode, a base station acts as a wireless access point hub, and nodes communicate through the hub. The hub usually, but not always, has a wired or fiber network connection, and may have permanent wireless connections to other nodes.

Wireless access points are usually fixed, and provide service to their client nodes within range.

Wireless clients, such as laptops, smartphones etc. connect to the access point to join the network.

Sometimes a network will have a multiple access points, with the same 'SSID' and security arrangement. In that case connecting to any access point on that network joins the client to the network. In that case, the client software will try to choose the access point to try to give the best service, such as the access point with the strongest signal.

Peer-to-peer

Peer-to-Peer / Ad-Hoc

Peer-to-Peer or ad hoc wireless LAN

An ad hoc network (not the same as a WiFi Direct network) is a network where stations communicate only peer to peer (P2P). There is no base and no one gives permission to talk. This is accomplished using the Independent Basic Service Set (IBSS).

A WiFi Direct network is another type of network where stations communicate peer to peer.

In a Wi-Fi P2P group, the group owner operates as an access point and all other devices are clients. There are two main methods to establish a group owner in the Wi-Fi Direct group. In one approach, the user sets up a P2P group owner manually. This method is also known as Autonomous Group Owner (autonomous GO). In the second method, also called negotiation-based group creation, two devices compete based on the group owner intent value. The device with higher intent value becomes a group owner and the second device becomes a client. Group owner intent value can depend on whether the wireless device performs a cross-connection between an infrastructure WLAN service and a P2P group, remaining power in the wireless device, whether the wireless device is already a group owner in another group and/or a received signal strength of the first wireless device.

A peer-to-peer network allows wireless devices to directly communicate with each other. Wireless devices within range of each other can discover and communicate directly without involving central access points. This method is typically used by two computers so that they can connect to each other to form a network. This can basically occur in devices within a closed range.

If a signal strength meter is used in this situation, it may not read the strength accurately and can be misleading, because it registers the strength of the strongest signal, which may be the closest computer.

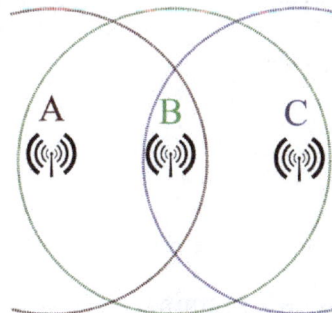

Hidden node problem: Devices A and C are both communicating with B, but are unaware of each other

IEEE 802.11 defines the physical layer (PHY) and MAC (Media Access Control) layers based on CSMA/CA (Carrier Sense Multiple Access with Collision Avoidance). The 802.11 specification includes provisions designed to minimize collisions, because two mobile units may both be in range of a common access point, but out of range of each other.

Bridge

"A bridge can be used to connect networks, typically of different types. A wireless Ethernet bridge allows the connection of devices on a wired Ethernet network to a wireless network. The bridge acts as the connection point to the Wireless LAN.

Wireless Distribution System

A Wireless Distribution System enables the wireless interconnection of access points in an IEEE 802.11 network. It allows a wireless network to be expanded using multiple access points without the need for a wired backbone to link them, as is traditionally required. The notable advantage of WDS over other solutions is that it preserves the MAC addresses of client packets across links between access points.

An access point can be either a main, relay or remote base station. A main base station is typically connected to the wired Ethernet. A relay base station relays data between remote base stations, wireless clients or other relay stations to either a main or another relay base station. A remote base station accepts connections from wireless clients and passes them to relay or main stations. Connections between "clients" are made using MAC addresses rather than by specifying IP assignments.

All base stations in a Wireless Distribution System must be configured to use the same radio channel, and share WEP keys or WPA keys if they are used. They can be configured to different service set identifiers. WDS also requires that every base station be configured to forward to others in the system as mentioned above.

WDS may also be referred to as repeater mode because it appears to bridge and accept wireless clients at the same time (unlike traditional bridging). It should be noted, however, that throughput in this method is halved for all clients connected wirelessly.

When it is difficult to connect all of the access points in a network by wires, it is also possible to put up access points as repeaters.

Roaming

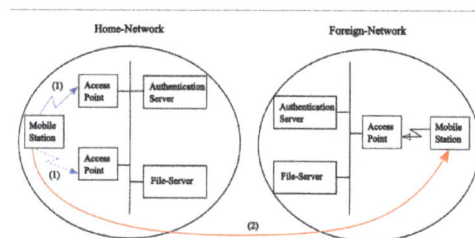

Roaming among Wireless Local Area Networks

There are two definitions for wireless LAN roaming:

- Internal Roaming: The Mobile Station (MS) moves from one access point (AP) to another AP within a home network because the signal strength is too weak. An authentication server (RADIUS) performs the re-authentication of MS via 802.1x (e.g. with PEAP). The billing of QoS is in the home network. A Mobile Station roaming from one access point to another often interrupts the flow of data among the Mobile Station and an application connected to the network. The Mobile Station, for instance, periodically monitors the presence of alternative access points (ones that will provide a better connection). At some point, based on proprietary mechanisms, the Mobile Station decides to re-associate with an access point having a stronger wireless signal. The Mobile Station, however, may lose a connection with an access point before associating with another access point. In order to provide reliable connections with applications, the Mobile Station must generally include software that provides session persistence.

- External Roaming: The MS (client) moves into a WLAN of another Wireless Internet Service Provider (WISP) and takes their services (Hotspot). The user can independently of his home network use another foreign network, if this is open for visitors. There must be special authentication and billing systems for mobile services in a foreign network.

Applications

Wireless LANs have a great deal of applications. Modern implementations of WLANs range from small in-home networks to large, campus-sized ones to completely mobile networks on airplanes and trains.

Users can access the Internet from WLAN hotspots in restaurants, hotels, and now with portable devices that connect to 3G or 4G networks. Oftentimes these types of public access points require no registration or password to join the network. Others can be accessed once registration has occurred and/or a fee is paid.

Existing Wireless LAN infrastructures can also be used to work as indoor positioning systems with no modification to the existing hardware.

Performance and Throughput

WLAN, organised in various layer 2 variants (IEEE 802.11), has different characteristics. Across all flavours of 802.11, maximum achievable throughputs are either given based on measurements under ideal conditions or in the layer 2 data rates. This, however, does not apply to typical deployments in which data are being transferred between two endpoints of which at least one is typically connected to a wired infrastructure and the other endpoint is connected to an infrastructure via a wireless link.

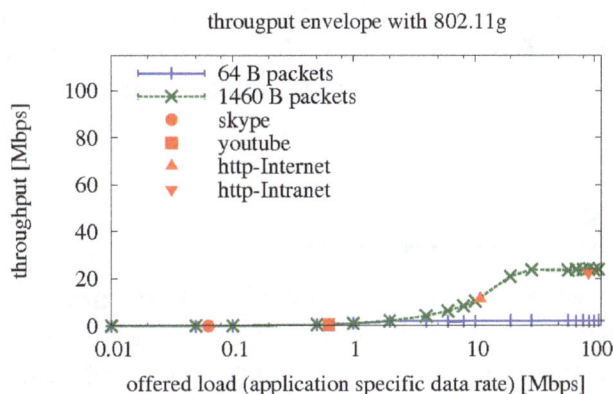

Graphical representation of Wi-Fi application specific (UDP) performance envelope 2.4 GHz band, with 802.11g

throughput envelope with 802.11n (40MHz Channelwidth)

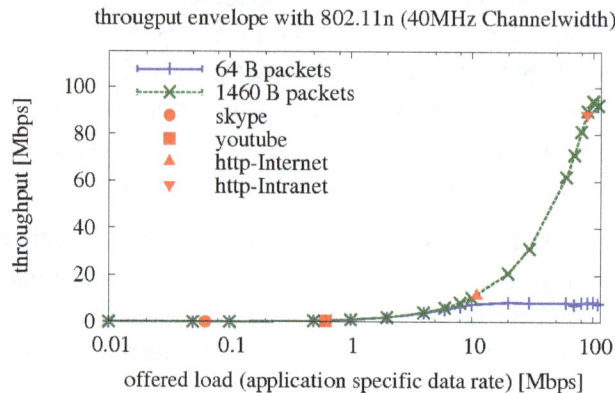

Graphical representation of Wi-Fi application specific (UDP) performance envelope 2.4 GHz band, with 802.11n with 40 MHz

This means that typically data frames pass an 802.11 (WLAN) medium and are being converted to 802.3 (Ethernet) or vice versa.

Due to the difference in the frame (header) lengths of these two media, the packet size of an application determines the speed of the data transfer. This means that an application which uses small packets (e.g. VoIP) creates a data flow with a high overhead traffic (e.g. a low goodput).

Other factors which contribute to the overall application data rate are the speed with which the application transmits the packets (i.e. the data rate) and the energy with which the wireless signal is received.

The latter is determined by distance and by the configured output power of the communicating devices.

Same references apply to the attached throughput graphs which show measurements of UDP throughput measurements. Each represents an average (UDP) throughput (the error bars are there, but barely visible due to the small variation) of 25 measurements.

Each is with a specific packet size (small or large) and with a specific data rate (10 kbit/s – 100 Mbit/s). Markers for traffic profiles of common applications are included as well. This text and measurements do not cover packet errors but information about this can be found at above references. The table below shows the maximum achievable (application specific) UDP throughput in the same scenarios (same references again) with various difference WLAN (802.11) flavours. The measurement hosts have been 25 meters apart from each other; loss is again ignored.

Wireless Access Point

In computer networking, a wireless access point (WAP) is a networking hardware device that allows a Wi-Fi compliant device to connect to a wired network. The WAP usually connects to a router (via a wired network) as a standalone device, but it can also be an integral component of the router itself. A WAP is differentiated from a hotspot, which is the physical location where Wi-Fi access to a WLAN is available.

Introduction

Linksys "WAP54G" 802.11g wireless access point

Prior to wireless networks, setting up a computer network in a business, home or school often required running many cables through walls and ceilings in order to deliver network access to all of the network-enabled devices in the building. With the creation of the wireless access point, network users are now able to add devices that access the network with few or no cables. A WAP normally connects directly to a wired Ethernet connection and the WAP then provides wireless connections using radio frequency links for other devices to utilize that wired connection. Most WAPs support the connection of multiple wireless devices to one wired connection. Modern WAPs are built to support a standard for sending and receiving data using these radio frequencies. Those standards, and the frequencies they use are defined by the IEEE. Most APs use IEEE 802.11 standards.

Common AP Applications

Typical corporate use involves attaching several WAPs to a wired network and then providing wireless access to the office LAN. The wireless access points are managed by a WLAN Controller which handles automatic adjustments to RF power, channels, authentication, and security. Furthermore, controllers can be combined to form a wireless mobility group to allow inter-controller roaming. The controllers can be part of a mobility domain to allow clients access throughout large or regional office locations. This saves the clients time and administrators overhead because it can automatically re-associate or re-authenticate.

A hotspot is a common public application of WAPs, where wireless clients can connect to the Internet without regard for the particular networks to which they have attached for the moment. The concept has become common in large cities, where a combination of coffeehouses, libraries, as well as privately owned open access points, allow clients to stay more or less continuously connected to the Internet, while moving around. A collection of connected hotspots can be referred to as a lily pad network.

WAPs are commonly used in home wireless networks. Home networks generally have only one AP to connect all the computers in a home. Most are wireless routers, meaning converged devices that include the WAP, a router, and, often, an Ethernet switch. Many also include a broadband modem. In places where most homes have their own WAP within range of the neighbours' AP, it's possible for technically savvy people to turn off their encryption and set up a wireless community network, creating an intra-city communication network although this does not negate the requirement for a wired network.

A WAP may also act as the network's arbitrator, negotiating when each nearby client device can transmit. However, the vast majority of currently installed IEEE 802.11 networks do not implement this, using a distributed pseudo-random algorithm called CSMA/CA instead.

Wireless Access Point vs. Ad Hoc Network

Some people confuse wireless access points with wireless ad hoc networks. An ad hoc network uses a connection between two or more devices without using a wireless access point: the devices communicate directly when in range. An ad hoc network is used in situations such as a quick data exchange or a multiplayer LAN game because setup is easy and does not require an access point. Due to its peer-to-peer layout, ad hoc connections are similar to Bluetooth ones.

But ad hoc connections are generally not recommended for a permanent installation. The reason is that Internet access via ad hoc networks, using features like Windows' Internet Connection Sharing, may work well with a small number of devices that are close to each other, but ad hoc networks don't scale well. Internet traffic will converge to the nodes with direct internet connection, potentially congesting these nodes. For internet-enabled nodes, access points have a clear advantage, with the possibility of having multiple access points connected by a wired LAN.

Limitations

One IEEE 802.11 AP can typically communicate with 30 client systems located within a radius of 103 metres. However, the actual range of communication can vary significantly, depending on such variables as indoor or outdoor placement, height above ground, nearby obstructions, other electronic devices that might actively interfere with the signal by broadcasting on the same frequency, type of antenna, the current weather, operating radio frequency, and the power output of devices. Network designers can extend the range of APs through the use of repeaters and reflectors, which can bounce or amplify radio signals that ordinarily would go un-received. In experimental conditions, wireless networking has operated over distances of several hundred kilometers.

Most jurisdictions have only a limited number of frequencies legally available for use by wireless networks. Usually, adjacent WAPs will use different frequencies (Channels) to communicate with their clients in order to avoid interference between the two nearby systems. Wireless devices can "listen" for data traffic on other frequencies, and can rapidly switch from one frequency to another to achieve better reception. However, the limited number of frequencies becomes problematic in crowded downtown areas with tall buildings using multiple WAPs. In such an environment, signal overlap becomes an issue causing interference, which results in signal droppage and data errors.

Wireless networking lags wired networking in terms of increasing bandwidth and throughput. While (as of 2013) high-density 256-QAM (TurboQAM) modulation, 3-antenna wireless devices for the consumer market can reach sustained real-world speeds of some 240 Mbit/s at 13 m behind two standing walls (NLOS) depending on their nature or 360 Mbit/s at 10 m line of sight or 380 Mbit/s at 2 m line of sight (IEEE 802.11ac) or 20 to 25 Mbit/s at 2 m line of sight (IEEE 802.11g), wired hardware of similar cost reaches somewhat less than 1000 Mbit/s up to specified distance of 100 m with twisted-pair cabling (Cat5, Cat5e, Cat6, or Cat7) (Gigabit Ethernet). One impediment to increasing the speed of wireless communications comes from Wi-Fi's use of a shared communications medium: Thus, two stations in infrastructure mode that are communicating with each other even over the same AP must have each and every frame transmitted twice: from the sender to the AP, then from the AP to the receiver. This approximately halves the effective bandwidth, so an AP is only able to use somewhat less than half the actual over-the-air rate for data throughput. Thus a typical 54 Mbit/s wireless connection actually carries TCP/IP data at 20 to 25 Mbit/s. Users of legacy wired networks expect faster speeds, and people using wireless connections keenly want to the wireless networks catch up.

By 2012, 802.11n based access points and client devices have already taken a fair share of the marketplace and with the finalization of the 802.11n standard in 2009 inherent problems integrating products from different vendors are less prevalent.

Security

Wireless access has special security considerations. Many wired networks base the security on physical access control, trusting all the users on the local network, but if wireless access points are connected to the network, anybody within range of the AP (which typically extends farther than the intended area) can attach to the network.

The most common solution is wireless traffic encryption. Modern access points come with built-in encryption. The first generation encryption scheme 'WEP' proved easy to crack; the second and third generation schemes, WPA and WPA2, are considered secure if a strong enough password or passphrase is used.

Some APs support hotspot style authentication using RADIUS and other authentication servers.

Opinions about wireless network security vary widely. Bruce Schneier asserted the net benefits of open Wi-Fi without passwords outweigh the risks, a position supported in 2014 by Peter Eckersley of the Electronic Frontier Foundation.

The opposite position was taken by Nick Mediati in an article for *PC World*, in which he takes the position that every wireless access point should be locked down with a password.

Wireless WAN

A wireless wide area network (WWAN), is a form of wireless network. The larger size of a wide area network compared to a local area network requires differences in technology. Wireless networks of all sizes deliver data in the form of telephone calls, web pages, and streaming video.

A WWAN often differs from wireless local area network (WLAN) by using mobile telecommunication cellular network technologies such as LTE, WiMAX (often called a wireless metropolitan area network or WMAN), UMTS, CDMA2000, GSM, cellular digital packet data (CDPD) and Mobitex to transfer data. It can also use Local Multipoint Distribution Service (LMDS) or Wi-Fi to provide Internet access. These technologies are offered regionally, nationwide, or even globally and are provided by a wireless service provider. WWAN connectivity allows a user with a laptop and a WWAN card to surf the web, check email, or connect to a virtual private network (VPN) from anywhere within the regional boundaries of cellular service. Various computers can have integrated WWAN capabilities.

A WWAN may also be a closed network that covers a large geographic area. For example, a mesh network or MANET with nodes on building, tower, trucks, and planes could also considered a WWAN.

Since radio communications systems do not provide a physically secure connection path, WWANs typically incorporate encryption and authentication methods to make them more secure. Unfortunately some of the early GSM encryption techniques were flawed, and security experts have issued warnings that cellular communication, including WWAN, is no longer secure. UMTS (3G) encryption was developed later and has yet to be broken.

Wireless Mesh Network

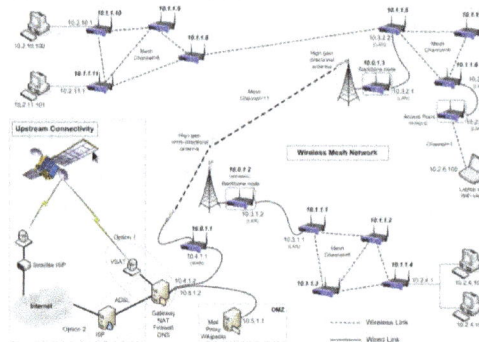

Diagram showing a possible configuration for a wireless mesh network, connected upstream via a VSAT link (click to enlarge)

A wireless mesh network (WMN) is a communications network made up of radio nodes organized in a mesh topology. It is also a form of wireless ad hoc network. Wireless mesh networks often consist of mesh clients, mesh routers and gateways. The mesh clients are often laptops, cell phones and other wireless devices while the mesh routers forward traffic to and from the gateways which may, but need not, be connected to the Internet. The coverage area of the radio nodes working as a single network is sometimes called a mesh cloud. Access to this mesh cloud is dependent on the radio nodes working in harmony with each other to create a radio network. A mesh network is reliable and offers redundancy. When one node can no longer operate, the rest of the nodes can still communicate with each other, directly or through one or more intermediate nodes. Wireless mesh networks can self form and self heal. Wireless mesh networks can be implemented with various wireless technologies including 802.11, 802.15, 802.16, cellular technologies and need not be restricted to any one technology or protocol.

History

Architecture

Wireless mesh architecture is a first step towards providing cost effective and dynamic high-bandwidth networks over a specific coverage area. Wireless mesh infrastructure is, in effect, a network of routers minus the cabling between nodes. It's built of peer radio devices that don't have to be cabled to a wired port like traditional WLAN access points (AP) do. Mesh infrastructure carries data over large distances by splitting the distance into a series of short hops. Intermediate nodes not only boost the signal, but cooperatively pass data from point A to point B by making forwarding decisions based on their knowledge of the network, i.e. perform routing. Such an architecture may, with careful design, provide high bandwidth, spectral efficiency, and economic advantage over the coverage area.

Wireless mesh networks have a relatively stable topology except for the occasional failure of nodes or addition of new nodes. The path of traffic, being aggregated from a large number of end users, changes infrequently. Practically all the traffic in an infrastructure mesh network is either forwarded to or from a gateway, while in ad hoc networks or client mesh networks the traffic flows between arbitrary pairs of nodes.

Management

This type of infrastructure can be decentralized (with no central server) or centrally managed (with a central server). Both are relatively inexpensive, and can be very reliable and resilient, as each node needs only transmit as far as the next node. Nodes act as routers to transmit data from nearby nodes to peers that are too far away to reach in a single hop, resulting in a network that can span larger distances. The topology of a mesh network is also reliable, as each node is connected to several other nodes. If one node drops out of the network, due to hardware failure or any other reason, its neighbors can quickly find another route using a routing protocol.

Applications

Mesh networks may involve either fixed or mobile devices. The solutions are as diverse as communication needs, for example in difficult environments such as emergency situations, tunnels, oil rigs, battlefield surveillance, high-speed mobile-video applications on board public transport or real-time racing-car telemetry. An important possible application for wireless mesh networks is VoIP. By using a Quality of Service scheme, the wireless mesh may support local telephone calls to be routed through the mesh.

Some current applications:

- U.S. military forces are now using wireless mesh networking to connect their computers, mainly ruggedized laptops, in field operations.

- Electric meters now being deployed on residences transfer their readings from one to another and eventually to the central office for billing without the need for human meter readers or the need to connect the meters with cables.

- The laptops in the One Laptop per Child program use wireless mesh networking to enable students to exchange files and get on the Internet even though they lack wired or cell phone or other physical connections in their area.

- The 66-satellite Iridium constellation operates as a mesh network, with wireless links between adjacent satellites. Calls between two satellite phones are routed through the mesh, from one satellite to another across the constellation, without having to go through an earth station. This makes for a smaller travel distance for the signal, reducing latency, and also allows for the constellation to operate with far fewer earth stations than would be required for 66 traditional communications satellites.

Operation

The principle is similar to the way packets travel around the wired Internet – data will hop from one device to another until it eventually reaches its destination. Dynamic routing algorithms implemented in each device allow this to happen. To implement such dynamic routing protocols, each device needs to communicate routing information to other devices in the network. Each device then determines what to do with the data it receives – either pass it on to the next device or keep it, depending on the protocol. The routing algorithm used should attempt to always ensure that the data takes the most appropriate (fastest) route to its destination.

Multi-radio Mesh

Multi-radio mesh refers to a unique pair of dedicated radios on each end of the link. This means there is a unique frequency used for each wireless hop and thus a dedicated CSMA collision domain. This is a true mesh link where you can achieve maximum performance without bandwidth degradation in the mesh and without adding latency. Thus voice and video applications work just as they would on a wired Ethernet network. In true 802.11 networks, there is no concept of a mesh. There are only APs and Stations. A multi-radio wireless mesh node will dedicate one of the radios to act as a station, and connect to a neighbor node AP radio.

Research Topics

One of the more often cited papers on Wireless Mesh Networks identified the following areas as open research problems in 2005

- New modulation scheme

    - In order to achieve higher transmission rate, new wideband transmission schemes other than OFDM and UWB are needed.

- Advanced antenna processing

    - Advanced antenna processing including directional, smart and multiple antenna technologies is further investigated, since their complexity and cost are still too high for wide commercialization.

- Flexible spectrum management

    - Tremendous efforts on research of frequency-agile techniques are being performed for increased efficiency.

- Cross-layer optimization

    - Cross-layer research is a popular current research topic where information is shared between different communications layers in order to increase the knowledge and current state of the network. This could enable new and more efficient protocols to be developed. A joint protocol which combines various design problems like routing, scheduling, channel assignment etc. can achieve higher performance since it is proven that these problems are strongly co-related. It is important to note that careless cross-layer design could lead to code which is difficult to maintain and extend.

- Software-defined wireless networking

  - Centralized, distributed, or hybrid? - In a new SDN architecture for WDNs is explored that eliminates the need for multi-hop flooding of route information and therefore enables WDNs to easily expand. The key idea is to split network control and data forwarding by using two separate frequency bands. The forwarding nodes and the SDN controller exchange link-state information and other network control signaling in one of the bands, while actual data forwarding takes place in the other band.

Protocols

Routing Protocols

There are more than 70 competing schemes for routing packets across mesh networks. Some of these include:

- AODV (Ad hoc On-Demand Distance Vector)

- B.A.T.M.A.N. (Better Approach To Mobile Adhoc Networking)

- Babel (protocol) (a distance-vector routing protocol for IPv6 and IPv4 with fast convergence properties)

- DNVR (Dynamic NIx-Vector Routing)

- DSDV (Destination-Sequenced Distance-Vector Routing)

- DSR (Dynamic Source Routing)

- HSLS (Hazy-Sighted Link State)

- HWMP (Hybrid Wireless Mesh Protocol)

- IWMP (Infrastructure Wireless Mesh Protocol) for Infrastructure Mesh Networks by GRECO UFPB-Brazil

- Wireless mesh networks routing protocol (MRP) by Jangeun Jun and Mihail L. Sichitiu

- OLSR (Optimized Link State Routing protocol)

- OORP (OrderOne Routing Protocol) (OrderOne Networks Routing Protocol)

- OSPF (Open Shortest Path First Routing)

- Routing Protocol for Low-Power and Lossy Networks (IETF ROLL RPL protocol, RFC 6550)

- PWRP (Predictive Wireless Routing Protocol)

- TORA (Temporally-Ordered Routing Algorithm)

- ZRP (Zone Routing Protocol)

The IEEE is developing a set of standards under the title 802.11s to define an architecture and protocol for ESS Mesh Networking.

A less thorough list can be found at Ad hoc routing protocol list.

Autoconfiguration Protocols

Standard autoconfiguration protocols, such as DHCP or IPv6 stateless autoconfiguration may be used over mesh networks.

Mesh network specific autoconfiguration protocols include:

- Ad Hoc Configuration Protocol (AHCP)

- Proactive Autoconfiguration (Proactive Autoconfiguration Protocol)

- Dynamic WMN Configuration Protocol (DWCP)

Communities and Providers

- CUWiN

- Freifunk (DE) / FunkFeuer (AT) / OpenWireless (CH)

- Firetide

- Guifi.net

- Netsukuku

- Ninux (IT)

- Senceive

Hotspot (Wi-Fi)

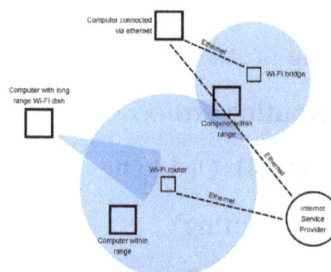

A diagram showing a Wi-Fi network

A hotspot is a physical location where people may obtain Internet access, typically using Wi-Fi technology, via a wireless local area network (WLAN) using a router connected to an internet service provider.

Public hotspots may be found in an increasing number of businesses for use of customers in many developed urban areas throughout the world, such as coffee shops. Many hotels offer wifi access to guests, either in guest rooms or in the lobby. Hotspots differ from wireless access points, which are the hardware devices used to provide a wireless network service. Private hotspots allow Internet access to a device (such as a tablet) via another device which may have data access via say a mobile device.

History

Public park in Brooklyn, New York, has free Wi-Fi from a local corporation

Public access wireless local area networks (LANs) were first proposed by Henrik Sjödin at the NetWorld+Interop conference in The Moscone Center in San Francisco in August 1993. Sjödin did not use the term hotspot but referred to publicly accessible wireless LANs.

The first commercial venture to attempt to create a public local area access network was a firm founded in Richardson, Texas known as PLANCOM (Public Local Area Network Communications). The founders of the venture, Mark Goode, Greg Jackson, and Brett Stewart dissolved the firm in 1998, while Goode and Jackson created MobileStar Networks. The firm was one of the first to sign such public access locations as Starbucks, American Airlines, and Hilton Hotels. The company was sold to Deutsche Telecom in 2001, who then converted the name of the firm into "T-Mobile Hotspot." It was then that the term "hotspot" entered the popular vernacular as a reference to a location where a publicly accessible wireless LAN is available.

ABI Research reported there was a total of 4.9 million global Wi-Fi hotspots in 2012 and projected that number would surpass 6.3 million by the end of 2013. The latest Wireless Broadband Alliance (WBA) Industry Report outlines a positive scenario for the Wi-Fi market: a steady annual increase from 5.2m public hotspots in 2012 to 10.5m public hotspots in 2018. Collectively, WBA operator members serve more than 1 billion subscribers and operate more than 15 million hotspots globally.

Uses

The public can use a laptop or other suitable portable device to access the wireless connection (usually Wi-Fi) provided. Of the estimated 150 million laptops, 14 million PDAs, and other emerging Wi-Fi devices sold per year for the last few years, most include the Wi-Fi feature.

For venues that have broadband Internet access, offering wireless access is as simple as configuring one access point (AP), in conjunction with a router and connecting the AP to the Internet connection. A single wireless router combining these functions may suffice.

The iPass 2014 interactive map, that shows data provided by the analysts Maravedis Rethink, shows that in December 2014 there are 46,000,000 hotspots worldwide and more than 22,000,000 roamable hotspots. More than 10,900 hotspots are on trains, planes and airports (Wi-Fi in motion) and more than 8,500,000 are "branded" hotspots (retail, cafés, hotels). The region with the largest number of public hotspots is Europe, followed by North America and Asia.

Security

Security is a serious concern in connection with Hotspots. There are three possible attack vectors. First, there is the wireless connection between the client and the access point. This needs to be encrypted, so that the connection cannot be eavesdropped or attacked by a man-in-the-middle-attack. Second, there is the Hotspot itself. The WLAN encryption ends at the interface, then travels its network stack unencrypted and then travels over the wired connection up to the BRAS of the ISP. Third, there is the connection from the Access Point to the BRAS of the ISP.

The safest method when accessing the Internet over a Hotspot, with unknown security measures, is end-to-end encryption. Examples of strong end-to-end encryption are HTTPS and SSH.

Locations

Hotspots are often found at airports, bookstores, coffee shops, department stores, fuel stations, hotels, hospitals, libraries, public pay phones, restaurants, RV parks and campgrounds, supermarkets, train stations, and other public places. Additionally, many schools and universities have wireless networks in their campuses.

Types

Free hotspots operate in two ways:

- Using an open public network is the easiest way to create a free hotspot. All that is needed is a Wi-Fi router. Similarly, when users of private wireless routers turn off their authentication requirements, opening their connection, intentionally or not, they permit piggybacking (sharing) by anyone in range.

- Closed public networks use a HotSpot Management System to control access to hotspots. This software runs on the router itself or an external computer allowing operators to authorize only specific users to access the Internet. Providers of such hotspots often associate the free access with a menu, membership, or purchase limit. Operators may also limit each user's available bandwidth (upload and download speed) to ensure that everyone gets a good quality service. Often this is done through service-level agreements.

Commercial Hotspots

A commercial hotspot may feature:

- A captive portal / login screen / splash page that users are redirected to for authentication and/or payment. The captive portal / splash page sometimes includes the social login buttons.

- A payment option using a credit card, iPass, PayPal, or another payment service (voucher-based Wi-Fi)

- A walled garden feature that allows free access to certain sites

- Service-oriented provisioning to allow for improved revenue

- Data analytics and data capture tools, to analyze and export data from Wi-Fi clients

Many services provide payment services to hotspot providers, for a monthly fee or commission from the end-user income. For example, Amazingports can be used to set up hotspots that intend to offer both fee-based and free internet access, and ZoneCD is a Linux distribution that provides payment services for hotspot providers who wish to deploy their own service.

Major airports and business hotels are more likely to charge for service, though most hotels provide free service to guests; and increasingly, small airports and airline lounges offer free service.. Retail shops, public venues and offices usually provide a free Wi-Fi SSID for their guests and visitors.

Roaming services are expanding among major hotspot service providers. With roaming service the users of a commercial provider can have access to other providers' hotspots, either free of charge or for extra fees, which users will usually be charged on an access-per-minute basis.

Software Hotspots

Many Wi-Fi adapters built into or easily added to consumer computers and mobile devices include the functionality to operate as private or mobile hotspots, sometimes referred to as "mi-fi". The use of a private hotspot to enable other personal devices to access the WAN (usually but not always the Internet) is a form of bridging, and known as tethering. Manufacturers and firmware creators can enable this functionality in Wi-Fi devices on many Wi-Fi devices, depending upon the capabilities of the hardware, and most modern consumer operating systems, including Android, Apple OS X 10.6 and later, Windows mobile, and Linux include features to support this. Additionally wireless chipset manufacturers such as Atheros, Broadcom, Intel and others, may add the capability for certain Wi-Fi NICs, usually used in a client role, to also be used for hotspot purposes. However, some service providers, such as AT&T, Sprint, and T-Mobile charge users for this service or prohibit and disconnect user connections if tethering is detected.

Third-party software vendors offer applications to allow users to operate their own hotspot, whether to access the Internet when on the go, share an existing connection, or extend the range of another hotspot. Third party implementations of software hotspots include:

- AmazingPorts Hotspot software

- Antamedia HotSpot software

- Connectify Hotspot

- Jaze Hotspot Gateway by Jaze Networks

- Hot Spot Network Manager (HSNM)

- Virtual Router

- Tanaza

- Start Hotspot software

Hotspot 2.0

Hotspot 2.0, also known as HS2 and Wi-Fi Certified Passpoint, is an approach to public access Wi-Fi by the Wi-Fi Alliance. The idea is for mobile devices to automatically join a Wi-Fi subscriber service whenever the user enters a Hotspot 2.0 area, in order to provide better bandwidth and services-on-demand to end-users and relieve carrier infrastructure of some traffic.

Hotspot 2.0 is based on the IEEE 802.11u standard, which is a set of protocols published in 2011 to enable cellular-like roaming. If the device supports 802.11u and is subscribed to a Hotspot 2.0 service it will automatically connect and roam.

Supported Devices

- Some Chinese tablet computers

- Some THL smartphones

- Apple mobile devices running iOS 7 and up

- Some Samsung Galaxy smartphones

- Windows 10 devices have full support for network discovery and connection

- Windows 8 and Windows 8.1 lack network discovery, but supports connecting to a network when the credentials are known

Billing

EDCF User-Priority-List

		Net traffic					
		low			high		
		Audio	Video	Data	Audio	Video	Data
User needs	time-critical	7	5	0	6	4	0
	not time-critical	-	-	2	-	-	2

The so-called "User-Fairness-Model " is a dynamic billing model, which allows a volume-based billing, charged only by the amount of payload (data, video, audio). Moreover, the tariff is classified by net traffic and user needs (Pommer, p. 116ff).

If the net traffic increases, then the user has to pay the next higher tariff class. By the way the user is asked for if he still wishes the session also by a higher traffic class.[dubious – discuss] Moreover, in time-critical applications (video, audio) a higher class fare is charged, than for non time-critical applications (such as reading Web pages, e-mail).

Tariff classes of the User-Fairness-Model

		Net traffic	
		low	high
User needs	time-critical	standard	exclusive
	not time-critical	low priced	standard

The "User-fairness model" can be implemented with the help of EDCF (IEEE 802.11e). A EDCF user priority list shares the traffic in 3 access categories (data, video, audio) and user priorities (UP) (Pommer, p. 117):

- Data [UP 0|2]
- Video [UP 5|4]
- Audio [UP 7|6]

Service-oriented provisioning for viable implementations

Security Concerns

Some hotspots authenticate users; however, this does not prevent users from viewing network traffic using packet sniffers.

Some vendors provide a download option that deploys WPA support. This conflicts with enterprise configurations that have solutions specific to their internal WLAN.

In order to provide robust security to hotspot users, the Wi-Fi Alliance is developing a new hotspot program that aims to encrypt hotspot traffic with WPA2 security. The program was scheduled to launch in the first half of 2012.

Legal Concerns

Depending upon the location, providers of public hotspot access may have legal obligations, related to privacy requirements and liability for use for unlawful purposes. In countries where the internet is regulated or freedom of speech more restricted, there may be requirements such as licensing, logging, or recording of user information. Concerns may also relate to child safety, and social issues such as exposure to objectionable content, protection against cyberbullying and illegal behaviours, and prevention of perpetration of such behaviors by hotspot users themselves.

European Union

- Data Retention Directive Hotspot owners must retain key user statistics for 12 months.

- Directive on Privacy and Electronic Communications

United Kingdom

- Data Protection Act 1998 The hotspot owner must retain individual's information within the confines of the law.

- Digital Economy Act 2010 Deals with, among other things, copyright infringement, and imposes fines of up to £250,000 for contravention.

Li-Fi

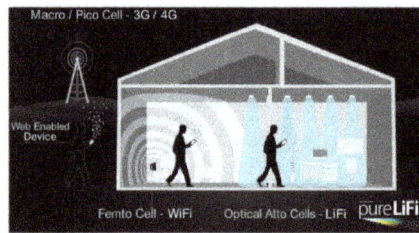

LiFi works in complement with existing and emerging wireless systems.

Light Fidelity (Li-Fi) is a bidirectional, high-speed and fully networked wireless communication technology similar to Wi-Fi. The term was coined by Harald Haas and is a form of visible light communication and a subset of optical wireless communications (OWC) and could be a complement to RF communication (Wi-Fi or cellular networks), or even a replacement in contexts of data broadcasting.

It is wireless and uses visible-light communication or infrared and near-ultraviolet instead of radio-frequency spectrum, part of optical wireless communications technology, which carries much more information, and has been proposed as a solution to the RF-bandwidth limitations.

Technology Details

This OWC technology uses light from light-emitting diodes (LEDs) as a medium to deliver networked, mobile, high-speed communication in a similar manner to Wi-Fi. The Li-Fi market is projected to have a compound annual growth rate of 82% from 2013 to 2018 and to be worth over $6 billion per year by 2018.

Visible light communications (VLC) works by switching the current to the LEDs off and on at a very high rate, too quick to be noticed by the human eye. Although Li-Fi LEDs would have to be kept on to transmit data, they could be dimmed to below human visibility while still emitting enough light to carry data. The light waves cannot penetrate walls which makes a much shorter range, though more secure from hacking, relative to Wi-Fi. Direct line of sight is not necessary for Li-Fi to transmit a signal; light reflected off the walls can achieve 70 Mbit/s.

Li-Fi has the advantage of being useful in electromagnetic sensitive areas such as in aircraft cabins, hospitals and nuclear power plants without causing electromagnetic interference. Both Wi-Fi and Li-Fi transmit data over the electromagnetic spectrum, but whereas Wi-Fi utilizes radio waves, Li-Fi uses visible light. While the US Federal Communications Commission has warned of a potential spectrum crisis because Wi-Fi is close to full capacity, Li-Fi has almost no limitations on capacity. The visible light spectrum is 10,000 times larger than the entire radio frequency spectrum. Researchers have reached data rates of over 10 Gbit/s, which is much faster than typical fast broadband in 2013. Li-Fi is expected to be ten times cheaper than Wi-Fi. Short range, low reliability and high installation costs are the potential downsides.

PureLiFi demonstrated the first commercially available Li-Fi system, the Li-1st, at the 2014 Mobile World Congress in Barcelona.

Bg-Fi is a Li-Fi system consisting of an application for a mobile device, and a simple consumer product, like an IoT (Internet of Things) device, with color sensor, microcontroller, and embedded software. Light from the mobile device display communicates to the color sensor on the consumer product, which converts the light into digital information. Light emitting diodes enable the consumer product to communicate synchronously with the mobile device.

History

Harald Haas, who teaches at the University of Edinburgh in Scotland, coined the term "Li-Fi" at his TED Global Talk where he introduced the idea of "Wireless data from every light". He is Chairman of Mobile Communications at the University of Edinburgh and co-founder of pureLiFi.

The general term visible light communication (VLC), whose history dates back to the 1880s, includes any use of the visible light portion of the electromagnetic spectrum to transmit information. The D-Light project at Edinburgh's Institute for Digital Communications was funded from January 2010 to January 2012. Haas promoted this technology in his 2011 TED Global talk and helped start a company to market it. PureLiFi, formerly pureVLC, is an original equipment manufacturer (OEM) firm set up to commercialize Li-Fi products for integration with existing LED-lighting systems.

In October 2011, companies and industry groups formed the Li-Fi Consortium, to promote high-speed optical wireless systems and to overcome the limited amount of radio-based wireless spectrum available by exploiting a completely different part of the electromagnetic spectrum.

A number of companies offer uni-directional VLC products, which is not the same as Li-Fi - a term defined by the IEEE 802.15.7r1 standardization committee.

VLC technology was exhibited in 2012 using Li-Fi. By August 2013, data rates of over 1.6 Gbit/s were demonstrated over a single color LED. In September 2013, a press release said that Li-Fi, or VLC systems in general, do not require line-of-sight conditions. In October 2013, it was reported Chinese manufacturers were working on Li-Fi development kits.

In April 2014, the Russian company Stins Coman announced the development of a Li-Fi wireless local network called BeamCaster. Their current module transfers data at 1.25 gigabytes per second but they fore boosting speeds up to 5 GB/second in the near future. In 2014 a new record was established by Sisoft (a Mexican company) that was able to transfer data at speeds of up to 10 Gbit/s across a light spectrum emitted by LED lamps.

Standards

Like Wi-Fi, Li-Fi is wireless and uses similar 802.11 protocols; but it uses visible light communication (instead of radio frequency waves), which has much wider bandwidth.

One part of VLC is modeled after communication protocols established by the IEEE 802 workgroup. However, the IEEE 802.15.7 standard is out-of-date, it fails to consider the latest technological developments in the field of optical wireless communications, specifically with the introduction of optical orthogonal frequency-division multiplexing (O-OFDM) modulation methods which have been optimized for data rates, multiple-access and energy efficiency. The introduction of O-OFDM means that a new drive for standardization of optical wireless communications is required.

Nonetheless, the IEEE 802.15.7 standard defines the physical layer (PHY) and media access control (MAC) layer. The standard is able to deliver enough data rates to transmit audio, video and multimedia services. It takes into account optical transmission mobility, its compatibility with artificial lighting present in infrastructures, and the interference which may be generated by ambient lighting. The MAC layer permits using the link with the other layers as with the TCP/IP protocol.

The standard defines three PHY layers with different rates:

- The PHY I was established for outdoor application and works from 11.67 kbit/s to 267.6 kbit/s.

- The PHY II layer permits reaching data rates from 1.25 Mbit/s to 96 Mbit/s.

- The PHY III is used for many emissions sources with a particular modulation method called color shift keying (CSK). PHY III can deliver rates from 12 Mbit/s to 96 Mbit/s.

The modulation formats recognized for PHY I and PHY II are on-off keying (OOK) and variable pulse position modulation (VPPM). The Manchester coding used for the PHY I and PHY II layers includes the clock inside the transmitted data by representing a logic 0 with an OOK symbol "01" and a logic 1 with an OOK symbol "10", all with a DC component. The DC component avoids light extinction in case of an extended run of logic 0's.

The first VLC smartphone prototype was presented at the Consumer Electronics Show in Las Vegas from January 7–10 in 2014. The phone uses SunPartner's Wysips CONNECT, a technique that converts light waves into usable energy, making the phone capable of receiving and decoding signals without drawing on its battery. A clear thin layer of crystal glass can be added to small screens like watches and smartphones that make them solar powered. Smartphones could gain 15% more battery life during a typical day. This first smartphones using this technology should arrive in 2015. This screen can also receive VLC signals as well as the smartphone camera. The cost of these screens per smartphone is between $2 and $3, much cheaper than most new technology.

Philips lighting company has developed a VLC system for shoppers at stores. They have to download an app on their smartphone and then their smartphone works with the LEDs in the store. The

LEDs can pinpoint where they are located in the store and give them corresponding coupons and information based on which aisle they are on and what they are looking at.

Bluetooth

Bluetooth is a wireless technology standard for exchanging data over short distances (using short-wavelength UHF radio waves in the ISM band from 2.4 to 2.485 GHz) from fixed and mobile devices, and building personal area networks (PANs). Invented by telecom vendor Ericsson in 1994, it was originally conceived as a wireless alternative to RS-232 data cables. It can connect several devices, overcoming problems of synchronization.

Bluetooth is managed by the Bluetooth Special Interest Group (SIG), which has more than 25,000 member companies in the areas of telecommunication, computing, networking, and consumer electronics. The IEEE standardized Bluetooth as IEEE 802.15.1, but no longer maintains the standard. The Bluetooth SIG overs development of the specification, manages the qualification program, and protects the trademarks. A manufacturer must make a device meet Bluetooth SIG standards to market it as a Bluetooth device. A network of patents apply to the technology, which are licensed to individual qualifying devices.

Origin

The development of the "short-link" radio technology, later named Bluetooth, was initiated in 1989 by Dr. Nils Rydbeck, CTO at Ericsson Mobile in Lund, and Dr. Johan Ullman. The purpose was to develop wireless headsets, according to two inventions by Johan Ullman, SE 8902098-6, issued 1989-06-12 and SE 9202239, issued 1992-07-24. Nils Rydbeck tasked Tord Wingren with specifying and Jaap Haartsen and Sven Mattisson with developing. Both were working for Ericsson in Lund, Sweden. The specification is based on frequency-hopping spread spectrum technology.

Name and Logo

The name "Bluetooth" is an Anglicised version of the Scandinavian *Blåtand/Blåtann* (Old Norse *blát☐nn*), the epithet of the tenth-century king Harald Bluetooth who united dissonant Danish tribes into a single kingdom and, according to legend, introduced Christianity as well. The idea of this name was proposed in 1997 by Jim Kardach who developed a system that would allow mobile phones to communicate with computers. At the time of this proposal he was reading Frans G. Bengtsson's historical novel *The Long Ships* about Vikings and King Harald Bluetooth. The implication is that Bluetooth does the same with communications protocols, uniting them into one universal standard.

The Bluetooth logo is a bind rune merging the Younger Futhark runes ᚼ (Hagall) and ᛒ (Bjarkan) , Harald's initials.

Implementation

Bluetooth operates at frequencies between 2402 and 2480 MHz, or 2400 and 2483.5 MHz including guard bands 2 MHz wide at the bottom end and 3.5 MHz wide at the top. This is in the globally unlicensed (but not unregulated) Industrial, Scientific and Medical (ISM) 2.4 GHz short-range radio frequency band. Bluetooth uses a radio technology called frequency-hopping spread spectrum. Bluetooth divides transmitted data into packets, and transmits each packet on one of 79 designated Bluetooth channels. Each channel has a bandwidth of 1 MHz. It usually performs 800 hops per second, with Adaptive Frequency-Hopping (AFH) enabled. Bluetooth low energy uses 2 MHz spacing, which accommodates 40 channels.

Originally, Gaussian frequency-shift keying (GFSK) modulation was the only modulation scheme available. Since the introduction of Bluetooth 2.0+EDR, $\pi/4$-DQPSK (Differential Quadrature Phase Shift Keying) and 8DPSK modulation may also be used between compatible devices. Devices functioning with GFSK are said to be operating in basic rate (BR) mode where an instantaneous data rate of 1 Mbit/s is possible. The term Enhanced Data Rate (EDR) is used to describe $\pi/4$-DPSK and 8DPSK schemes, each giving 2 and 3 Mbit/s respectively. The combination of these (BR and EDR) modes in Bluetooth radio technology is classified as a "BR/EDR radio".

Bluetooth is a packet-based protocol with a master-slave structure. One master may communicate with up to seven slaves in a piconet. All devices share the master's clock. Packet exchange is based on the basic clock, defined by the master, which ticks at 312.5 μs intervals. Two clock ticks make up a slot of 625 μs, and two slots make up a slot pair of 1250 μs. In the simple case of single-slot packets the master transmits in even slots and receives in odd slots. The slave, conversely, receives in even slots and transmits in odd slots. Packets may be 1, 3 or 5 slots long, but in all cases the master's transmission begins in even slots and the slave's in odd slots.

The above is valid for "classic" BT. Bluetooth Low Energy, introduced in the 4.0 specification, uses the same spectrum but somewhat differently; Bluetooth low energy#Radio interface.

Communication and Connection

A master Bluetooth device can communicate with a maximum of seven devices in a piconet (an ad-hoc computer network using Bluetooth technology), though not all devices reach this maximum. The devices can switch roles, by agreement, and the slave can become the master (for example, a headset initiating a connection to a phone necessarily begins as master—as initiator of the connection—but may subsequently operate as slave).

The Bluetooth Core Specification provides for the connection of two or more piconets to form a scatternet, in which certain devices simultaneously play the master role in one piconet and the slave role in another.

At any given time, data can be transferred between the master and one other device (except for the little-used broadcast mode.) The master chooses which slave device to address; typically, it switches rapidly from one device to another in a round-robin fashion. Since it is the master that chooses which slave to address, whereas a slave is (in theory) supposed to listen in each receive slot, being a master is a lighter burden than being a slave. Being a master of seven slaves is possible; being a slave of more than one master is difficult. The specification is vague as to required behavior in scatternets.

Uses

Class	Max. permitted power		Typ. range (m)
	(mW)	(dBm)	
1	100	20	~100
2	2.5	4	~10
3	1	0	~1
4	0.5	-3	~0.5

Bluetooth is a standard wire-replacement communications protocol primarily designed for low-power consumption, with a short range based on low-cost transceiver microchips in each device. Because the devices use a radio (broadcast) communications system, they do not have to be in visual line of sight of each other, however a *quasi optical* wireless path must be viable. Range is power-class-dependent, but effective ranges vary in practice; the table on the right.

Officially Class 3 radios have a range of up to 1 metre (3 ft), Class 2, most commonly found in mobile devices, 10 metres (33 ft), and Class 1, primarily for industrial use cases,100 metres (300 ft). Bluetooth Marketing qualifies that Class 1 range is in most cases 20–30 metres (66–98 ft), and Class 2 range 5–10 metres (16–33 ft).

Bluetooth Version	Maximum Speed	Maximum Range
3.0	25 Mbit/s	
4.0	25 Mbit/s	200 feet(60.96 m)
5.0	50 Mbit/s	800 feet(243.84 m)

The effective range varies due to propagation conditions, material coverage, production sample variations, antenna configurations and battery conditions. Most Bluetooth applications are for indoor conditions, where attenuation of walls and signal fading due to signal reflections make the range far lower than specified line-of-sight ranges of the Bluetooth products. Most Bluetooth applications are battery powered Class 2 devices, with little difference in range whether the other end of the link is a Class 1 or Class 2 device as the lower powered device tends to set the range limit. In some cases the effective range of the data link can be extended when a Class 2 device is connecting to a Class 1 transceiver with both higher sensitivity and transmission power than a typical Class 2 device. Mostly, however, the Class 1 devices have a similar sensitivity to Class 2 devices. Connecting two Class 1 devices with both high sensitivity and high power can allow ranges far in excess of the typical 100m, depending on the throughput required by the application. Some such devices allow open field ranges of up to 1 km and beyond between two similar devices without exceeding legal emission limits.

The Bluetooth Core Specification mandates a range of not less than 10 metres (33 ft), but there is no upper limit on actual range. Manufacturers' implementations can be tuned to provide the range needed for each case.

Bluetooth Profiles

To use Bluetooth wireless technology, a device must be able to interpret certain Bluetooth profiles, which are definitions of possible applications and specify general behaviours that Bluetooth-enabled devices use to communicate with other Bluetooth devices. These profiles include settings to parametrize and to control the communication from start. Adherence to profiles saves the time for transmitting the parameters anew before the bi-directional link becomes effective. There are a wide range of Bluetooth profiles that describe many different types of applications or use cases for devices.

List of Applications

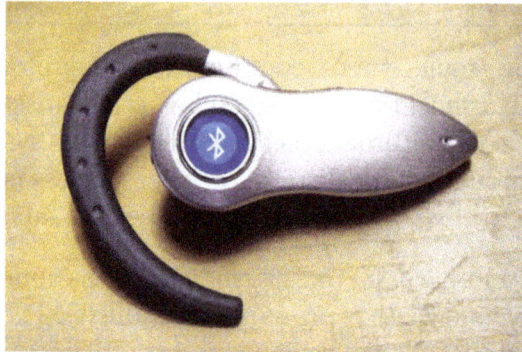

A typical Bluetooth mobile phone headset.

- Wireless control of and communication between a mobile phone and a handsfree headset. This was one of the earliest applications to become popular.

- Wireless control of and communication between a mobile phone and a Bluetooth compatible car stereo system.

- Wireless control of and communication with iOS and Android device phones, tablets and portable wireless speakers.

- Wireless Bluetooth headset and Intercom. Idiomatically, a headset is sometimes called "a Bluetooth".

- Wireless streaming of audio to headphones with or without communication capabilities.

- Wireless streaming of data collected by Bluetooth-enabled fitness devices to phone or PC.

- Wireless networking between PCs in a confined space and where little bandwidth is required.

- Wireless communication with PC input and output devices, the most common being the mouse, keyboard and printer.

- Transfer of files, contact details, calendar appointments, and reminders between devices with OBEX.

- Replacement of previous wired RS-232 serial communications in test equipment, GPS receivers, medical equipment, bar code scanners, and traffic control devices.

- For controls where infrared was often used.

- For low bandwidth applications where higher USB bandwidth is not required and cable-free connection desired.

- Sending small advertisements from Bluetooth-enabled advertising hoardings to other, discoverable, Bluetooth devices.

- Wireless bridge between two Industrial Ethernet (*e.g.*, PROFINET) networks.

- Seventh and eighth generation game consoles such as Nintendo's Wii, and Sony's PlayStation 3 use Bluetooth for their respective wireless controllers.

- Dial-up internet access on personal computers or PDAs using a data-capable mobile phone as a wireless modem.

- Short range transmission of health sensor data from medical devices to mobile phone, settop box or dedicated telehealth devices.

- Allowing a DECT phone to ring and answer calls on behalf of a nearby mobile phone.

- Real-time location systems (RTLS), are used to track and identify the location of objects in real-time using "Nodes" or "tags" attached to, or embedded in the objects tracked, and "Readers" that receive and process the wireless signals from these tags to determine their locations.

- Personal security application on mobile phones for prevention of theft or loss of items. The protected item has a Bluetooth marker (*e.g.*, a tag) that is in constant communication with the phone. If the connection is broken (the marker is out of range of the phone) then an alarm is raised. This can also be used as a man overboard alarm. A product using this technology has been available since 2009.

- Calgary, Alberta, Canada's Roads Traffic division uses data collected from travelers' Bluetooth devices to predict travel times and road congestion for motorists.

- Wireless transmission of audio (a more reliable alternative to FM transmitters)

Bluetooth vs. Wi-Fi (IEEE 802.11)

Bluetooth and Wi-Fi (the brand name for products using IEEE 802.11 standards) have some similar applications: setting up networks, printing, or transferring files. Wi-Fi is intended as a replacement for high speed cabling for general local area network access in work areas or home. This category of applications is sometimes called wireless local area networks (WLAN). Bluetooth was intended for portable equipment and its applications. The category of applications is outlined as the wireless personal area network (WPAN). Bluetooth is a replacement for cabling in a variety of personally carried applications in any setting, and also works for fixed location applications such as smart energy functionality in the home (thermostats, etc.).

Wi-Fi and Bluetooth are to some extent complementary in their applications and usage. Wi-Fi is usually access point-centered, with an asymmetrical client-server connection with all traffic routed through the access point, while Bluetooth is usually symmetrical, between two Bluetooth devices. Bluetooth serves well in simple applications where two devices need to connect with minimal configuration like a button press, as in headsets and remote controls, while Wi-Fi suits better in applications where some degree of client configuration is possible and high speeds are required, especially for network access through an access node. However, Bluetooth access points do exist and ad-hoc connections are possible with Wi-Fi though not as simply as with Bluetooth. Wi-Fi Direct was recently developed to add a more Bluetooth-like ad-hoc functionality to Wi-Fi.

Devices

A Bluetooth USB dongle with a 100 m range.

Bluetooth exists in many products, such as telephones, tablets, media players, robotics systems, handheld, laptops and console gaming equipment, and some high definition headsets, modems, and watches. The technology is useful when transferring information between two or more devices that are near each other in low-bandwidth situations. Bluetooth is commonly used to transfer sound data with telephones (i.e., with a Bluetooth headset) or byte data with hand-held computers (transferring files).

Bluetooth protocols simplify the discovery and setup of services between devices. Bluetooth devices can advertise all of the services they provide. This makes using services easier, because more of the security, network address and permission configuration can be automated than with many other network types.

Computer Requirements

A typical Bluetooth USB dongle.

A personal computer that does not have embedded Bluetooth can use a Bluetooth adapter that enables the PC to communicate with Bluetooth devices. While some desktop computers and most recent laptops come with a built-in Bluetooth radio, others require an external adapter, typically in the form of a small USB "dongle."

An internal notebook Bluetooth card (14×36×4 mm).

Unlike its predecessor, IrDA, which requires a separate adapter for each device, Bluetooth lets multiple devices communicate with a computer over a single adapter.

Operating System Implementation

For Microsoft platforms, Windows XP Service Pack 2 and SP3 releases work natively with Bluetooth v1.1, v2.0 and v2.0+EDR. Previous versions required users to install their Bluetooth adapter's own drivers, which were not directly supported by Microsoft. Microsoft's own Bluetooth dongles (packaged with their Bluetooth computer devices) have no external drivers and thus require at least Windows XP Service Pack 2. Windows Vista RTM/SP1 with the Feature Pack for Wireless or Windows Vista SP2 work with Bluetooth v2.1+EDR. Windows 7 works with Bluetooth v2.1+EDR and Extended Inquiry Response (EIR).

The Windows XP and Windows Vista/Windows 7 Bluetooth stacks support the following Bluetooth profiles natively: PAN, SPP, DUN, HID, HCRP. The Windows XP stack can be replaced by a third party stack that supports more profiles or newer Bluetooth versions. The Windows Vista/Windows 7 Bluetooth stack supports vendor-supplied additional profiles without requiring that the Microsoft stack be replaced.

Apple products have worked with Bluetooth since Mac OS X v10.2, which was released in 2002.

Linux has two popular Bluetooth stacks, BlueZ and Affix. The BlueZ stack is included with most Linux kernels and was originally developed by Qualcomm. The Affix stack was developed by Nokia.

FreeBSD features Bluetooth since its v5.0 release.

NetBSD features Bluetooth since its v4.0 release. Its Bluetooth stack has been ported to OpenBSD as well.

Specifications and Features

The specifications were formalized by the Bluetooth Special Interest Group (SIG). The SIG was formally announced on 20 May 1998. Today it has a membership of over 30,000 companies worldwide. It was established by Ericsson, IBM, Intel, Toshiba and Nokia, and later joined by many other companies.

All versions of the Bluetooth standards support downward compatibility. That lets the latest standard cover all older versions.

The Bluetooth Core Specification Working Group (CSWG) produces mainly 4 kinds of specifications

- The Bluetooth Core Specification, release cycle is typically a few years in between

- Core Specification Addendum (CSA), release cycle can be as tight as a few times per year

- Core Specification Supplements (CSS), can be released very quickly

- Errata

Bluetooth v1.0 and v1.0B

Versions 1.0 and 1.0B had many problems and manufacturers had difficulty making their products interoperable. Versions 1.0 and 1.0B also included mandatory Bluetooth hardware device address (BD_ADDR) transmission in the Connecting process (rendering anonymity impossible at the protocol level), which was a major setback for certain services planned for use in Bluetooth environments.

Bluetooth v1.1

- Ratified as IEEE Standard 802.15.1–2002

- Many errors found in the v1.0B specifications were fixed.

- Added possibility of non-encrypted channels.

- Received Signal Strength Indicator (RSSI).

Bluetooth v1.2

Major enhancements include the following:

- Faster Connection and Discovery

- *Adaptive frequency-hopping spread spectrum (AFH)*, which improves resistance to radio frequency interference by avoiding the use of crowded frequencies in the hopping sequence.

- Higher transmission speeds in practice, up to 721 kbit/s, than in v1.1.

- Extended Synchronous Connections (eSCO), which improve voice quality of audio links by allowing retransmissions of corrupted packets, and may optionally increase audio latency to provide better concurrent data transfer.

- Host Controller Interface (HCI) operation with three-wire UART.

- Ratified as IEEE Standard 802.15.1–2005

- Introduced Flow Control and Retransmission Modes for L2CAP.

Bluetooth v2.0 + EDR

This version of the Bluetooth Core Specification was released in 2004. The main difference is the introduction of an Enhanced Data Rate (EDR) for faster data transfer. The nominal rate of EDR is about 3 Mbit/s, although the practical data transfer rate is 2.1 Mbit/s. EDR uses a combination of GFSK and Phase Shift Keying modulation (PSK) with two variants, π/4-DQPSK and 8DPSK. EDR can provide a lower power consumption through a reduced duty cycle.

The specification is published as *Bluetooth v2.0 + EDR*, which implies that EDR is an optional feature. Aside from EDR, the v2.0 specification contains other minor improvements, and products may claim compliance to "Bluetooth v2.0" without supporting the higher data rate. At least one commercial device states "Bluetooth v2.0 without EDR" on its data sheet.

Bluetooth v2.1 + EDR

Bluetooth Core Specification Version 2.1 + EDR was adopted by the Bluetooth SIG on 26 July 2007.

The headline feature of v2.1 is secure simple pairing (SSP): this improves the pairing experience for Bluetooth devices, while increasing the use and strength of security. the section on Pairing below for more details.

Version 2.1 allows various other improvements, including "Extended inquiry response" (EIR), which provides more information during the inquiry procedure to allow better filtering of devices before connection; and sniff subrating, which reduces the power consumption in low-power mode.

Bluetooth v3.0 + HS

Version 3.0 + HS of the Bluetooth Core Specification was adopted by the Bluetooth SIG on 21 April 2009. Bluetooth v3.0 + HS provides theoretical data transfer speeds of up to 24 Mbit/s, though not over the Bluetooth link itself. Instead, the Bluetooth link is used for negotiation and establishment, and the high data rate traffic is carried over a colocated 802.11 link.

The main new feature is AMP (Alternative MAC/PHY), the addition of 802.11 as a high speed transport. The High-Speed part of the specification is not mandatory, and hence only devices that display the "+HS" logo actually support Bluetooth over 802.11 high-speed data transfer. A Bluetooth v3.0 device without the "+HS" suffix is only required to support features introduced in Core Specification Version 3.0 or earlier Core Specification Addendum 1.

L2CAP Enhanced Modes

Enhanced Retransmission Mode (ERTM) implements reliable L2CAP channel, while Streaming Mode (SM) implements unreliable channel with no retransmission or flow control. Introduced in Core Specification Addendum 1.

Alternative MAC/PHY

Enables the use of alternative MAC and PHYs for transporting Bluetooth profile data. The Bluetooth radio is still used for device discovery, initial connection and profile configuration. However, when large quantities of data must be sent, the high speed alternative MAC PHY 802.11 (typically associated with Wi-Fi) transports the data. This means that Bluetooth uses proven low power connection models when the system is idle, and the faster radio when it must send large quantities of data. AMP links require enhanced L2CAP modes.

Unicast Connectionless Data

Permits sending service data without establishing an explicit L2CAP channel. It is intended for use by applications that require low latency between user action and reconnection/transmission of data. This is only appropriate for small amounts of data.

Enhanced Power Control

Updates the power control feature to remove the open loop power control, and also to clarify ambiguities in power control introduced by the new modulation schemes added for EDR. Enhanced power control removes the ambiguities by specifying the behaviour that is expected. The feature also adds closed loop power control, meaning RSSI filtering can start as the response is received. Additionally, a "go straight to maximum power" request has been introduced. This is expected to deal with the headset link loss issue typically observed when a user puts their phone into a pocket on the opposite side to the headset.

Ultra-wideband

The high speed (AMP) feature of Bluetooth v3.0 was originally intended for UWB, but the WiMedia Alliance, the body responsible for the flavor of UWB intended for Bluetooth, announced in March 2009 that it was disbanding, and ultimately UWB was omitted from the Core v3.0 specification.

On 16 March 2009, the WiMedia Alliance announced it was entering into technology transfer agreements for the WiMedia Ultra-wideband (UWB) specifications. WiMedia has transferred all current and future specifications, including work on future high speed and power optimized implementations, to the Bluetooth Special Interest Group (SIG), Wireless USB Promoter Group and the USB Implementers Forum. After successful completion of the technology transfer, marketing, and related administrative items, the WiMedia Alliance ceased operations.

In October 2009 the Bluetooth Special Interest Group suspended development of UWB as part of the alternative MAC/PHY, Bluetooth v3.0 + HS solution. A small, but significant, number of former WiMedia members had not and would not sign up to the necessary agreements for the IP transfer. The Bluetooth SIG is now in the process of evaluating other options for its longer term roadmap.

Bluetooth v4.0

The Bluetooth SIG completed the Bluetooth Core Specification version 4.0 (called Bluetooth Smart) and has been adopted as of 30 June 2010. It includes *Classic Bluetooth*, *Bluetooth high speed* and *Bluetooth low energy* protocols. Bluetooth high speed is based on Wi-Fi, and Classic Bluetooth consists of legacy Bluetooth protocols.

Bluetooth low energy, previously known as Wibree, is a subset of Bluetooth v4.0 with an entirely new protocol stack for rapid build-up of simple links. As an alternative to the Bluetooth standard protocols that were introduced in Bluetooth v1.0 to v3.0, it is aimed at very low power applications running off a coin cell. Chip designs allow for two types of implementation, dual-mode, single-mode and enhanced past versions. The provisional names *Wibree* and *Bluetooth ULP* (Ultra Low Power) were abandoned and the BLE name was used for a while. In late 2011, new logos "Bluetooth Smart Ready" for hosts and "Bluetooth Smart" for sensors were introduced as the general-public face of BLE.

- In a single-mode implementation, only the low energy protocol stack is implemented. ST-Microelectronics, AMICCOM, CSR, Nordic Semiconductor and Texas Instruments have released single mode Bluetooth low energy solutions.

- In a dual-mode implementation, Bluetooth Smart functionality is integrated into an existing Classic Bluetooth controller. As of March 2011, the following semiconductor companies have announced the availability of chips meeting the standard: Qualcomm-Atheros, CSR, Broadcom and Texas Instruments. The compliant architecture shares all of Classic Bluetooth's existing radio and functionality resulting in a negligible cost increase compared to Classic Bluetooth.

Cost-reduced single-mode chips, which enable highly integrated and compact devices, feature a lightweight Link Layer providing ultra-low power idle mode operation, simple device discovery, and reliable point-to-multipoint data transfer with advanced power-save and secure encrypted connections at the lowest possible cost.

General improvements in version 4.0 include the changes necessary to facilitate BLE modes, as well the Generic Attribute Profile (GATT) and Security Manager (SM) services with AES Encryption.

Core Specification Addendum 2 was unveiled in December 2011; it contains improvements to the audio Host Controller Interface and to the High Speed (802.11) Protocol Adaptation Layer.

Core Specification Addendum 3 revision 2 has an adoption date of 24 July 2012.

Core Specification Addendum 4 has an adoption date of 12 February 2013.

Bluetooth v4.1

The Bluetooth SIG announced formal adoption of the Bluetooth v4.1 specification on 4 December 2013. This specification is an incremental software update to Bluetooth Specification v4.0, and not a hardware update. The update incorporates Bluetooth Core Specification Addenda (CSA 1, 2, 3 & 4) and adds new features that improve consumer usability. These include increased co-existence support for LTE, bulk data exchange rates—and aid developer innovation by allowing devices to support multiple roles simultaneously.

New features of this specification include:

- Mobile Wireless Service Coexistence Signaling

- Train Nudging and Generalized Interlaced Scanning

- Low Duty Cycle Directed Advertising

- L2CAP Connection Oriented and Dedicated Channels with Credit Based Flow Control

- Dual Mode and Topology

- LE Link Layer Topology

- 802.11n PAL

- Audio Architecture Updates for Wide Band Speech

- Fast Data Advertising Interval

- Limited Discovery Time

Notice that some features were already available in a Core Specification Addendum (CSA) before the release of v4.1.

Bluetooth v4.2

Bluetooth v4.2 was released on December 2, 2014. It Introduces some key features for IoT. Some features, such as Data Length Extension, require a hardware update. But some older Bluetooth hardware may receive some Bluetooth v4.2 features, such as privacy updates via firmware.

The major areas of improvement are:

- LE Data Packet Length Extension

- LE Secure Connections

- Link Layer Privacy

- Link Layer Extended Scanner Filter Policies

- IP connectivity for Bluetooth Smart devices to become available soon after the introduction of BT v4.2 via the new Internet Protocol Support Profile (IPSP).

- IPSP adds an IPv6 connection option for Bluetooth Smart, to support connected home and other IoT implementations.

Bluetooth v5

Bluetooth 5 was announced in June 2016. It will quadruple the range, double the speed, and provide an eight-fold increase in data broadcasting capacity of low energy Bluetooth connections, in addition to adding functionality for connectionless services like location-relevant information and navigation.

It is mainly focused on Internet of Things emerging technology. The release of products is scheduled for late 2016 to early 2017.

Technical Information

Bluetooth Protocol Stack

Bluetooth is defined as a layer protocol architecture consisting of core protocols, cable replacement protocols, telephony control protocols, and adopted protocols. Mandatory protocols for all

Bluetooth stacks are: LMP, L2CAP and SDP. In addition, devices that communicate with Bluetooth almost universally can use these protocols: HCI and RFCOMM.

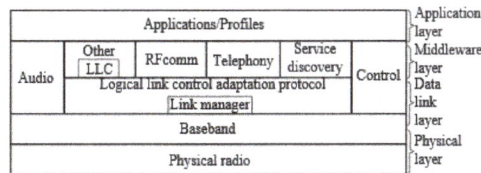

Bluetooth Protocol Stack

LMP

The *Link Management Protocol* (LMP) is used for set-up and control of the radio link between two devices. Implemented on the controller.

L2CAP

The *Logical Link Control and Adaptation Protocol* (L2CAP) is used to multiplex multiple logical connections between two devices using different higher level protocols. Provides segmentation and reassembly of on-air packets.

In *Basic* mode, L2CAP provides packets with a payload configurable up to 64 kB, with 672 bytes as the default MTU, and 48 bytes as the minimum mandatory supported MTU.

In *Retransmission and Flow Control* modes, L2CAP can be configured either for isochronous data or reliable data per channel by performing retransmissions and CRC checks.

Bluetooth Core Specification Addendum 1 adds two additional L2CAP modes to the core specification. These modes effectively deprecate original Retransmission and Flow Control modes:

- Enhanced Retransmission Mode (ERTM): This mode is an improved version of the original retransmission mode. This mode provides a reliable L2CAP channel.

- Streaming Mode (SM): This is a very simple mode, with no retransmission or flow control. This mode provides an unreliable L2CAP channel.

Reliability in any of these modes is optionally and/or additionally guaranteed by the lower layer Bluetooth BDR/EDR air interface by configuring the number of retransmissions and flush timeout (time after which the radio flushes packets). In-order sequencing is guaranteed by the lower layer.

Only L2CAP channels configured in ERTM or SM may be operated over AMP logical links.

SDP

The *Service Discovery Protocol* (SDP) allows a device to discover services offered by other devices, and their associated parameters. For example, when you use a mobile phone with a Bluetooth headset, the phone uses SDP to determine which Bluetooth profiles the headset can use (Headset Profile, Hands Free Profile, Advanced Audio Distribution Profile (A2DP) etc.) and the protocol multiplexer settings needed for the phone to connect to the headset using each of them. Each service is identified by a Universally Unique Identifier (UUID), with official services (Bluetooth profiles) assigned a short form UUID (16 bits rather than the full 128).

RFCOMM

Radio Frequency Communications (RFCOMM) is a cable replacement protocol used to generate a virtual serial data stream. RFCOMM provides for binary data transport and emulates EIA-232 (formerly RS-232) control signals over the Bluetooth baseband layer, i.e. it is a serial port emulation.

RFCOMM provides a simple reliable data stream to the user, similar to TCP. It is used directly by many telephony related profiles as a carrier for AT commands, as well as being a transport layer for OBEX over Bluetooth.

Many Bluetooth applications use RFCOMM because of its widespread support and publicly available API on most operating systems. Additionally, applications that used a serial port to communicate can be quickly ported to use RFCOMM.

BNEP

The *Bluetooth Network Encapsulation Protocol* (BNEP) is used for transferring another protocol stack's data via an L2CAP channel. Its main purpose is the transmission of IP packets in the Personal Area Networking Profile. BNEP performs a similar function to SNAP in Wireless LAN.

AVCTP

The *Audio/Video Control Transport Protocol* (AVCTP) is used by the remote control profile to transfer AV/C commands over an L2CAP channel. The music control buttons on a stereo headset use this protocol to control the music player.

AVDTP

The *Audio/Video Distribution Transport Protocol* (AVDTP) is used by the advanced audio distribution profile to stream music to stereo headsets over an L2CAP channel intended for video distribution profile in the Bluetooth transmission.

TCS

The *Telephony Control Protocol – Binary* (TCS BIN) is the bit-oriented protocol that defines the call control signaling for the establishment of voice and data calls between Bluetooth devices. Additionally, "TCS BIN defines mobility management procedures for handling groups of Bluetooth TCS devices."

TCS-BIN is only used by the cordless telephony profile, which failed to attract implementers. As such it is only of historical interest.

Adopted Protocols

Adopted protocols are defined by other standards-making organizations and incorporated into

Bluetooth's protocol stack, allowing Bluetooth to code protocols only when necessary. The adopted protocols include:

- Point-to-Point Protocol (PPP): Internet standard protocol for transporting IP datagrams over a point-to-point link.

- TCP/IP/UDP: Foundation Protocols for TCP/IP protocol suite

- Object Exchange Protocol (OBEX): Session-layer protocol for the exchange of objects, providing a model for object and operation representation

- Wireless Application Environment/Wireless Application Protocol (WAE/WAP): WAE specifies an application framework for wireless devices and WAP is an open standard to provide mobile users access to telephony and information services.

Baseband Error Correction

Depending on packet type, individual packets may be protected by error correction, either 1/3 rate forward error correction (FEC) or 2/3 rate. In addition, packets with CRC will be retransmitted until acknowledged by automatic repeat request (ARQ).

Setting Up Connections

Any Bluetooth device in *discoverable mode* transmits the following information on demand:

- Device name

- Device class

- List of services

- Technical information (for example: device features, manufacturer, Bluetooth specification used, clock offset)

Any device may perform an inquiry to find other devices to connect to, and any device can be configured to respond to such inquiries. However, if the device trying to connect knows the address of the device, it always responds to direct connection requests and transmits the information shown in the list above if requested. Use of a device's services may require pairing or acceptance by its owner, but the connection itself can be initiated by any device and held until it goes out of range. Some devices can be connected to only one device at a time, and connecting to them prevents them from connecting to other devices and appearing in inquiries until they disconnect from the other device.

Every device has a unique 48-bit address. However, these addresses are generally not shown in inquiries. Instead, friendly Bluetooth names are used, which can be set by the user. This name appears when another user scans for devices and in lists of paired devices.

Most cellular phones have the Bluetooth name set to the manufacturer and model of the phone by default. Most cellular phones and laptops show only the Bluetooth names and special programs are required to get additional information about remote devices. This can be confusing as, for example, there could be several cellular phones in range named T610.

Pairing and Bonding

Motivation

Many services offered over Bluetooth can expose private data or let a connecting party control the Bluetooth device. Security reasons make it necessary to recognize specific devices, and thus enable control over which devices can connect to a given Bluetooth device. At the same time, it is useful for Bluetooth devices to be able to establish a connection without user intervention (for example, as soon as in range).

To resolve this conflict, Bluetooth uses a process called *bonding*, and a bond is generated through a process called *pairing*. The pairing process is triggered either by a specific request from a user to generate a bond (for example, the user explicitly requests to "Add a Bluetooth device"), or it is triggered automatically when connecting to a service where (for the first time) the identity of a device is required for security purposes. These two cases are referred to as dedicated bonding and general bonding respectively.

Pairing often involves some level of user interaction. This user interaction confirms the identity of the devices. When pairing successfully completes, a bond forms between the two devices, enabling those two devices to connect to each other in the future without repeating the pairing process to confirm device identities. When desired, the user can remove the bonding relationship.

Implementation

During pairing, the two devices establish a relationship by creating a shared secret known as a *link key*. If both devices store the same link key, they are said to be *paired* or *bonded*. A device that wants to communicate only with a bonded device can cryptographically authenticate the identity of the other device, ensuring it is the same device it previously paired with. Once a link key is generated, an authenticated Asynchronous Connection-Less (ACL) link between the devices may be encrypted to protect exchanged data against eavesdropping. Users can delete link keys from either device, which removes the bond between the devices—so it is possible for one device to have a stored link key for a device it is no longer paired with.

Bluetooth services generally require either encryption or authentication and as such require pairing before they let a remote device connect. Some services, such as the Object Push Profile, elect not to explicitly require authentication or encryption so that pairing does not interfere with the user experience associated with the service use-cases.

Pairing Mechanisms

Pairing mechanisms changed significantly with the introduction of Secure Simple Pairing in Bluetooth v2.1. The following summarizes the pairing mechanisms:

- *Legacy pairing*: This is the only method available in Bluetooth v2.0 and before. Each device must enter a PIN code; pairing is only successful if both devices enter the same PIN code. Any 16-byte UTF-8 string may be used as a PIN code; however, not all devices may be capable of entering all possible PIN codes.

  - *Limited input devices*: The obvious example of this class of device is a Bluetooth

Hands-free headset, which generally have few inputs. These devices usually have a *fixed PIN*, for example "0000" or "1234", that are hard-coded into the device.

- *Numeric input devices*: Mobile phones are classic examples of these devices. They allow a user to enter a numeric value up to 16 digits in length.

- *Alpha-numeric input devices*: PCs and smartphones are examples of these devices. They allow a user to enter full UTF-8 text as a PIN code. If pairing with a less capable device the user must be aware of the input limitations on the other device, there is no mechanism available for a capable device to determine how it should limit the available input a user may use.

- *Secure Simple Pairing* (SSP): This is required by Bluetooth v2.1, although a Bluetooth v2.1 device may only use legacy pairing to interoperate with a v2.0 or earlier device. Secure Simple Pairing uses a form of public key cryptography, and some types can help protect against man in the middle, or MITM attacks. SSP has the following authentication mechanisms:

  - *Just works*: As the name implies, this method just works, with no user interaction. However, a device may prompt the user to confirm the pairing process. This method is typically used by headsets with very limited IO capabilities, and is more secure than the fixed PIN mechanism this limited set of devices uses for legacy pairing. This method provides no man-in-the-middle (MITM) protection.

  - *Numeric comparison*: If both devices have a display, and at least one can accept a binary yes/no user input, they may use Numeric Comparison. This method displays a 6-digit numeric code on each device. The user should compare the numbers to ensure they are identical. If the comparison succeeds, the user(s) should confirm pairing on the device(s) that can accept an input. This method provides MITM protection, assuming the user confirms on both devices and actually performs the comparison properly.

  - *Passkey Entry*: This method may be used between a device with a display and a device with numeric keypad entry (such as a keyboard), or two devices with numeric keypad entry. In the first case, the display is used to show a 6-digit numeric code to the user, who then enters the code on the keypad. In the second case, the user of each device enters the same 6-digit number. Both of these cases provide MITM protection.

  - *Out of band* (OOB): This method uses an external means of communication, such as Near Field Communication (NFC) to exchange some information used in the pairing process. Pairing is completed using the Bluetooth radio, but requires information from the OOB mechanism. This provides only the level of MITM protection that is present in the OOB mechanism.

SSP is considered simple for the following reasons:

- In most cases, it does not require a user to generate a passkey.
- For use-cases not requiring MITM protection, user interaction can be eliminated.

- For *numeric comparison*, MITM protection can be achieved with a simple equality comparison by the user.

- Using OOB with NFC enables pairing when devices simply get close, rather than requiring a lengthy discovery process.

Security Concerns

Prior to Bluetooth v2.1, encryption is not required and can be turned off at any time. Moreover, the encryption key is only good for approximately 23.5 hours; using a single encryption key longer than this time allows simple XOR attacks to retrieve the encryption key.

- Turning off encryption is required for several normal operations, so it is problematic to detect if encryption is disabled for a valid reason or for a security attack.

- Bluetooth v2.1 addresses this in the following ways:

- Encryption is required for all non-SDP (Service Discovery Protocol) connections

- A new Encryption Pause and Resume feature is used for all normal operations that require that encryption be disabled. This enables easy identification of normal operation from security attacks.

- The encryption key must be refreshed before it expires.

Link keys may be stored on the device file system, not on the Bluetooth chip itself. Many Bluetooth chip manufacturers let link keys be stored on the device—however, if the device is removable, this means that the link key moves with the device.

Security

Overview

Bluetooth implements confidentiality, authentication and key derivation with custom algorithms based on the SAFER+ block cipher. Bluetooth key generation is generally based on a Bluetooth PIN, which must be entered into both devices. This procedure might be modified if one of the devices has a fixed PIN (e.g., for headsets or similar devices with a restricted user interface). During pairing, an initialization key or master key is generated, using the E22 algorithm. The E0 stream cipher is used for encrypting packets, granting confidentiality, and is based on a shared cryptographic secret, namely a previously generated link key or master key. Those keys, used for subsequent encryption of data sent via the air interface, rely on the Bluetooth PIN, which has been entered into one or both devices.

An overview of Bluetooth vulnerabilities exploits was published in 2007 by Andreas Becker.

In September 2008, the National Institute of Standards and Technology (NIST) published a Guide to Bluetooth Security as a reference for organizations. It describes Bluetooth security capabilities and how to secure Bluetooth technologies effectively. While Bluetooth has its benefits, it is susceptible to denial-of-service attacks, eavesdropping, man-in-the-middle attacks, message modification, and resource misappropriation. Users and organizations must evaluate their acceptable level of risk and incorporate security into the lifecycle of Bluetooth devices. To help mitigate risks, included in the NIST document are security checklists with guidelines and recommendations for creating and maintaining secure Bluetooth piconets, headsets, and smart card readers.

Bluetooth v2.1 – finalized in 2007 with consumer devices first appearing in 2009 – makes significant changes to Bluetooth's security, including pairing. the pairing mechanisms section for more about these changes.

Bluejacking

Bluejacking is the sending of either a picture or a message from one user to an unsuspecting user through *Bluetooth* wireless technology. Common applications include short messages, *e.g.*, "You've just been bluejacked!". Bluejacking does not involve the removal or alteration of any data from the device. Bluejacking can also involve taking control of a mobile device wirelessly and phoning a premium rate line, owned by the bluejacker. Security advances have alleviated this issue.

History of Security Concerns

2001–2004

In 2001, Jakobsson and Wetzel from Bell Laboratories discovered flaws in the Bluetooth pairing protocol and also pointed to vulnerabilities in the encryption scheme. In 2003, Ben and Adam Laurie from A.L. Digital Ltd. discovered that serious flaws in some poor implementations of Bluetooth security may lead to disclosure of personal data. In a subsequent experiment, Martin Herfurt from the trifinite.group was able to do a field-trial at the CeBIT fairgrounds, showing the importance of the problem to the world. A new attack called BlueBug was used for this experiment. In 2004 the first purported virus using Bluetooth to spread itself among mobile phones appeared on the Symbian OS. The virus was first described by Kaspersky Lab and requires users to confirm the installation of unknown software before it can propagate. The virus was written as a proof-of-concept by a group of virus writers known as "29A" and sent to anti-virus groups. Thus, it should be regarded as a potential (but not real) security threat to Bluetooth technology or Symbian OS since the virus has never spread outside of this system. In August 2004, a world-record-setting experiment showed that the range of Class 2 Bluetooth radios could be extended to 1.78 km (1.11 mi) with directional antennas and signal amplifiers. This poses a potential security threat because it enables attackers to access vulnerable Bluetooth devices from a distance beyond expectation. The attacker must also be able to receive information from the victim to set up a connection. No attack can be made against a Bluetooth device unless the attacker knows its Bluetooth address and which channels to transmit on, although these can be deduced within a few minutes if the device is in use.

2005

In January 2005, a mobile malware worm known as *John Cena.* began targeting mobile phones using Symbian OS (Series 60 platform) using Bluetooth enabled devices to replicate itself and spread to other devices. The worm is self-installing and begins once the mobile user approves the transfer of the file (velasco.sis) from another device. Once installed, the worm begins looking for other Bluetooth enabled devices to infect. Additionally, the worm infects other .SIS files on the device, allowing replication to another device through use of removable media (Secure Digital, Compact Flash, etc.). The worm can render the mobile device unstable.

In April 2005, Cambridge University security researchers published results of their actual implementation of passive attacks against the PIN-based pairing between commercial Bluetooth devices. They confirmed that attacks are practically fast, and the Bluetooth symmetric key establishment method is vulnerable. To rectify this vulnerability, they designed an implementation that showed that stronger, asymmetric key establishment is feasible for certain classes of devices, such as mobile phones.

In June 2005, Yaniv Shaked and Avishai Wool published a paper describing both passive and active methods for obtaining the PIN for a Bluetooth link. The passive attack allows a suitably equipped attacker to eavesdrop on communications and spoof, if the attacker was present at the time of initial pairing. The active method makes use of a specially constructed message that must be inserted at a specific point in the protocol, to make the master and slave repeat the pairing process. After that, the first method can be used to crack the PIN. This attack's major weakness is that it requires the user of the devices under attack to re-enter the PIN during the attack when the device prompts them to. Also, this active attack probably requires custom hardware, since most commercially available Bluetooth devices are not capable of the timing necessary.

In August 2005, police in Cambridgeshire, England, issued warnings about thieves using Bluetooth enabled phones to track other devices left in cars. Police are advising users to ensure that any mobile networking connections are de-activated if laptops and other devices are left in this way.

2006

In April 2006, researchers from Secure Network and F-Secure published a report that warns of the large number of devices left in a visible state, and issued statistics on the spread of various Bluetooth services and the ease of spread of an eventual Bluetooth worm.

2007

In October 2007, at the Luxemburgish Hack.lu Security Conference, Kevin Finistere and Thierry Zoller demonstrated and released a remote root shell via Bluetooth on Mac OS X v10.3.9 and v10.4. They also demonstrated the first Bluetooth PIN and Linkkeys cracker, which is based on the research of Wool and Shaked.

Mitigation

Options to mitigate against Bluetooth security attacks include:>

- Enable Bluetooth only when required
- Enable Bluetooth discovery only when necessary, and disable discovery when finished
- Do not enter link keys or PINs when unexpectedly prompted to do so
- Remove paired devices when not in use
- Regularly update firmware on Bluetooth-enabled devices

Health Concerns

Bluetooth uses the microwave radio frequency spectrum in the 2.402 GHz to 2.480 GHz range. Maximum power output from a Bluetooth radio is 100 mW for class 1, 2.5 mW for class 2, and 1 mW for class 3 devices. Even the maximum power output of class 1 is a lower level than the lowest powered mobile phones. UMTS & W-CDMA outputs 250 mW, GSM1800/1900 outputs 1000 mW, and GSM850/900 outputs 2000 mW.

Interference Caused by USB 3.0

USB 3.0 devices, ports and cables have been proven to interfere with Bluetooth devices due to the electronic noise they release falling over the same operating band as Bluetooth. The close proximity of Bluetooth and USB 3.0 devices can result in a drop in throughput or complete connection loss of the Bluetooth device/s connected to a computer.

Various strategies can be applied to resolve the problem, ranging from simple solutions such as increasing the distance of USB 3.0 devices from any Bluetooth devices or purchasing better shielded USB cables. Other solutions include applying additional shielding to the internal USB components of a computer.

Bluetooth Award Programs

The Bluetooth Innovation World Cup, a marketing initiative of the Bluetooth Special Interest Group (SIG), was an international competition that encouraged the development of innovations for applications leveraging Bluetooth technology in sports, fitness and health care products. The aim of the competition was to stimulate new markets.

The Bluetooth Innovation World Cup morphed into the Bluetooth Breakthrough Awards in 2013. The Breakthrough Awards Bluetooth program highlights the most innovative products and applications available today, prototypes coming soon, and student-led projects in the making.

References

- G. Miao, J. Zander, K-W Sung, and B. Slimane, Fundamentals of Mobile Data Networks, Cambridge University Press, ISBN 1107143217, 2016.

- Sherman, Joshua (30 October 2013). "How LED Light Bulbs could replace Wi-Fi". Digital Trends. Retrieved 29 November 2015.

- "Global Visible Light Communication (VLC)/Li-Fi Technology Market worth $6,138.02 Million by 2018". MarketsandMarkets. 10 January 2013. Retrieved 29 November 2015.

- Coetzee, Jacques (13 January 2013). "LiFi beats Wi-Fi with 1Gb wireless speeds over pulsing LEDs". Gearburn. Retrieved 29 November 2015.

- Vincent, James (29 October 2013). "Li-Fi revolution: internet connections using light bulbs are 250 times". The Independent. Retrieved 29 November 2015.

- "pureLiFi to demonstrate first ever Li-Fi system at Mobile World Congress". Virtual-Strategy Magazine. 19 February 2014. Retrieved 29 November 2015.

- Vega, Anna (14 July 2014). "Li-fi record data transmission of 10Gbps set using LED lights". Engineering and Technology Magazine. Retrieved 29 November 2015.

- Van Camp, Jeffrey (19 January 2014). "Wysips Solar Charging Screen Could Eliminate Chargers and Wi-Fi". Digital Trends. Retrieved 29 November 2015.

- "Towards Energy-Awareness in Managing Wireless LAN Applications". IEEE/IFIP NOMS 2012: IEEE/IFIP Network Operations and Management Symposium. Retrieved 2014-08-11.

- "Application Level Energy and Performance Measurements in a Wireless LAN". The 2011 IEEE/ACM International Conference on Green Computing and Communications. Retrieved 2014-08-11.

- Rigg, Jamie (January 11, 2014). "Smartphone concept incorporates LiFi sensor for receiving light-based data". Engadget. Retrieved January 16, 2014.

- Scarfone, K. & Padgette, J. (September 2008). "Guide to Bluetooth Security" (PDF). National Institute of Standards and Technology. Retrieved 3 July 2013.

- Vilorio, Dennis. "You're a what? Tower Climber" (PDF). Occupational Outlook Quarterly. Archived (PDF) from the original on February 3, 2013. Retrieved December 6, 2013.

- "American Airlines and MobileStar Network to Deliver Wireless Internet Connectivity to American's Passengers". PR Newswire. 11 May 2000. Retrieved 13 April 2013.

- "MobileStar Network to Supply U.S. Hilton Hotels With Wireless High-Speed Internet Access". 28 October 1998. Retrieved 13 April 2013.

- Brownlee, John (2013-06-12). "iOS 7 Will Make It Possible To Roam Between Open Wi-Fi Networks Without Your Data Ever Dropping". Cult of Mac. Retrieved 2013-09-16.

- "Visible-light communication: Tripping the light fantastic: A fast and cheap optical version of Wi-Fi is coming". The Economist. 28 January 2012. Retrieved 22 October 2013.

- Povey,, Gordon. "About Visible Light Communications". pureVLC. Archived from the original on 18 August 2013. Retrieved 22 October 2013.

- Povey, Gordon (19 October 2011). "Li-Fi Consortium is Launched". D-Light Project. Archived from the original on 18 August 2013. Retrieved 22 October 2013.

- pureVLC (6 August 2012). "pureVLC Demonstrates Li-Fi Streaming along with Research Supporting World's Fastest Li-Fi Speeds up to 6 Gbit/s". Press release (Edinburgh). Retrieved 22 October 2013.

- Thomson, Iain (18 October 2013). "Forget Wi-Fi, boffins get 150Mbps Li-Fi connection from lightbulbs: Many (Chinese) hands make light work". The Register. Retrieved 22 October 2013.

- Moser, Max; Schrödel, Philipp (2007-12-05). "27Mhz Wireless Keyboard Analysis Report aka "We know what you typed last summer"" (PDF). Retrieved 6 February 2012

- Ermanno Pietrosemoli. "Setting Long Distance WiFi Records: Proofing Solutions for Rural Connectivity". Fundación Escuela Latinoamericana de Redes University of the Andes (Venezuela). Retrieved March 17, 2012.

Wireless Security and Concerns

The prevention of unauthorized access or damage to computers is wireless security. The most common types of wireless security are wireless security, wired equivalent privacy, Wi-Fi protected access, mobile and computer security. This chapter explains to the reader the importance of wireless security and its various aspects.

Wireless Security

Wireless security is the prevention of unauthorized access or damage to computers using wireless networks. The most common types of wireless security are Wired Equivalent Privacy (WEP) and Wi-Fi Protected Access (WPA). WEP is a notoriously weak security standard. The password it uses can often be cracked in a few minutes with a basic laptop computer and widely available software tools. WEP is an old IEEE 802.11 standard from 1999, which was outdated in 2003 by WPA, or Wi-Fi Protected Access. WPA was a quick alternative to improve security over WEP. The current standard is WPA2; some hardware cannot support WPA2 without firmware upgrade or replacement. WPA2 uses an encryption device that encrypts the network with a 256-bit key; the longer key length improves security over WEP.

An example wireless router, that can implement wireless security features

Many laptop computers have wireless cards pre-installed. The ability to enter a network while mobile has great benefits. However, wireless networking is prone to some security issues. Hackers have found wireless networks relatively easy to break into, and even use wireless technology to hack into wired networks. As a result, it is very important that enterprises define effective wireless security policies that guard against unauthorized access to important resources. Wireless Intrusion Prevention Systems (WIPS) or Wireless Intrusion Detection Systems (WIDS) are commonly used to enforce wireless security policies.

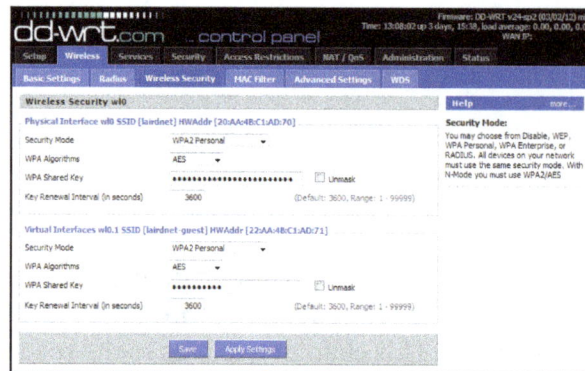

Security settings panel for a DD-WRT router

The risks to users of wireless technology have increased as the service has become more popular. There were relatively few dangers when wireless technology was first introduced. Hackers had not yet had time to latch on to the new technology, and wireless networks were not commonly found in the work place. However, there are many security risks associated with the current wireless protocols and encryption methods, and in the carelessness and ignorance that exists at the user and corporate IT level. Hacking methods have become much more sophisticated and innovative with wireless access. Hacking has also become much easier and more accessible with easy-to-use Windows- or Linux-based tools being made available on the web at no charge.

Some organizations that have no wireless access points installed do not feel that they need to address wireless security concerns. In-Stat MDR and META Group have estimated that 95% of all corporate laptop computers that were planned to be purchased in 2005 were equipped with wireless cards. Issues can arise in a supposedly non-wireless organization when a wireless laptop is plugged into the corporate network. A hacker could sit out in the parking lot and gather information from it through laptops and/or other devices, or even break in through this wireless card–equipped laptop and gain access to the wired network.

Background

Anyone within the geographical network range of an open, unencrypted wireless network can "sniff", or capture and record, the traffic, gain unauthorized access to internal network resources as well as to the internet, and then use the information and resources to perform disruptive or illegal acts. Such security breaches have become important concerns for both enterprise and home networks.

If router security is not activated or if the owner deactivates it for convenience, it creates a free hotspot. Since most 21st-century laptop PCs have wireless networking built in, they don't need a third-party adapter such as a PCMCIA Card or USB dongle. Built-in wireless networking might be enabled by default, without the owner realizing it, thus broadcasting the laptop's accessibility to any computer nearby.

Modern operating systems such as Linux, Mac OS, or Microsoft Windows make it fairly easy to set up a PC as a wireless LAN "base station" using Internet Connection Sharing, thus allowing all the PCs in the home to access the Internet through the "base" PC. However, lack of knowledge among users about the security issues inherent in setting up such systems often may allow others nearby access to the connection. Such "piggybacking" is usually achieved without the wireless network operator's knowledge; it may even be without the knowledge of the intruding user if their computer automatically selects a nearby unsecured wireless network to use as an access point.

The Threat Situation

Wireless security is just an aspect of computer security; however, organizations may be particularly vulnerable to security breaches caused by rogue access points.

If an employee (trusted entity) brings in a wireless router and plugs it into an unsecured switchport, the entire network can be exposed to anyone within range of the signals. Similarly, if an employee adds a wireless interface to a networked computer using an open USB port, they may create a breach in network security that would allow access to confidential materials. However, there are effective countermeasures (like disabling open switchports during switch configuration and VLAN configuration to limit network access) that are available to protect both the network and the information it contains, but such countermeasures must be applied uniformly to all network devices.

Threats and Vulnerabilites in an Industrial (M2M) Context

Due to its availability and low cost, the use of wireless communication technologies increases in domains beyond the originally intended usage areas, e.g. M2M communication in industrial applications. Such industrial applications often have specific security requirements. Hence, it is important to understand the characteristics of such applications and evaluate the vulnerabilities bearing the highest risk in this context. Evaluation of these vulnerabilities and the resulting vulnerability catalogs in an industrial context when considering WLAN, NFC and ZigBee are available.

The Mobility Advantage

Wireless networks are very common, both for organizations and individuals. Many laptop computers have wireless cards pre-installed. The ability to enter a network while mobile has great benefits. However, wireless networking is prone to some security issues. Hackers have found wireless networks relatively easy to break into, and even use wireless technology to hack into wired networks. As a result, it is very important that enterprises define effective wireless security policies that guard against unauthorized access to important resources. Wireless Intrusion Prevention Systems (WIPS) or Wireless Intrusion Detection Systems (WIDS) are commonly used to enforce wireless security policies.

The Air Interface and Link Corruption Risk

There were relatively few dangers when wireless technology was first introduced, as the effort to maintain the communication was high and the effort to intrude is always higher. The variety of risks to users of wireless technology have increased as the service has become more popular and the technology more commonly available. Today there are a great number of security risks associated with the current wireless protocols and encryption methods, as carelessness and ignorance exists at the user and corporate IT level. Hacking methods have become much more sophisticated and innovative with wireless.

Modes of Unauthorized Access

The modes of unauthorised access to links, to functions and to data is as variable as the respective entities make use of program code. There does not exist a full scope model of such threat. To some extent the prevention relies on known modes and methods of attack and relevant methods for suppression of the applied methods. However, each new mode of operation will create new options of threatening. Hence prevention requires a steady drive for improvement. The described modes of attack are just a snapshot of typical methods and scenarios where to apply.

Accidental Association

Violation of the security perimeter of a corporate network can come from a number of different methods and intents. One of these methods is referred to as "accidental association". When a user turns on a computer and it latches on to a wireless access point from a neighboring company's overlapping network, the user may not even know that this has occurred. However, it is a security breach in that proprietary company information is exposed and now there could exist a link from one company to the other. This is especially true if the laptop is also hooked to a wired network.

Accidental association is a case of wireless vulnerability called as "mis-association". Mis-association can be accidental, deliberate (for example, done to bypass corporate firewall) or it can result from deliberate attempts on wireless clients to lure them into connecting to attacker's APs.

Malicious Association

"Malicious associations" are when wireless devices can be actively made by attackers to connect to a company network through their laptop instead of a company access point (AP). These types of laptops are known as "soft APs" and are created when a cyber criminal runs some software that makes his/her wireless network card look like a legitimate access point. Once the thief has gained access, he/she can steal passwords, launch attacks on the wired network, or plant trojans. Since wireless networks operate at the Layer 2 level, Layer 3 protections such as network authentication and virtual private networks (VPNs) offer no barrier. Wireless 802.1x authentications do help with some protection but are still vulnerable to hacking. The idea behind this type of attack may not be to break into a VPN or other security measures. Most likely the criminal is just trying to take over the client at the Layer 2 level.

Ad Hoc Networks

Ad hoc networks can pose a security threat. Ad hoc networks are defined as [peer to peer] networks between wireless computers that do not have an access point in between them. While these types of networks usually have little protection, encryption methods can be used to provide security.

The security hole provided by Ad hoc networking is not the Ad hoc network itself but the bridge it provides into other networks, usually in the corporate environment, and the unfortunate default settings in most versions of Microsoft Windows to have this feature turned on unless explicitly disabled. Thus the user may not even know they have an unsecured Ad hoc network in operation on their computer. If they are also using a wired or wireless infrastructure network at the same time, they are providing a bridge to the secured organizational network through the unsecured Ad hoc connection. Bridging is in two forms. A direct bridge, which requires the user actually configure a bridge between the two connections and is thus unlikely to be initiated unless explicitly desired, and an indirect bridge which is the shared resources on the user computer. The indirect bridge may expose private data that is shared from the user's computer to LAN connections, such as shared folders or private Network Attached Storage, making no distinction between authenticated or private connections and unauthenticated Ad-Hoc networks. This presents no threats not already familiar to open/public or unsecured wifi access points, but firewall rules may be circumvented in the case of poorly configured operating systems or local settings.

Non-traditional Networks

Non-traditional networks such as personal network Bluetooth devices are not safe from hacking and should be regarded as a security risk. Even barcode readers, handheld PDAs, and wireless printers and copiers should be secured. These non-traditional networks can be easily overlooked by IT personnel who have narrowly focused on laptops and access points.

Identity Theft (MAC Spoofing)

Identity theft (or MAC spoofing) occurs when a hacker is able to listen in on network traffic and identify the MAC address of a computer with network privileges. Most wireless systems allow some kind of MAC filtering to allow only authorized computers with specific MAC IDs to gain access and utilize the network. However, programs exist that have network "sniffing" capabilities. Combine these programs with other software that allow a computer to pretend it has any MAC address that the hacker desires, and the hacker can easily get around that hurdle.

MAC filtering is effective only for small residential (SOHO) networks, since it provides protection only when the wireless device is "off the air". Any 802.11 device "on the air" freely transmits its unencrypted MAC address in its 802.11 headers, and it requires no special equipment or software to detect it. Anyone with an 802.11 receiver (laptop and wireless adapter) and a freeware wireless packet analyzer can obtain the MAC address of any transmitting 802.11 within range. In an organizational environment, where most wireless devices are "on the air" throughout the active working shift, MAC filtering provides only a false sense of security since it prevents only "casual" or unintended connections to the organizational infrastructure and does nothing to prevent a directed attack.

Man-in-the-middle Attacks

A man-in-the-middle attacker entices computers to log in to a computer which is set up as a soft AP (Access Point). Once this is done, the hacker connects to a real access point through another wireless card offering a steady flow of traffic through the transparent hacking computer to the real network. The hacker can then sniff the traffic. One type of man-in-the-middle attack relies on security faults in challenge and handshake protocols to execute a "de-authentication attack". This attack forces AP-connected computers to drop their connections and reconnect with the hacker's soft AP (disconnects the user from the modem so they have to connect again using their password which one can extract from the recording of the event). Man-in-the-middle attacks are enhanced by software such as LANjack and AirJack which automate multiple steps of the process, meaning what once required some skill can now be done by script kiddies. Hotspots are particularly vulnerable to any attack since there is little to no security on these networks..

Denial of Service

A Denial-of-Service attack (DoS) occurs when an attacker continually bombards a targeted AP (Access Point) or network with bogus requests, premature successful connection messages, failure messages, and/or other commands. These cause legitimate users to not be able to get on the network and may even cause the network to crash. These attacks rely on the abuse of protocols such as the Extensible Authentication Protocol (EAP).

The DoS attack in itself does little to expose organizational data to a malicious attacker, since the interruption of the network prevents the flow of data and actually indirectly protects data by preventing it from being transmitted. The usual reason for performing a DoS attack is to observe the recovery of the wireless network, during which all of the initial handshake codes are re-transmitted by all devices, providing an opportunity for the malicious attacker to record these codes and use various cracking tools to analyze security weaknesses and exploit them to gain unauthorized access to the system. This works best on weakly encrypted systems such as WEP, where there are a number of tools available which can launch a dictionary style attack of "possibly accepted" security keys based on the "model" security key captured during the network recovery.

Network Injection

In a network injection attack, a hacker can make use of access points that are exposed to non-filtered network traffic, specifically broadcasting network traffic such as "Spanning Tree" (802.1D), OSPF, RIP, and HSRP. The hacker injects bogus networking re-configuration commands that affect routers, switches, and intelligent hubs. A whole network can be brought down in this manner and require rebooting or even reprogramming of all intelligent networking devices.

Caffe Latte Attack

The Caffe Latte attack is another way to defeat WEP. It is not necessary for the attacker to be in the area of the network using this exploit. By using a process that targets the Windows wireless stack, it is possible to obtain the WEP key from a remote client. By sending a flood of encrypted ARP requests, the assailant takes advantage of the shared key authentication and the message modification flaws in 802.11 WEP. The attacker uses the ARP responses to obtain the WEP key in less than 6 minutes.

Wireless Intrusion Prevention Concepts

There are three principal ways to secure a wireless network.

- For closed networks (like home users and organizations) the most common way is to configure access restrictions in the access points. Those restrictions may include encryption and checks on MAC address. Wireless Intrusion Prevention Systems can be used to provide wireless LAN security in this network model.

- For commercial providers, hotspots, and large organizations, the preferred solution is often to have an open and unencrypted, but completely isolated wireless network. The users will at first have no access to the Internet nor to any local network resources. Commercial providers usually forward all web traffic to a captive portal which provides for payment and/or authorization. Another solution is to require the users to connect securely to a privileged network using VPN.

- Wireless networks are less secure than wired ones; in many offices intruders can easily visit and hook up their own computer to the wired network without problems, gaining access to the network, and it is also often possible for remote intruders to gain access to the network through backdoors like Back Orifice. One general solution may be end-to-end encryption, with independent authentication on all resources that shouldn't be available to the public.

There is no ready designed system to prevent from fraudulent usage of wireless communication or to protect data and functions with wirelessly communicating computers and other entities. However, there is a system of qualifying the taken measures as a whole according to a common understanding what shall be seen as state of the art. The system of qualifying is an international consensus as specified in ISO/IEC 15408.

A Wireless Intrusion Prevention System

A Wireless Intrusion Prevention System (WIPS) is a concept for the most robust way to counteract wireless security risks. However such WIPS does not exist as a ready designed solution to implement as a software package. A WIPS is typically implemented as an overlay to an existing Wireless LAN infrastructure, although it may be deployed standalone to enforce no-wireless policies within an organization. WIPS is considered so important to wireless security that in July 2009, the Payment Card Industry Security Standards Council published wireless guidelines for PCI DSS recommending the use of WIPS to automate wireless scanning and protection for large organizations.

Security Measures

There are a range of wireless security measures, of varying effectiveness and practicality.

SSID Hiding

A simple but ineffective method to attempt to secure a wireless network is to hide the SSID (Service Set Identifier). This provides very little protection against anything but the most casual intrusion efforts.

MAC ID Filtering

One of the simplest techniques is to only allow access from known, pre-approved MAC addresses. Most wireless access points contain some type of MAC ID filtering. However, an attacker can simply sniff the MAC address of an authorized client and spoof this address.

Static IP Addressing

Typical wireless access points provide IP addresses to clients via DHCP. Requiring clients to set their own addresses makes it more difficult for a casual or unsophisticated intruder to log onto the network, but provides little protection against a sophisticated attacker.

802.11 Security

IEEE 802.1X is the IEEE Standard authentication mechanisms to devices wishing to attach to a Wireless LAN.

Regular WEP

The Wired Equivalent Privacy (WEP) encryption standard was the original encryption standard for wireless, but since 2004 with the ratification WPA2 the IEEE has declared it "deprecated", and while often supported, it is seldom or never the default on modern equipment.

Concerns were raised about its security as early as 2001, dramatically demonstrated in 2005 by the FBI, yet in 2007 T.J. Maxx admitted a massive security breach due in part to a reliance on WEP and the Payment Card Industry took until 2008 to prohibit its use - and even then allowed existing use to continue until June 2010.

WPAv1

The Wi-Fi Protected Access (WPA and WPA2) security protocols were later created to address the problems with WEP. If a weak password, such as a dictionary word or short character string is used, WPA and WPA2 can be cracked. Using a long enough random password (e.g. 14 random letters) or passphrase (e.g. 5 randomly chosen words) makes pre-shared key WPA virtually uncrackable. The second generation of the WPA security protocol (WPA2) is based on the final IEEE 802.11i amendment to the 802.11 standard and is eligible for FIPS 140-2 compliance. With all those encryption schemes, any client in the network that knows the keys can read all the traffic.

Wi-Fi Protected Access (WPA) is a software/firmware improvement over WEP. All regular WLAN-equipment that worked with WEP are able to be simply upgraded and no new equipment needs to be bought. WPA is a trimmed-down version of the 802.11i security standard that was developed by the IEEE 802.11 to replace WEP. The TKIP encryption algorithm was developed for WPA to provide improvements to WEP that could be fielded as firmware upgrades to existing 802.11 devices. The WPA profile also provides optional support for the AES-CCMP algorithm that is the preferred algorithm in 802.11i and WPA2.

WPA Enterprise provides RADIUS based authentication using 802.1x. WPA Personal uses a pre-shared Shared Key (PSK) to establish the security using an 8 to 63 character passphrase. The PSK may also be entered as a 64 character hexadecimal string. Weak PSK passphrases can be broken using off-line dictionary attacks by capturing the messages in the four-way exchange when the client reconnects after being deauthenticated. Wireless suites such as aircrack-ng can crack a weak passphrase in less than a minute. Other WEP/WPA crackers are AirSnort and Auditor Security Collection. Still, WPA Personal is secure when used with 'good' passphrases or a full 64-character hexadecimal key.

There was information, however, that Erik Tews (the man who created the fragmentation attack against WEP) was going to reveal a way of breaking the WPA TKIP implementation at Tokyo's PacSec security conference in November 2008, cracking the encryption on a packet in between 12–15 minutes. Still, the announcement of this 'crack' was somewhat overblown by the media, because as of August, 2009, the best attack on WPA (the Beck-Tews attack) is only partially successful in that it only works on short data packets, it cannot decipher the WPA key, and it requires very specific WPA implementations in order to work.

Additions to WPAv1

In addition to WPAv1, TKIP, WIDS and EAP may be added alongside. Also, VPN-networks (non-continuous secure network connections) may be set up under the 802.11-standard. VPN implementations include PPTP, L2TP, IPsec and SSH. However, this extra layer of security may also be cracked with tools such as Anger, Deceit and Ettercap for PPTP; and ike-scan, IKEProbe, ipsectrace, and IKEcrack for IPsec-connections.

TKIP

This stands for Temporal Key Integrity Protocol and the acronym is pronounced as tee-kip. This is part of the IEEE 802.11i standard. TKIP implements per-packet key mixing with a re-keying system and also provides a message integrity check. These avoid the problems of WEP.

EAP

The WPA-improvement over the IEEE 802.1X standard already improved the authentication and authorization for access of wireless and wired LANs. In addition to this, extra measures such as the Extensible Authentication Protocol (EAP) have initiated an even greater amount of security. This, as EAP uses a central authentication server. Unfortunately, during 2002 a Maryland professor discovered some shortcomings. Over the next few years these shortcomings were addressed with the use of TLS and other enhancements. This new version of EAP is now called Extended EAP and is available in several versions; these include: EAP-MD5, PEAPv0, PEAPv1, EAP-MSCHAPv2, LEAP, EAP-FAST, EAP-TLS, EAP-TTLS, MSCHAPv2, and EAP-SIM.

EAP-versions

EAP-versions include LEAP, PEAP and other EAP's.

LEAP

This stands for the Lightweight Extensible Authentication Protocol. This protocol is based on 802.1X and helps minimize the original security flaws by using WEP and a sophisticated key management system. This EAP-version is safer than EAP-MD5. This also uses MAC address authentication. LEAP is not secure; THC-LeapCracker can be used to break Cisco's version of LEAP and be used against computers connected to an access point in the form of a dictionary attack. Anwrap and asleap finally are other crackers capable of breaking LEAP.

PEAP

This stands for Protected Extensible Authentication Protocol. This protocol allows for a secure transport of data, passwords, and encryption keys without the need of a certificate server. This was developed by Cisco, Microsoft, and RSA Security.

Other EAPs There are other types of Extensible Authentication Protocol implementations that are based on the EAP framework. The framework that was established supports existing EAP types as well as future authentication methods. EAP-TLS offers very good protection because of its mutual authentication. Both the client and the network are authenticated using certificates and per-session WEP keys. EAP-FAST also offers good protection. EAP-TTLS is another alternative made by Certicom and Funk Software. It is more convenient as one does not need to distribute certificates to users, yet offers slightly less protection than EAP-TLS.

Restricted Access Networks

Solutions include a newer system for authentication, IEEE 802.1x, that promises to enhance security on both wired and wireless networks. Wireless access points that incorporate technologies like these often also have routers built in, thus becoming wireless gateways.

End-to-end Encryption

One can argue that both layer 2 and layer 3 encryption methods are not good enough for protecting valuable data like passwords and personal emails. Those technologies add encryption only to parts of the communication path, still allowing people to spy on the traffic if they have gained access to the wired network somehow. The solution may be encryption and authorization in the application layer, using technologies like SSL, SSH, GnuPG, PGP and similar.

The disadvantage with the end-to-end method is, it may fail to cover all traffic. With encryption on the router level or VPN, a single switch encrypts all traffic, even UDP and DNS lookups. With end-to-end encryption on the other hand, each service to be secured must have its encryption "turned on", and often every connection must also be "turned on" separately. For sending emails, every recipient must support the encryption method, and must exchange keys correctly. For Web, not all web sites offer https, and even if they do, the browser sends out IP addresses in clear text.

The most prized resource is often access to Internet. An office LAN owner seeking to restrict such access will face the nontrivial enforcement task of having each user authenticate themselves for the router.

802.11i Security

The newest and most rigorous security to implement into WLAN's today is the 802.11i RSN-standard. This full-fledged 802.11i standard (which uses WPAv2) however does require the newest hardware (unlike WPAv1), thus potentially requiring the purchase of new equipment. This new hardware required may be either AES-WRAP (an early version of 802.11i) or the newer and better AES-CCMP-equipment. One should make sure one needs WRAP or CCMP-equipment, as the 2 hardware standards are not compatible.

WPAv2

WPA2 is a WiFi Alliance branded version of the final 802.11i standard. The primary enhancement over WPA is the inclusion of the AES-CCMP algorithm as a mandatory feature. Both WPA and WPA2 support EAP authentication methods using RADIUS servers and preshared key (PSK).

The number of WPA and WPA2 networks are increasing, while the number of WEP networks are decreasing, because of the security vulnerabilities in WEP.

WPA2 has been found to have at least one security vulnerability, nicknamed Hole196. The vulnerability uses the WPA2 Group Temporal Key (GTK), which is a shared key among all users of the same BSSID, to launch attacks on other users of the same BSSID. It is named after page 196 of the IEEE 802.11i specification, where the vulnerability is discussed. In order for this exploit to be performed, the GTK must be known by the attacker.

Additions to WPAv2

Unlike 802.1X, 802.11i already has most other additional security-services such as TKIP. Just as with WPAv1, WPAv2 may work in cooperation with EAP and a WIDS.

WAPI

This stands for WLAN Authentication and Privacy Infrastructure. This is a wireless security standard defined by the Chinese government.

Smart Cards, USB Tokens, and Software Tokens

This is a very strong form of security. When combined with some server software, the hardware or software card or token will use its internal identity code combined with a user entered PIN to create a powerful algorithm that will very frequently generate a new encryption code. The server will be time synced to the card or token. This is a very secure way to conduct wireless transmissions. Companies in this area make USB tokens, software tokens, and smart cards. They even make hardware versions that double as an employee picture badge. Currently the safest security measures are the smart cards / USB tokens. However, these are expensive. The next safest methods are WPA2 or WPA with a RADIUS server. Any one of the three will provide a good base foundation for security. The third item on the list is to educate both employees and contractors on security risks and personal preventive measures. It is also IT's task to keep the company workers' knowledge base up-to-date on any new dangers that they should be cautious about. If the employees are educated, there will be a much lower chance that anyone will accidentally cause a breach in security by not locking down their laptop or bring in a wide open home access point to extend their mobile range. Employees need to be made aware that company laptop security extends to outside of their site walls as well. This includes places such as coffee houses where workers can be at their most vulnerable. The last item on the list deals with 24/7 active defense measures to ensure that the company network is secure and compliant. This can take the form of regularly looking at access point, server, and firewall logs to try to detect any unusual activity. For instance, if any large files went through an access point in the early hours of the morning, a serious investigation into the incident would be called for. There are a number of software and hardware devices that can be used to supplement the usual logs and usual other safety measures.

RF Shielding

It's practical in some cases to apply specialized wall paint and window film to a room or building to significantly attenuate wireless signals, which keeps the signals from propagating outside a facility. This can significantly improve wireless security because it's difficult for hackers to receive the signals beyond the controlled area of an enterprise, such as within parking lots.

Despite security measures as encryption, hackers may still be able to crack them. This is done using several techniques and tools. An overview of them can be found at the Network encryption cracking article, to understand what we are dealing with. Understanding the mindset/techniques of the hacker allows one to better protect their system.

For closed networks (like home users and organizations) the most common way is to configure access restrictions in the access points. Those restrictions may include encryption and checks on MAC address. Another option is to disable ESSID broadcasting, making the access point difficult for outsiders to detect. Wireless Intrusion Prevention Systems can be used to provide wireless LAN security in this network model.

For commercial providers, hotspots, and large organizations, the preferred solution is often to have an open and unencrypted, but completely isolated wireless network. The users will at first have no access to the Internet nor to any local network resources. Commercial providers usually forward all web traffic to a captive portal which provides for payment and/or authorization. Another solution is to require the users to connect securely to a privileged network using VPN. Wireless networks are less secure than wired ones; in many offices intruders can easily visit and hook up their own computer to the wired network without problems, gaining access to the network, and it is also often possible for remote intruders to gain access to the network through backdoors like Back Orifice. One general solution may be end-to-end encryption, with independent authentication on all resources that shouldn't be available to the public.

Denial of Service Defense

Most DoS attacks are easy to detect. However, a lot of them are difficult to stop even after detection. Here are three of the most common ways to stop a DoS attack.

- Black Holing

- Validating the Handshake

- Rate Limiting

Black Holing is one possible way of stopping a DoS attack. This is a situation where we drop all IP packets from an attacker. This is not a very good long-term strategy because attackers can change their source address very quickly.

This may have negative effects if done automatically. An attacker could knowingly spoof attack packets with the IP address of a corporate partner. Automated defenses could block legitimate traffic from that partner and cause additional problems.

Validating the Handshake involves creating false opens, and not setting aside resources until the sender acknowledges. Some firewalls address SYN floods by pre-validating the TCP handshake. This is done by creating false opens. Whenever a SYN segment arrives, the firewall sends back a SYN/ACK segment, without passing the SYN segment on to the target server.

Only when the firewall gets back an ACK, which would happen only in a legitimate connection, would the firewall send the original SYN segment on to the server for which it was originally intended. The firewall doesn't set aside resources for a connection when a SYN segment arrives, so handling a large number of false SYN segments is only a small burden.

Rate limiting can be used to reduce a certain type of traffic down to an amount the can be reasonably dealt with. Broadcasting to the internal network could still be used, but only at a limited rate for example. This is for more subtle DoS attacks. This is good if an attack is aimed at a single server because it keeps transmission lines at least partially open for other communication.

Rate limiting frustrates both the attacker, and the legitimate users. This helps but does not fully solve the problem. Once DoS traffic clogs the access line going to the internet, there is nothing a border firewall can do to help the situation. Most DoS attacks are problems of the community which can only be stopped with the help of ISP's and organizations whose computers are taken over as bots and used to attack other firms.

Mobile Devices

With increasing number of mobile devices with 802.1x interfaces, security of such mobile devices becomes a concern. While open standards such as Kismet are targeted towards securing laptops, access points solutions should extend towards covering mobile devices also. Host based solutions for mobile handsets and PDA's with 802.1x interface.

Security within mobile devices fall under three categories:

1. Protecting against ad hoc networks

2. Connecting to rogue access points

3. Mutual authentication schemes such as WPA2 as described above

Wireless IPS solutions now offer wireless security for mobile devices.

Mobile patient monitoring devices are becoming an integral part of healthcare industry and these devices will eventually become the method of choice for accessing and implementing health checks for patients located in remote areas. For these types of patient monitoring systems, security and reliability are critical, because they can influence the condition of patients, and could leave medical professionals in the dark about the condition of the patient if compromised.

Implementing Network Encryption

In order to implement 802.11i, one must first make sure both that the router/access point(s), as well as all client devices are indeed equipped to support the network encryption. If this is done, a server such as RADIUS, ADS, NDS, or LDAP needs to be integrated. This server can be a computer on the local network, an access point / router with integrated authentication server, or a remote server. AP's/routers with integrated authentication servers are often very expensive and specifically an option for commercial usage like hot spots. Hosted 802.1X servers via the Internet require a monthly fee; running a private server is free yet has the disadvantage that one must set it up and that the server needs to be on continuously.

To set up a server, server and client software must be installed. Server software required is an enterprise authentication server such as RADIUS, ADS, NDS, or LDAP. The required software can be picked from various suppliers as Microsoft, Cisco, Funk Software, Meetinghouse Data, and from some open-source projects. Software includes:

- Aradial RADIUS Server

- Cisco Secure Access Control Software

- freeRADIUS (open-source)

- Funk Software Steel Belted RADIUS (Odyssey)

- Microsoft Internet Authentication Service

- Meetinghouse Data EAGIS

- SkyFriendz (free cloud solution based on freeRADIUS)

Client software comes built-in with Windows XP and may be integrated into other OS's using any of following software:

- AEGIS-client

- Cisco ACU-client

- Intel PROSet/Wireless Software

- Odyssey client

- Xsupplicant (open1X)-project

RADIUS

Remote Authentication Dial In User Service (RADIUS) is an AAA (authentication, authorization and accounting) protocol used for remote network access. RADIUS was originally proprietary but was later published under ISOC documents RFC 2138 and RFC 2139. The idea is to have an inside server act as a gatekeeper by verifying identities through a username and password that is already pre-determined by the user. A RADIUS server can also be configured to enforce user policies and restrictions as well as record accounting information such as connection time for purposes such as billing.

Open Access Points

Today, there is almost full wireless network coverage in many urban areas - the infrastructure for the wireless community network (which some consider to be the future of the internet) is already in place. One could roam around and always be connected to Internet if the nodes were open to the public, but due to security concerns, most nodes are encrypted and the users don't know how to disable encryption. Many people consider it proper etiquette to leave access points open to the public, allowing free access to Internet. Others think the default encryption provides substantial protection at small inconvenience, against dangers of open access that they fear may be substantial even on a home DSL router.

The density of access points can even be a problem - there are a limited number of channels available, and they partly overlap. Each channel can handle multiple networks, but places with many private wireless networks (for example, apartment complexes), the limited number of Wi-Fi radio channels might cause slowness and other problems.

According to the advocates of Open Access Points, it shouldn't involve any significant risks to open up wireless networks for the public:

- The wireless network is after all confined to a small geographical area. A computer connected to the Internet and having improper configurations or other security problems can be exploited by anyone from anywhere in the world, while only clients in a small geographical range can exploit an open wireless access point. Thus the exposure is low with an open wireless access point, and the risks with having an open wireless network are small. However, one should be aware that an open wireless router will give access to the local network, often including access to file shares and printers.

- The only way to keep communication truly secure is to use end-to-end encryption. For example, when accessing an internet bank, one would almost always use strong encryption from the web browser and all the way to the bank - thus it shouldn't be risky to do banking over an unencrypted wireless network. The argument is that anyone can sniff the traffic applies to wired networks too, where system administrators and possible hackers have access to the links and can read the traffic. Also, anyone knowing the keys for an encrypted wireless network can gain access to the data being transferred over the network.

- If services like file shares, access to printers etc. are available on the local net, it is advisable to have authentication (i.e. by password) for accessing it (one should never assume that the private network is not accessible from the outside). Correctly set up, it should be safe to allow access to the local network to outsiders.

- With the most popular encryption algorithms today, a sniffer will usually be able to compute the network key in a few minutes.

- It is very common to pay a fixed monthly fee for the Internet connection, and not for the traffic - thus extra traffic will not be detrimental.

- Where Internet connections are plentiful and cheap, freeloaders will seldom be a prominent nuisance.

On the other hand, in some countries including Germany, persons providing an open access point may be made (partially) liable for any illegal activity conducted via this access point. Also, many contracts with ISPs specify that the connection may not be shared with other persons.

Wired Equivalent Privacy

Wired Equivalent Privacy (WEP) is a security algorithm for IEEE 802.11 wireless networks. Introduced as part of the original 802.11 standard ratified in 1997, its intention was to provide data confidentiality comparable to that of a traditional wired network. WEP, recognizable by the key of

10 or 26 hexadecimal digits, was at one time widely in use and was often the first security choice presented to users by router configuration tools.

In 2003 the Wi-Fi Alliance announced that WEP had been superseded by Wi-Fi Protected Access (WPA). In 2004, with the ratification of the full 802.11i standard (i.e. WPA2), the IEEE declared that both WEP-40 and WEP-104 have been deprecated.

Encryption Details

WEP was included as the privacy component of the original IEEE 802.11 standard ratified in 1997. WEP uses the stream cipher RC4 for confidentiality, and the CRC-32 checksum for integrity. It was deprecated in 2004 and is documented in the current standard.

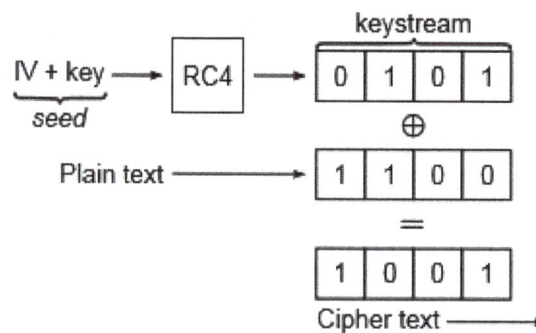

Basic WEP encryption: RC4 keystream XORed with plaintext

Standard 64-bit WEP uses a 40 bit key (also known as WEP-40), which is concatenated with a 24-bit initialization vector (IV) to form the RC4 key. At the time that the original WEP standard was drafted, the U.S. Government's export restrictions on cryptographic technology limited the key size. Once the restrictions were lifted, manufacturers of access points implemented an extended 128-bit WEP protocol using a 104-bit key size (WEP-104).

A 64-bit WEP key is usually entered as a string of 10 hexadecimal (base 16) characters (0–9 and A–F). Each character represents 4 bits, 10 digits of 4 bits each gives 40 bits; adding the 24-bit IV produces the complete 64-bit WEP key (4 bits × 10 + 24 bits IV = 64 bits of WEP key). Most devices also allow the user to enter the key as 5 ASCII characters (0–9, a–z, A–Z), each of which is turned into 8 bits using the character's byte value in ASCII (8 bits × 5 + 24 bits IV = 64 bits of WEP key); however, this restricts each byte to be a printable ASCII character, which is only a small fraction of possible byte values, greatly reducing the space of possible keys.

A 128-bit WEP key is usually entered as a string of 26 hexadecimal characters. 26 digits of 4 bits each gives 104 bits; adding the 24-bit IV produces the complete 128-bit WEP key (4 bits × 26 + 24 bits IV = 128 bits of WEP key). Most devices also allow the user to enter it as 13 ASCII characters (8 bits × 13 + 24 bits IV = 128 bits of WEP key).

A 152-bit and a 256-bit WEP systems are available from some vendors. As with the other WEP variants, 24 bits of that is for the IV, leaving 128 or 232 bits for actual protection. These 128 or 232 bits are typically entered as 32 or 58 hexadecimal characters (4 bits × 32 + 24 bits IV = 152 bits of WEP key, 4 bits × 58 + 24 bits IV = 256 bits of WEP key). Most devices also allow the user to enter it as 16 or 29 ASCII characters (8 bits × 16 + 24 bits IV = 152 bits of WEP key, 8 bits × 29 + 24 bits IV = 256 bits of WEP key).

Authentication

Two methods of authentication can be used with WEP: Open System authentication and Shared Key authentication.

For the sake of clarity, we discuss WEP authentication in the Infrastructure mode (that is, between a WLAN client and an Access Point). The discussion applies to the ad hoc mode as well.

In Open System authentication, the WLAN client need not provide its credentials to the Access Point during authentication. Any client can authenticate with the Access Point and then attempt to associate. In effect, no authentication occurs. Subsequently, WEP keys can be used for encrypting data frames. At this point, the client must have the correct keys.

In Shared Key authentication, the WEP key is used for authentication in a four-step challenge-response handshake:

1. The client sends an authentication request to the Access Point.

2. The Access Point replies with a clear-text challenge.

3. The client encrypts the challenge-text using the configured WEP key and sends it back in another authentication request.

4. The Access Point decrypts the response. If this matches the challenge text, the Access Point sends back a positive reply.

After the authentication and association, the pre-shared WEP key is also used for encrypting the data frames using RC4.

At first glance, it might seem as though Shared Key authentication is more secure than Open System authentication, since the latter offers no real authentication. However, it is quite the reverse. It is possible to derive the keystream used for the handshake by capturing the challenge frames in Shared Key authentication. Therefore, data can be more easily intercepted and decrypted with Shared Key authentication than with Open System authentication. If privacy is a primary concern, it is more advisable to use Open System authentication for WEP authentication, rather than Shared Key authentication; however, this also means that any WLAN client can connect to the AP. (Both authentication mechanisms are weak; Shared Key WEP is deprecated in favor of WPA/WPA2.)

Security Details

Because RC4 is a stream cipher, the same traffic key must never be used twice. The purpose of an IV, which is transmitted as plain text, is to prevent any repetition, but a 24-bit IV is not long enough to ensure this on a busy network. The way the IV was used also opened WEP to a related key attack. For a 24-bit IV, there is a 50% probability the same IV will repeat after 5000 packets.

In August 2001, Scott Fluhrer, Itsik Mantin, and Adi Shamir published a cryptanalysis of WEP that exploits the way the RC4 ciphers and IV are used in WEP, resulting in a passive attack that can recover the RC4 key after eavesdropping on the network. Depending on the amount of network traffic, and thus the number of packets available for inspection, a successful key recovery could take as little as one minute. If an insufficient number of packets are being sent, there are ways for an attacker to send packets on the network and thereby stimulate reply packets which can then be inspected to find the key. The attack was soon implemented, and automated tools have since been released. It is possible to perform the attack with a personal computer, off-the-shelf hardware and freely available software such as aircrack-ng to crack *any* WEP key in minutes.

Cam-Winget et al. surveyed a variety of shortcomings in WEP. They write *"Experiments in the field show that, with proper equipment, it is practical to eavesdrop on WEP-protected networks from distances of a mile or more from the target."* They also reported two generic weaknesses:

- the use of WEP was optional, resulting in many installations never even activating it, and

- by default, WEP relies on a single shared key among users, which leads to practical problems in handling compromises, which often leads to ignoring compromises.

In 2005, a group from the U.S. Federal Bureau of Investigation gave a demonstration where they cracked a WEP-protected network in 3 minutes using publicly available tools. Andreas Klein presented another analysis of the RC4 stream cipher. Klein showed that there are more correlations between the RC4 keystream and the key than the ones found by Fluhrer, Mantin and Shamir which can additionally be used to break WEP in WEP-like usage modes.

In 2006, Bittau, Handley, and Lackey showed that the 802.11 protocol itself can be used against WEP to enable earlier attacks that were previously thought impractical. After eavesdropping a single packet, an attacker can rapidly bootstrap to be able to transmit arbitrary data. The eavesdropped packet can then be decrypted one byte at a time (by transmitting about 128 packets per byte to decrypt) to discover the local network IP addresses. Finally, if the 802.11 network is connected to the Internet, the attacker can use 802.11 fragmentation to replay eavesdropped packets while crafting a new IP header onto them. The access point can then be used to decrypt these packets and relay them on to a buddy on the Internet, allowing real-time decryption of WEP traffic within a minute of eavesdropping the first packet.

In 2007, Erik Tews, Andrei Pychkine, and Ralf-Philipp Weinmann were able to extend Klein's 2005 attack and optimize it for usage against WEP. With the new attack it is possible to recover a 104-bit WEP key with probability 50% using only 40,000 captured packets. For 60,000 available data packets, the success probability is about 80% and for 85,000 data packets about 95%. Using active techniques like deauth and ARP re-injection, 40,000 packets can be captured in less than one minute under good conditions. The actual computation takes about 3 seconds and 3 MB of main memory on a Pentium-M 1.7 GHz and can additionally be optimized for devices with slower CPUs. The same attack can be used for 40-bit keys with an even higher success probability.

In 2008, Payment Card Industry (PCI) Security Standards Council's latest update of the Data Security Standard (DSS), prohibits use of the WEP as part of any credit-card processing after 30 June

2010, and prohibits any new system from being installed that uses WEP after 31 March 2009. The use of WEP contributed to the T.J. Maxx parent company network invasion.

Remedies

Use of encrypted tunneling protocols (e.g. IPSec, Secure Shell) can provide secure data transmission over an insecure network. However, replacements for WEP have been developed with the goal of restoring security to the wireless network itself.

802.11i (WPA and WPA2)

The recommended solution to WEP security problems is to switch to WPA2. WPA was an intermediate solution for hardware that could not support WPA2. Both WPA and WPA2 are much more secure than WEP. To add support for WPA or WPA2, some old Wi-Fi access points might need to be replaced or have their firmware upgraded. WPA was designed as an interim software-implementable solution for WEP that could forestall immediate deployment of new hardware. However, TKIP (the basis of WPA) has reached the end of its designed lifetime, has been partially broken, and had been officially deprecated with the release of the 802.11-2012 standard.

Implemented Non-standard Fixes

WEP2

This stopgap enhancement to WEP was present in some of the early 802.11i drafts. It was implementable on *some* (not all) hardware not able to handle WPA or WPA2, and extended both the IV and the key values to 128 bits. It was hoped to eliminate the duplicate IV deficiency as well as stop brute force key attacks.

After it became clear that the overall WEP algorithm was deficient (and not just the IV and key sizes) and would require even more fixes, both the WEP2 name and original algorithm were dropped. The two extended key lengths remained in what eventually became WPA's TKIP.

WEPplus

WEPplus, also known as WEP+, is a proprietary enhancement to WEP by Agere Systems (formerly a subsidiary of Lucent Technologies) that enhances WEP security by avoiding "weak IVs". It is only completely effective when WEPplus is used at *both ends* of the wireless connection. As this cannot easily be enforced, it remains a serious limitation. It also does not necessarily prevent replay attacks, and is ineffective against later statistical attacks that do not rely on weak IVs.

Dynamic Wep

Dynamic WEP refers to the combination of 802.1x technology and the Extensible Authentication Protocol. Dynamic WEP changes WEP keys dynamically. It is a vendor-specific feature provided by several vendors such as 3Com.

The dynamic change idea made it into 802.11i as part of TKIP, but not for the actual WEP algorithm.

Wi-Fi Protected Access

An example of a Wi-Fi Protected Access label found on a consumer device

Wi-Fi Protected Access (WPA) and Wi-Fi Protected Access II (WPA2) are two security protocols and security certification programs developed by the Wi-Fi Alliance to secure wireless computer networks. The Alliance defined these in response to serious weaknesses researchers had found in the previous system, Wired Equivalent Privacy (WEP).

WPA (sometimes referred to as the *draft IEEE 802.11i* standard) became available in 2003. The Wi-Fi Alliance intended it as an intermediate measure in anticipation of the availability of the more secure and complex WPA2. WPA2 became available in 2004 and is a common shorthand for the full IEEE 802.11i (or IEEE 802.11i-2004) standard.

A flaw in a feature added to Wi-Fi, called Wi-Fi Protected Setup, allows WPA and WPA2 security to be bypassed and effectively broken in many situations. WPA and WPA2 security implemented without using the Wi-Fi Protected Setup feature are unaffected by the security vulnerability.

WPA

The Wi-Fi Alliance intended WPA as an intermediate measure to take the place of WEP pending the availability of the full IEEE 802.11i standard. WPA could be implemented through firmware upgrades on wireless network interface cards designed for WEP that began shipping as far back as 1999. However, since the changes required in the wireless access points (APs) were more extensive than those needed on the network cards, most pre-2003 APs could not be upgraded to support WPA.

The WPA protocol implements much of the IEEE 802.11i standard. Specifically, the Temporal Key Integrity Protocol (TKIP) was adopted for WPA. WEP used a 64-bit or 128-bit encryption key that must be manually entered on wireless access points and devices and does not change. TKIP employs a per-packet key, meaning that it dynamically generates a new 128-bit key for each packet and thus prevents the types of attacks that compromised WEP.

WPA also includes a message integrity check, which is designed to prevent an attacker from altering and resending data packets. This replaces the cyclic redundancy check (CRC) that was used by the WEP standard. CRC's main flaw was that it did not provide a sufficiently strong data integrity guarantee for the packets it handled. Well tested message authentication codes existed to solve these problems, but they required too much computation to be used on old network cards. WPA uses a message integrity check algorithm called *Michael* to verify the integrity of the packets. Michael is much stronger than a CRC, but not as strong as the algorithm used in WPA2. Researchers have since discovered a flaw in WPA that relied on older weaknesses in WEP and the limitations of Michael to retrieve the keystream from short packets to use for re-injection and spoofing.

WPA2

WPA2 replaced WPA. WPA2, which requires testing and certification by the Wi-Fi Alliance, implements the mandatory elements of IEEE 802.11i. In particular, it includes mandatory support for CCMP, an AES-based encryption mode with strong security. Certification began in September, 2004; from March 13, 2006, WPA2 certification is mandatory for all new devices to bear the Wi-Fi trademark.

Hardware Support

WPA has been designed specifically to work with wireless hardware produced prior to the introduction of WPA protocol, which provides inadequate security through WEP. Some of these devices support WPA only after applying firmware upgrades, which are not available for some legacy devices.

Wi-Fi devices certified since 2006 support both the WPA and WPA2 security protocols. WPA2 may not work with some older network cards.

WPA Terminology

Different WPA versions and protection mechanisms can be distinguished based on the target end-user (according to the method of authentication key distribution), and the encryption protocol used.

Target Users (Authentication Key Distribution)

WPA-Personal

Also referred to as *WPA-PSK* (pre-shared key) mode, this is designed for home and small office networks and doesn't require an authentication server. Each wireless network device encrypts the network traffic using a 256 bit key. This key may be entered either as a string of 64 hexadecimal digits, or as a passphrase of 8 to 63 printable ASCII characters. If ASCII characters are used, the 256 bit key is calculated by applying the PBKDF2 key derivation function to the passphrase, using the SSID as the salt and 4096 iterations of HMAC-SHA1. WPA-Personal mode is available with both WPA and WPA2.

WPA-Enterprise

Also referred to as *WPA-802.1X mode*, and sometimes just *WPA* (as opposed to WPA-PSK), this is designed for enterprise networks and requires a RADIUS authentication server. This requires a more complicated setup, but provides additional security (e.g. protection against dictionary attacks on short passwords). Various kinds of the Extensible Authentication Protocol (EAP) are used for authentication. WPA-Enterprise mode is available with both WPA and WPA2.

Wi-Fi Protected Setup(wifi PS)

This is an alternative authentication key distribution method intended to simplify and strengthen the process, but which, as widely implemented, creates a major security hole via WPS PIN recovery.

Encryption Protocol

TKIP (Temporal Key Integrity Protocol)

> The RC4 stream cipher is used with a 128-bit per-packet key, meaning that it dynamically generates a new key for each packet. Used by WPA.

CCMP (*CTR* mode with *CBC-MAC* Protocol)

> The protocol used by WPA2, based on the Advanced Encryption Standard (AES) cipher along with strong message authenticity and integrity checking that is significantly stronger in protection for both privacy and integrity than the RC4-based TKIP used by WPA. Among informal names are "AES" and "AES-CCMP". According to the 802.11n specification, this encryption protocol must be used to achieve the fast 802.11n high bitrate schemes, though not all implementations enforce this. Otherwise, the data rate will not exceed 54 MBit/s.

EAP Extensions Under WPA and WPA2 Enterprise

Originally, only EAP-TLS (Extensible Authentication Protocol - Transport Layer Security) was certified by the Wi-Fi alliance. In April 2010, the Wi-Fi Alliance announced the inclusion of additional EAP types to its WPA- and WPA2- Enterprise certification programs. This was to ensure that WPA-Enterprise certified products can interoperate with one another.

As of 2010 the certification program includes the following EAP types:

- EAP-TLS (previously tested)

- EAP-TTLS/MSCHAPv2 (April 2005)

- PEAPv0/EAP-MSCHAPv2 (April 2005)

- PEAPv1/EAP-GTC (April 2005)

- PEAP-TLS

- EAP-SIM (April 2005)

- EAP-AKA (April 2009)

- EAP-FAST (April 2009)

802.1X clients and servers developed by specific firms may support other EAP types. This certification is an attempt for popular EAP types to interoperate; their failure to do so as of 2013 is one of the major issues preventing rollout of 802.1X on heterogeneous networks.

Commercial 802.1X servers include Microsoft Internet Authentication Service and Juniper Networks Steelbelted RADIUS as well as Aradial Radius server. FreeRADIUS is an open source 802.1X server.

Security Issues

Weak Password

Pre-shared key WPA and WPA2 remain vulnerable to password cracking attacks if users rely on a weak password or passphrase. To protect against a brute force attack, a truly random passphrase of 20 characters (selected from the set of 95 permitted characters) is probably sufficient.

Brute forcing of simple passwords can be attempted using the Aircrack Suite starting from the four-way authentication handshake exchanged during association or periodic re-authentication.

To further protect against intrusion, the network's SSID should not match any entry in the top 1,000 SSIDs as downloadable rainbow tables have been pre-generated for them and a multitude of common passwords.

WPA Packet Spoofing and Decryption

Mathy Vanhoef and Frank Piessens significantly improved upon the WPA-TKIP attacks of Erik Tews and Martin Beck. They demonstrated how to inject an arbitrary amount of packets, with each packet containing at most 112 bytes of payload. This was demonstrated by implementing a port scanner, which can be executed against any client using WPA-TKIP. Additionally they showed how to decrypt arbitrary packets sent to a client. They mentioned this can be used to hijack a TCP connection, allowing an attacker to inject malicious JavaScript when the victim visits a website. In contrast, the Beck-Tews attack could only decrypt short packets with mostly known content, such as ARP messages, and only allowed injection of 3 to 7 packets of at most 28 bytes. The Beck-Tews attack also requires Quality of Service (as defined in 802.11e) to be enabled, while the Vanhoef-Piessens attack does not. Both attacks do not lead to recovery of the shared session key between the client and Access Point. The authors say using a short rekeying interval can prevent some attacks but not all, and strongly recommend switching from TKIP to AES-based CCMP.

Halvorsen and others show how to modify the Beck-Tews attack to allow injection of 3 to 7 packets having a size of at most 596 bytes. The downside is that their attack requires substantially more time to execute: approximately 18 minutes and 25 seconds. In other work Vanhoef and Piessens showed that, when WPA is used to encrypt broadcast packets, their original attack can also be executed. This is an important extension, as substantially more networks use WPA to protect broadcast packets, than to protect unicast packets. The execution time of this attack is on average around 7 minutes, compared to the 14 minutes of the original Vanhoef-Piessens and Beck-Tews attack.

The vulnerabilities of TKIP are significant in that WPA-TKIP had been held to be an extremely safe combination; indeed, WPA-TKIP is still a configuration option upon a wide variety of wireless routing devices provided by many hardware vendors. A survey in 2013 showed that 71% still allow usage of TKIP, and 19% exclusively support TKIP.

WPS PIN Recovery

A more serious security flaw was revealed in December 2011 by Stefan Viehböck that affects wireless routers with the Wi-Fi Protected Setup (WPS) feature, regardless of which encryption method they use. Most recent models have this feature and enable it by default. Many consumer Wi-Fi device manufacturers had taken steps to eliminate the potential of weak passphrase choices by promoting alternative methods of automatically generating and distributing strong keys when users add a new wireless adapter or appliance to a network. These methods include pushing buttons on the devices or entering an 8-digit PIN.

The Wi-Fi Alliance standardized these methods as Wi-Fi Protected Setup; however the PIN feature as widely implemented introduced a major new security flaw. The flaw allows a remote attacker to recover the WPS PIN and, with it, the router's WPA/WPA2 password in a few hours. Users have been urged to turn off the WPS feature, although this may not be possible on some router models. Also note that the PIN is written on a label on most Wi-Fi routers with WPS, and cannot be changed if compromised.

MS-CHAPv2

Several weaknesses have been found in MS-CHAPv2, some of which severely reduce the complexity of brute-force attacks making them feasible with modern hardware. In 2012 the complexity of breaking MS-CHAPv2 was reduced to that of breaking a single DES key, work by Moxie Marlinspike and Marsh Ray. Moxie advised: "Enterprises who are depending on the mutual authentication properties of MS-CHAPv2 for connection to their WPA2 Radius servers should immediately start migrating to something else."

Hole196

Hole196 is a vulnerability in the WPA2 protocol that abuses the shared Group Temporal Key (GTK). It can be used to conduct man-in-the-middle and denial-of-service attacks. However, it assumes that the attacker is already authenticated against Access Point and thus in possession of the GTK.

Wireless Intrusion Prevention System

In computing, a wireless intrusion prevention system (WIPS) is a network device that monitors the radio spectrum for the presence of unauthorized access points *(intrusion detection)*, and can automatically take countermeasures *(intrusion prevention)*.

Purpose

The primary purpose of a WIPS is to prevent unauthorized network access to local area networks and other information assets by wireless devices. These systems are typically implemented as an overlay to an existing Wireless LAN infrastructure, although they may be deployed standalone to enforce no-wireless policies within an organization. Some advanced wireless infrastructure has integrated WIPS capabilities.

Large organizations with many employees are particularly vulnerable to security breaches caused by rogue access points. If an employee (trusted entity) in a location brings in an easily available wireless router, the entire network can be exposed to anyone within range of the signals.

In July 2009, the PCI Security Standards Council published wireless guidelines for PCI DSS recommending the use of WIPS to automate wireless scanning for large organizations.

Intrusion Detection

A wireless intrusion detection system (WIDS) monitors the radio spectrum for the presence of unauthorized, rogue access points and the use of wireless attack tools. The system monitors the radio spectrum used by wireless LANs, and immediately alerts a systems administrator whenever a rogue access point is detected. Conventionally it is achieved by comparing the MAC address of the participating wireless devices.

Rogue devices can spoof MAC address of an authorized network device as their own. New research uses fingerprinting approach to weed out devices with spoofed MAC addresses. The idea is to compare the unique signatures exhibited by the signals emitted by each wireless device against the known signatures of pre-authorized, known wireless devices.

Intrusion Prevention

In addition to intrusion detection, a WIPS also includes features that prevent against the threat *automatically*. For automatic prevention, it is required that the WIPS is able to accurately detect and automatically classify a threat.

The following types of threats can be prevented by a good WIPS:

- Rogue access points – WIPS should understand the difference between rogue APs and external (neighbor's) APs
- Mis-configured AP
- Client mis-association
- Unauthorized association
- Man-in-the-middle attack
- *Ad hoc* networks
- MAC spoofing
- Honeypot / evil twin attack
- Denial-of-service attack

Implementation

WIPS configurations consist of three components:

- Sensors — These devices contain antennas and radios that scan the wireless spectrum for packets and are installed throughout areas to be protected

- Server — The WIPS server centrally analyzes packets captured by sensors

- Console — The console provides the primary user interface into the system for administration and reporting

A simple intrusion detection system can be a single computer, connected to a wireless signal processing device, and antennas placed throughout the facility. For huge organizations, a Multi Network Controller provides central control of multiple WIPS servers, while for SOHO or SMB customers, all the functionality of WIPS is available in single box.

In a WIPS implementation, users first define the operating wireless policies in the WIPS. The WIPS sensors then analyze the traffic in the air and send this information to WIPS server. The WIPS server correlates the information, validates it against the defined policies, and classifies if it is a threat. The administrator of the WIPS is then notified of the threat, or, if a policy has been set accordingly, the WIPS takes automatic protection measures.

WIPS is configured as either a network implementation or a hosted implementation.

Network Implementation

In a network WIPS implementation, server, sensors and the console are all placed inside a private network and are not accessible from the Internet.

Sensors communicate with the server over a private network using a private port. Since the server resides on the private network, users can access the console only from within the private network.

A network implementation is suitable for organizations where all locations are within the private network.

Hosted Implementation

In a hosted WIPS implementation, sensors are installed inside a private network. However, the server is hosted in secure data center and is accessible on the Internet. Users can access the WIPS console from anywhere on the Internet. A hosted WIPS implementation is as secure as a network implementation because the data flow is encrypted between sensors and server, as well as between server and console. A hosted WIPS implementation requires very little configuration because the sensors are programmed to automatically look for the server on the Internet over a secure TLS connection.

For a large organization with locations that are not a part of a private network, a hosted WIPS implementation simplifies deployment significantly because sensors connect to the Server over the Internet without requiring any special configuration. Additionally, the Console can be accessed securely from anywhere on the Internet.

Hosted WIPS implementations are available in an on-demand, subscription-based software as a service model. Hosted implementations may be appropriate for organizations looking to fulfill the minimum scanning requirements of PCI DSS.

Computer Security

Computer security, also known as cybersecurity or IT security, is the protection of information systems from theft or damage to the hardware, the software, and to the information on them, as well as from disruption or misdirection of the services they provide.

It includes controlling physical access to the hardware, as well as protecting against harm that may come via network access, data and code injection, and due to malpractice by operators, whether intentional, accidental, or due to them being tricked into deviating from secure procedures.

The field is of growing importance due to the increasing reliance on computer systems and the Internet in most societies, wireless networks such as Bluetooth and Wi-Fi - and the growth of "smart" devices, including smartphones, televisions and tiny devices as part of the Internet of Things.

Vulnerabilities and Attacks

A vulnerability is a system susceptibility or flaw. Many vulnerabilities are documented in the Common Vulnerabilities and Exposures (CVE) database. An *exploitable* vulnerability is one for which at least one working attack or "exploit" exists.

To secure a computer system, it is important to understand the attacks that can be made against it, and these threats can typically be classified into one of the categories below:

Backdoors

A backdoor in a computer system, a cryptosystem or an algorithm, is any secret method of bypassing normal authentication or security controls. They may exist for a number of reasons, including by original design or from poor configuration. They may have been added by an authorized party to allow some legitimate access, or by an attacker for malicious reasons; but regardless of the motives for their existence, they create a vulnerability.

Denial-of-service Attack

Denial of service attacks are designed to make a machine or network resource unavailable to its intended users. Attackers can deny service to individual victims, such as by deliberately entering a wrong password enough consecutive times to cause the victim account to be locked, or they may overload the capabilities of a machine or network and block all users at once. While a network attack from a single IP address can be blocked by adding a new firewall rule, many forms of Distributed denial of service (DDoS) attacks are possible, where the attack comes from a large number of points – and defending is much more difficult. Such attacks can originate from the zombie computers of a botnet, but a range of other techniques are possible including reflection and amplification attacks, where innocent systems are fooled into sending traffic to the victim.

Direct-access Attacks

An unauthorized user gaining physical access to a computer is most likely able to directly copy data from it. They may also compromise security by making operating system modifications, installing software worms, keyloggers, covert listening devices or using wireless mice. Even when the system is protected by standard security measures, these may be able to be by-passed by booting another operating system or tool from a CD-ROM or other bootable media. Disk encryption and Trusted Platform Module are designed to prevent these attacks.

Eavesdropping

Eavesdropping is the act of surreptitiously listening to a private conversation, typically between hosts on a network. For instance, programs such as Carnivore and NarusInsight have been used by the FBI and NSA to eavesdrop on the systems of internet service providers. Even machines that operate as a closed system (i.e., with no contact to the outside world) can be eavesdropped upon via monitoring the faint electro-magnetic transmissions generated by the hardware; TEMPEST is a specification by the NSA referring to these attacks.

Spoofing

Spoofing, in general, is a fraudulent or malicious practice in which communication is sent from an unknown source disguised as a source known to the receiver. Spoofing is most prevalent in communication mechanisms that lack a high level of security.

Tampering

Tampering describes a malicious modification of products. So-called "Evil Maid" attacks and security services planting of surveillance capability into routers are examples.

Privilege Escalation

Privilege escalation describes a situation where an attacker with some level of restricted access is able to, without authorization, elevate their privileges or access level. So for example a standard computer user may be able to fool the system into giving them access to restricted data; or even to "become root" and have full unrestricted access to a system.

Phishing

Phishing is the attempt to acquire sensitive information such as usernames, passwords, and credit card details directly from users. Phishing is typically carried out by email spoofing or instant messaging, and it often directs users to enter details at a fake website whose look and feel are almost identical to the legitimate one. Preying on a victim's trusting, phishing can be classified as a form of social engineering.

Clickjacking

Clickjacking, also known as "UI redress attack or User Interface redress attack", is a malicious technique in which an attacker tricks a user into clicking on a button or link on another webpage while the user intended to click on the top level page. This is done using multiple transparent or opaque layers. The attacker is basically "hijacking" the clicks meant for the top level page and routing them to some other irrelevant page, most likely owned by someone else. A similar technique can be used to hijack keystrokes. Carefully drafting a combination of stylesheets, iframes, buttons and text boxes, a user can be led into believing that they are typing the password or other information on some authentic webpage while it is being channeled into an invisible frame controlled by the attacker.

Social Engineering

Social engineering aims to convince a user to disclose secrets such as passwords, card numbers, etc. by, for example, impersonating a bank, a contractor, or a customer.

A popular and profitable cyber scam involves fake CEO emails sent to accounting and finance departments. In early 2016, the FBI reported that the scam has cost US businesses more than $2bn in about two years.

In May 2016, the Milwaukee Bucks NBA team was the victim of this type of cyber scam with a perpetrator impersonating the team's president Peter Feigin, resulting in the handover of all the team's employees' 2015 W-2 tax forms.

Systems at Risk

Computer security is critical in almost any industry which uses computers. Currently, most electronic devices such as computers, laptops and cellphones come with built in firewall security software, but despite this, computers are not 100 percent accurate and dependable to protect our data (Smith, Grabosky & Urbas, 2004.) There are many different ways of hacking into computers. It can be done through a network system, clicking into unknown links, connecting to unfamiliar Wi-Fi, downloading software and files from unsafe sites, power consumption, electromagnetic radiation waves, and many more. However, computers can be protected through well built software and hardware. By having strong internal interactions of properties, software complexity can prevent software crash and security failure.

Financial Systems

Web sites and apps that accept or store credit card numbers, brokerage accounts, and bank account information are prominent hacking targets, because of the potential for immediate financial gain from transferring money, making purchases, or selling the information on the black market. In-store payment systems and ATMs have also been tampered with in order to gather customer account data and PINs.

Utilities and Industrial Equipment

Computers control functions at many utilities, including coordination of telecommunications, the power grid, nuclear power plants, and valve opening and closing in water and gas networks. The Internet is a potential attack vector for such machines if connected, but the Stuxnet worm demonstrated that even equipment controlled by computers not connected to the Internet can be vulnerable to physical damage caused by malicious commands sent to industrial equipment (in that case uranium enrichment centrifuges) which are infected via removable media. In 2014, the Computer Emergency Readiness Team, a division of the Department of Homeland Security, investigated 79 hacking incidents at energy companies. Vulnerabilities in smart meters (many of which use local radio or cellular communications) can cause problems with billing fraud.

Aviation

The aviation industry is very reliant on a series of complex system which could be attacked. A simple power outage at one airport can cause repercussions worldwide, much of the system relies on radio transmissions which could be disrupted, and controlling aircraft over oceans is especially dangerous because radar surveillance only extends 175 to 225 miles offshore. There is also potential for attack from within an aircraft.

The consequences of a successful attack range from loss of confidentiality to loss of system integrity, which may lead to more serious concerns such as exfiltration of data, network and air traffic control outages, which in turn can lead to airport closures, loss of aircraft, loss of passenger life, damages on the ground and to transportation infrastructure. A successful attack on a military aviation system that controls munitions could have even more serious consequences.

Consumer Devices

Desktop computers and laptops are commonly infected with malware either to gather passwords or financial account information, or to construct a botnet to attack another target. Smart phones, tablet computers, smart watches, and other mobile devices such as Quantified Self devices like activity trackers have also become targets and many of these have sensors such as cameras, microphones, GPS receivers, compasses, and accelerometers which could be exploited, and may collect personal information, including sensitive health information. Wifi, Bluetooth, and cell phone networks on any of these devices could be used as attack vectors, and sensors might be remotely activated after a successful breach.

Home automation devices such as the Nest thermostat are also potential targets.

Large Corporations

Large corporations are common targets. In many cases this is aimed at financial gain through identity theft and involves data breaches such as the loss of millions of clients' credit card details by Home Depot, Staples, and Target Corporation. Medical records have been targeted for use in general identify theft, health insurance fraud, and impersonating patients to obtain prescription drugs for recreational purposes or resale.

Not all attacks are financially motivated however; for example security firm HBGary Federal suffered a serious series of attacks in 2011 from hacktivist group Anonymous in retaliation for the firm's CEO claiming to have infiltrated their group, and Sony Pictures was attacked in 2014 where the motive appears to have been to embarrass with data leaks, and cripple the company by wiping workstations and servers.

Automobiles

If access is gained to a car's internal controller area network, it is possible to disable the brakes and turn the steering wheel. Computerized engine timing, cruise control, anti-lock brakes, seat belt tensioners, door locks, airbags and advanced driver assistance systems make these disruptions possible, and self-driving cars go even further. Connected cars may use wifi and bluetooth to communicate with onboard consumer devices, and the cell phone network to contact concierge and emergency assistance services or get navigational or entertainment information; each of these networks is a potential entry point for malware or an attacker. Researchers in 2011 were even able to use a malicious compact disc in a car's stereo system as a successful attack vector, and cars with built-in voice recognition or remote assistance features have onboard microphones which could be used for eavesdropping.

A 2015 report by U.S. Senator Edward Markey criticized manufacturers' security measures as inadequate, and also highlighted privacy concerns about driving, location, and diagnostic data being collected, which is vulnerable to abuse by both manufacturers and hackers.

Government

Government and military computer systems are commonly attacked by activists and foreign powers. Local and regional government infrastructure such as traffic light controls, police and intelligence agency communications, personnel records, student records, and financial systems are also potential targets as they are now all largely computerized. Passports and government ID cards that control access to facilities which use RFID can be vulnerable to cloning.

Internet of Things and Physical Vulnerabilities

The Internet of Things (IoT) is the network of physical objects such as devices, vehicles, buildings and that are embedded with electronics, software, sensors, and network connectivity that enables them to collect and exchange data - and concerns have been raised that this is being developed without appropriate consideration of the security challenges involved.

While the IoT creates opportunities for more direct integration of the physical world into computer-based systems, it also provides opportunities for misuse. In particular, as the Internet of Things spreads widely, cyber attacks are likely to become an increasingly physical (rather than simply virtual) threat. If a front door's lock is connected to the Internet, and can be locked/unlocked from a phone, then a criminal could enter the home at the press of a button from a stolen or hacked phone. People could stand to lose much more than their credit card numbers in a world controlled by IoT-enabled devices. Thieves have also used electronic means to circumvent non-Internet-connected hotel door locks.

Medical devices have either been successfully attacked or had potentially deadly vulnerabilities demonstrated, including both in-hospital diagnostic equipment and implanted devices including pacemakers and insulin pumps.

Impact of Security Breaches

Serious financial damage has been caused by security breaches, but because there is no standard model for estimating the cost of an incident, the only data available is that which is made public by the organizations involved. "Several computer security consulting firms produce estimates of total worldwide losses attributable to virus and worm attacks and to hostile digital acts in general.

The 2003 loss estimates by these firms range from $13 billion (worms and viruses only) to $226 billion (for all forms of covert attacks). The reliability of these estimates is often challenged; the underlying methodology is basically anecdotal."

However, reasonable estimates of the financial cost of security breaches can actually help organizations make rational investment decisions. According to the classic Gordon-Loeb Model analyzing the optimal investment level in information security, one can conclude that the amount a firm spends to protect information should generally be only a small fraction of the expected loss (i.e., the expected value of the loss resulting from a cyber/information security breach).

Attacker Motivation

As with physical security, the motivations for breaches of computer security vary between attackers. Some are thrill-seekers or vandals, others are activists or criminals looking for financial gain. State-sponsored attackers are now common and well resourced, but started with amateurs such as Markus Hess who hacked for the KGB, as recounted by Clifford Stoll, in *The Cuckoo's Egg*.

A standard part of threat modelling for any particular system is to identify what might motivate an attack on that system, and who might be motivated to breach it. The level and detail of precautions will vary depending on the system to be secured. A home personal computer, bank, and classified military network face very different threats, even when the underlying technologies in use are similar.

Computer Protection (Countermeasures)

In computer security a countermeasure is an action, device, procedure, or technique that reduces a threat, a vulnerability, or an attack by eliminating or preventing it, by minimizing the harm it can cause, or by discovering and reporting it so that corrective action can be taken.

Some common countermeasures are listed in the following sections:

Security by Design

Security by design, or alternately secure by design, means that the software has been designed from the ground up to be secure. In this case, security is considered as a main feature.

Some of the techniques in this approach include:

- The principle of least privilege, where each part of the system has only the privileges that are needed for its function. That way even if an attacker gains access to that part, they have only limited access to the whole system.

- Automated theorem proving to prove the correctness of crucial software subsystems.

- Code reviews and unit testing, approaches to make modules more secure where formal correctness proofs are not possible.

- Defense in depth, where the design is such that more than one subsystem needs to be violated to compromise the integrity of the system and the information it holds.

- Default secure settings, and design to "fail secure" rather than "fail insecure". Ideally, a secure system should require a deliberate, conscious, knowledgeable and free decision on the part of legitimate authorities in order to make it insecure.

- Audit trails tracking system activity, so that when a security breach occurs, the mechanism and extent of the breach can be determined. Storing audit trails remotely, where they can only be appended to, can keep intruders from covering their tracks.

- Full disclosure of all vulnerabilities, to ensure that the "window of vulnerability" is kept as short as possible when bugs are discovered.

Security Architecture

The Open Security Architecture organization defines IT security architecture as "the design artifacts that describe how the security controls (security countermeasures) are positioned, and how they relate to the overall information technology architecture. These controls serve the purpose to maintain the system's quality attributes: confidentiality, integrity, availability, accountability and assurance services".

Techopedia defines security architecture as "a unified security design that addresses the necessities and potential risks involved in a certain scenario or environment. It also specifies when and where to apply security controls. The design process is generally reproducible." The key attributes of security architecture are:

- the relationship of different components and how they depend on each other.

- the determination of controls based on risk assessment, good practice, finances, and legal matters.

- the standardization of controls.

Security Measures

A state of computer "security" is the conceptual ideal, attained by the use of the three processes: threat prevention, detection, and response. These processes are based on various policies and system components, which include the following:

- User account access controls and cryptography can protect systems files and data, respectively.

- Firewalls are by far the most common prevention systems from a network security perspective as they can (if properly configured) shield access to internal network services, and block certain kinds of attacks through packet filtering. Firewalls can be both hardware- or software-based.

- Intrusion Detection System (IDS) products are designed to detect network attacks in-progress and assist in post-attack forensics, while audit trails and logs serve a similar function for individual systems.

- "Response" is necessarily defined by the assessed security requirements of an individual system and may cover the range from simple upgrade of protections to notification of legal authorities, counter-attacks, and the like. In some special cases, a complete destruction of the compromised system is favored, as it may happen that not all the compromised resources are detected.

Today, computer security comprises mainly "preventive" measures, like firewalls or an exit procedure. A firewall can be defined as a way of filtering network data between a host or a network and another network, such as the Internet, and can be implemented as software running on the machine, hooking into the network stack (or, in the case of most UNIX-based operating systems such as Linux, built into the operating system kernel) to provide real time filtering and blocking. Another implementation is a so-called physical firewall which consists of a separate machine filtering network traffic. Firewalls are common amongst machines that are permanently connected to the Internet.

Some organizations are turning to big data platforms, such as Apache Hadoop, to extend data accessibility and machine learning to detect advanced persistent threats.

However, relatively few organisations maintain computer systems with effective detection systems, and fewer still have organised response mechanisms in place. As result, as Reuters points out: "Companies for the first time report they are losing more through electronic theft of data than physical stealing of assets". The primary obstacle to effective eradication of cyber crime could be traced to excessive reliance on firewalls and other automated "detection" systems. Yet it is basic evidence gathering by using packet capture appliances that puts criminals behind bars.

Vulnerability Management

Vulnerability management is the cycle of identifying, and remediating or mitigating vulnerabilities", especially in software and firmware. Vulnerability management is integral to computer security and network security.

Vulnerabilities can be discovered with a vulnerability scanner, which analyzes a computer system in search of known vulnerabilities, such as open ports, insecure software configuration, and susceptibility to malware

Beyond vulnerability scanning, many organisations contract outside security auditors to run regular penetration tests against their systems to identify vulnerabilities. In some sectors this is a contractual requirement.

Reducing Vulnerabilities

While formal verification of the correctness of computer systems is possible, it is not yet common. Operating systems formally verified include seL4, and SYSGO's PikeOS – but these make up a very small percentage of the market.

Cryptography properly implemented is now virtually impossible to directly break. Breaking them requires some non-cryptographic input, such as a stolen key, stolen plaintext (at either end of the transmission), or some other extra cryptanalytic information.

Two factor authentication is a method for mitigating unauthorized access to a system or sensitive information. It requires "something you know"; a password or PIN, and "something you have"; a card, dongle, cellphone, or other piece of hardware. This increases security as an unauthorized person needs both of these to gain access.

Social engineering and direct computer access (physical) attacks can only be prevented by non-computer means, which can be difficult to enforce, relative to the sensitivity of the information. Training is often involved to help mitigate this risk, but even in a highly disciplined environments (e.g. military organizations), social engineering attacks can still be difficult to foresee and prevent.

It is possible to reduce an attacker's chances by keeping systems up to date with security patches and updates, using a security scanner or/and hiring competent people responsible for security. The effects of data loss/damage can be reduced by careful backing up and insurance.

Hardware Protection Mechanisms

While hardware may be a source of insecurity, such as with microchip vulnerabilities maliciously introduced during the manufacturing process, hardware-based or assisted computer security also offers an alternative to software-only computer security. Using devices and methods such as dongles, trusted platform modules, intrusion-aware cases, drive locks, disabling USB ports, and mobile-enabled access may be considered more secure due to the physical access (or sophisticated backdoor access) required in order to be compromised. Each of these is covered in more detail below.

- USB dongles are typically used in software licensing schemes to unlock software capabilities, but they can also be seen as a way to prevent unauthorized access to a computer or other device's software. The dongle, or key, essentially creates a secure encrypted tunnel between the software application and the key. The principle is that an encryption scheme on the dongle, such as Advanced Encryption Standard (AES) provides a stronger measure of security, since it is harder to hack and replicate the dongle than to simply copy the native software to another machine and use it. Another security application for dongles is to use them for accessing web-based content such as cloud software or Virtual Private Networks (VPNs). In addition, a USB dongle can be configured to lock or unlock a computer.

- Trusted platform modules (TPMs) secure devices by integrating cryptographic capabilities onto access devices, through the use of microprocessors, or so-called computers-on-a-chip. TPMs used in conjunction with server-side software offer a way to detect and authenticate hardware devices, preventing unauthorized network and data access.

- Computer case intrusion detection refers to a push-button switch which is triggered when a computer case is opened. The firmware or BIOS is programmed to show an alert to the operator when the computer is booted up the next time.

- Drive locks are essentially software tools to encrypt hard drives, making them inaccessible to thieves. Tools exist specifically for encrypting external drives as well.

- Disabling USB ports is a security option for preventing unauthorized and malicious access to an otherwise secure computer. Infected USB dongles connected to a network from a computer inside the firewall are considered by the magazine Network World as the most common hardware threat facing computer networks.

- Mobile-enabled access devices are growing in popularity due to the ubiquitous nature of cell phones. Built-in capabilities such as Bluetooth, the newer Bluetooth low energy (LE), Near field communication (NFC) on non-iOS devices and biometric validation such as thumb print readers, as well as QR code reader software designed for mobile devices, offer new, secure ways for mobile phones to connect to access control systems. These control systems provide computer security and can also be used for controlling access to secure buildings.

Secure Operating Systems

One use of the term "computer security" refers to technology that is used to implement secure operating systems. In the 1980s the United States Department of Defense (DoD) used the "Orange Book" standards, but the current international standard ISO/IEC 15408, "Common Criteria" defines a number of progressively more stringent Evaluation Assurance Levels. Many common operating systems meet the EAL4 standard of being "Methodically Designed, Tested and Reviewed", but the formal verification required for the highest levels means that they are uncommon. An example of an EAL6 ("Semiformally Verified Design and Tested") system is Integrity-178B, which is used in the Airbus A380 and several military jets.

Secure Coding

In software engineering, secure coding aims to guard against the accidental introduction of security vulnerabilities. It is also possible to create software designed from the ground up to be secure. Such systems are "secure by design". Beyond this, formal verification aims to prove the correctness of the algorithms underlying a system; important for cryptographic protocols for example.

Capabilities and Access Control Lists

Within computer systems, two of many security models capable of enforcing privilege separation are access control lists (ACLs) and capability-based security. Using ACLs to confine programs has been proven to be insecure in many situations, such as if the host computer can be tricked into indirectly allowing restricted file access, an issue known as the confused deputy problem. It has also been shown that the promise of ACLs of giving access to an object to only one person can never be guaranteed in practice. Both of these problems are resolved by capabilities. This does not mean practical flaws exist in all ACL-based systems, but only that the designers of certain utilities must take responsibility to ensure that they do not introduce flaws.

Capabilities have been mostly restricted to research operating systems, while commercial OSs still use ACLs. Capabilities can, however, also be implemented at the language level, leading to a style of programming that is essentially a refinement of standard object-oriented design. An open source project in the area is the E language.

The most secure computers are those not connected to the Internet and shielded from any interference. In the real world, the most secure systems are operating systems where security is not an add-on.

Response to Breaches

Responding forcefully to attempted security breaches (in the manner that one would for attempted physical security breaches) is often very difficult for a variety of reasons:

- Identifying attackers is difficult, as they are often in a different jurisdiction to the systems they attempt to breach, and operate through proxies, temporary anonymous dial-up accounts, wireless connections, and other anonymising procedures which make backtracing difficult and are often located in yet another jurisdiction. If they successfully breach security, they are often able to delete logs to cover their tracks.

- The sheer number of attempted attacks is so large that organisations cannot spend time pursuing each attacker (a typical home user with a permanent (e.g., cable modem) connection will be attacked at least several times per day, so more attractive targets could be presumed). Note however, that most of the sheer bulk of these attacks are made by automated vulnerability scanners and computer worms.

- Law enforcement officers are often unfamiliar with information technology, and so lack the skills and interest in pursuing attackers. There are also budgetary constraints. It has been argued that the high cost of technology, such as DNA testing, and improved forensics mean less money for other kinds of law enforcement, so the overall rate of criminals not getting dealt with goes up as the cost of the technology increases. In addition, the identification of attackers across a network may require logs from various points in the network and in many countries, the release of these records to law enforcement (with the exception of being voluntarily surrendered by a network administrator or a system administrator) requires a search warrant and, depending on the circumstances, the legal proceedings required can be drawn out to the point where the records are either regularly destroyed, or the information is no longer relevant.

Notable Attacks and Breaches

Some illustrative examples of different types of computer security breaches are given below.

Robert Morris and the First Computer Worm

In 1988, only 60,000 computers were connected to the Internet, and most were mainframes, minicomputers and professional workstations. On November 2, 1988, many started to slow down, because they were running a malicious code that demanded processor time and that spread itself to other computers – the first internet "computer worm". The software was traced back to 23-year-old Cornell University graduate student Robert Tappan Morris, Jr. who said 'he wanted to count how many machines were connected to the Internet'.

Rome Laboratory

In 1994, over a hundred intrusions were made by unidentified crackers into the Rome Laboratory, the US Air Force's main command and research facility. Using trojan horses, hackers were able to obtain unrestricted access to Rome's networking systems and remove traces of their activities. The intruders were able to obtain classified files, such as air tasking order systems data and furthermore able to penetrate connected networks of National Aeronautics and Space Administration's Goddard Space Flight Center, Wright-Patterson Air Force Base, some Defense contractors, and other private sector organizations, by posing as a trusted Rome center user.

TJX Customer Credit Card Details

In early 2007, American apparel and home goods company TJX announced that it was the victim of an unauthorized computer systems intrusion and that the hackers had accessed a system that stored data on credit card, debit card, check, and merchandise return transactions.

Stuxnet Attack

The computer worm known as Stuxnet reportedly ruined almost one-fifth of Iran's nuclear centrifuges by disrupting industrial programmable logic controllers (PLCs) in a targeted attack generally believed to have been launched by Israel and the United States although neither has publicly acknowledged this.

Global Surveillance Disclosures

In early 2013, massive breaches of computer security by the NSA were revealed, including deliberately inserting a backdoor in a NIST standard for encryption and tapping the links between Google's data centres. These were disclosed by NSA contractor Edward Snowden.

Target and Home Depot Breaches

In 2013 and 2014, a Russian/Ukrainian hacking ring known as "Rescator" broke into Target Corporation computers in 2013, stealing roughly 40 million credit cards, and then Home Depot computers in 2014, stealing between 53 and 56 million credit card numbers. Warnings were delivered at both corporations, but ignored; physical security breaches using self checkout machines are believed to have played a large role. "The malware utilized is absolutely unsophisticated and uninteresting," says Jim Walter, director of threat intelligence operations at security technology company McAfee – meaning that the heists could have easily been stopped by existing antivirus software had administrators responded to the warnings. The size of the thefts has resulted in major attention from state and Federal United States authorities and the investigation is ongoing.

Ashley Madison Breach

In July 2015, a hacker group known as "The Impact Team" successfully breached the extramarital relationship website Ashley Madison. The group claimed that they had taken not only company data but user data as well. After the breach, The Impact Team dumped emails from the company's CEO, to prove their point, and threatened to dump customer data unless the website was taken down permanently. With this initial data release, the group stated "Avid Life Media has been instructed to take Ashley Madison and Established Men offline permanently in all forms, or we will release all customer records, including profiles with all the customers' secret sexual fantasies and matching credit card transactions, real names and addresses, and employee documents and emails. The other websites may stay online." When Avid Life Media, the parent company that created the Ashley Madison website, did not take the site offline, The Impact Group released two more compressed files, one 9.7GB and the second 20GB. After the second data dump, Avid Life Media CEO Noel Biderman resigned, but the website remained functional.

Legal Issues and Global Regulation

Conflict of laws in cyberspace has become a major cause of concern for computer security community. Some of the main challenges and complaints about the antivirus industry are the lack of global web regulations, a global base of common rules to judge, and eventually punish, cyber crimes and cyber criminals. There is no global cyber law and cybersecurity treaty that can be invoked for enforcing global cybersecurity issues.

International legal issues of cyber attacks are complicated in nature. Even if an antivirus firm locates the cyber criminal behind the creation of a particular virus or piece of malware or form of cyber attack, often the local authorities cannot take action due to lack of laws under which to prosecute. Authorship attribution for cyber crimes and cyber attacks is a major problem for all law enforcement agencies.

"[Computer viruses] switch from one country to another, from one jurisdiction to another – moving around the world, using the fact that we don't have the capability to globally police operations like this. So the Internet is as if someone [had] given free plane tickets to all the online criminals of the world." Use of dynamic DNS, fast flux and bullet proof servers have added own complexities to this situation.

Government

The role of the government is to make regulations to force companies and organizations to protect their systems, infrastructure and information from any cyber-attacks, but also to protect its own national infrastructure such as the national power-grid.

The question of whether the government should intervene or not in the regulation of the cyberspace is a very polemical one. Indeed, for as long as it has existed and by definition, the cyberspace is a virtual space free of any government intervention. Where everyone agree that an improvement on cybersecurity is more than vital, is the government the best actor to solve this issue? Many government officials and experts think that the government should step in and that there is a crucial need for regulation, mainly due to the failure of the private sector to solve efficiently the cybersecurity problem. R. Clarke said during a panel discussion at the RSA Security Conference in San Francisco, he believes that the "industry only responds when you threaten regulation. If industry doesn't respond (to the threat), you have to follow through." On the other hand, executives from the private sector agree that improvements are necessary, but think that the government intervention would affect their ability to innovate efficiently.

Actions and Teams in the US

Legislation

The 1986 18 U.S.C. § 1030, more commonly known as the Computer Fraud and Abuse Act is the key legislation. It prohibits unauthorized access or damage of "protected computers" as defined in 18 U.S.C. § 1030(e)(2).

Although various other measures have been proposed, such as the "Cybersecurity Act of 2010 – S. 773" in 2009, the "International Cybercrime Reporting and Cooperation Act – H.R.4962" and

"Protecting Cyberspace as a National Asset Act of 2010 – S.3480" in 2010 – none of these has succeeded.

Executive order 13636 *Improving Critical Infrastructure Cybersecurity* was signed February 12, 2013.

Agencies

The Department of Homeland Security has a dedicated division responsible for the response system, risk management program and requirements for cybersecurity in the United States called the National Cyber Security Division. The division is home to US-CERT operations and the National Cyber Alert System. The National Cybersecurity and Communications Integration Center brings together government organizations responsible for protecting computer networks and networked infrastructure.

The third priority of the Federal Bureau of Investigation (FBI) is to: *"Protect the United States against cyber-based attacks and high-technology crimes"*, and they, along with the National White Collar Crime Center (NW3C), and the Bureau of Justice Assistance (BJA) are part of the multi-agency task force, The Internet Crime Complaint Center, also known as IC3.

In addition to its own specific duties, the FBI participates alongside non-profit organizations such as InfraGard.

In the criminal division of the United States Department of Justice operates a section called the Computer Crime and Intellectual Property Section. The CCIPS is in charge of investigating computer crime and intellectual property crime and is specialized in the search and seizure of digital evidence in computers and networks.

The United States Cyber Command, also known as USCYBERCOM, is tasked with the defense of specified Department of Defense information networks and *"ensure US/Allied freedom of action in cyberspace and deny the same to our adversaries."* It has no role in the protection of civilian networks.

The U.S. Federal Communications Commission's role in cybersecurity is to strengthen the protection of critical communications infrastructure, to assist in maintaining the reliability of networks during disasters, to aid in swift recovery after, and to ensure that first responders have access to effective communications services.

The Food and Drug Administration has issued guidance for medical devices, and the National Highway Traffic Safety Administration is concerned with automotive cybersecurity. After being criticized by the Government Accountability Office, and following successful attacks on airports and claimed attacks on airplanes, the Federal Aviation Administration has devoted funding to securing systems on board the planes of private manufacturers, and the Aircraft Communications Addressing and Reporting System. Concerns have also been raised about the future Next Generation Air Transportation System.

Computer Emergency Readiness Team

"Computer emergency response team" is a name given to expert groups that handle computer security incidents. In the US, two distinct organization exist, although they do work closely together.

- US-CERT: part of the National Cyber Security Division of the United States Department of Homeland Security.

- CERT/CC: created by the Defense Advanced Research Projects Agency (DARPA) and run by the Software Engineering Institute (SEI).

International Actions

Many different teams and organisations exist, including:

- The Forum of Incident Response and Security Teams (FIRST) is the global association of CSIRTs. The US-CERT, AT&T, Apple, Cisco, McAfee, Microsoft are all members of this international team.

- The Council of Europe helps protect societies worldwide from the threat of cybercrime through the Convention on Cybercrime.

- The purpose of the Messaging Anti-Abuse Working Group (MAAWG) is to bring the messaging industry together to work collaboratively and to successfully address the various forms of messaging abuse, such as spam, viruses, denial-of-service attacks and other messaging exploitations. France Telecom, Facebook, AT&T, Apple, Cisco, Sprint are some of the members of the MAAWG.

- ENISA : The European Network and Information Security Agency (ENISA) is an agency of the European Union with the objective to improve network and information security in the European Union.

Europe

CSIRTs in Europe collaborate in the TERENA task force TF-CSIRT. TERENA's Trusted Introducer service provides an accreditation and certification scheme for CSIRTs in Europe. A full list of known CSIRTs in Europe is available from the Trusted Introducer website.

National Teams

Here are the main computer emergency response teams around the world. Most countries have their own team to protect network security.

Canada

On October 3, 2010, Public Safety Canada unveiled Canada's Cyber Security Strategy, following a Speech from the Throne commitment to boost the security of Canadian cyberspace. The aim of the strategy is to strengthen Canada's "cyber systems and critical infrastructure sectors, support economic growth and protect Canadians as they connect to each other and to the world." Three main pillars define the strategy: securing government systems, partnering to secure vital cyber systems outside the federal government, and helping Canadians to be secure online. The strategy involves multiple departments and agencies across the Government of Canada. The Cyber Incident Management Framework for Canada outlines these responsibilities, and provides a plan for coordinated response between government and other partners in the event of a cyber incident. The

Action Plan 2010–2015 for Canada's Cyber Security Strategy outlines the ongoing implementation of the strategy.

Public Safety Canada's Canadian Cyber Incident Response Centre (CCIRC) is responsible for mitigating and responding to threats to Canada's critical infrastructure and cyber systems. The CCIRC provides support to mitigate cyber threats, technical support to respond and recover from targeted cyber attacks, and provides online tools for members of Canada's critical infrastructure sectors. The CCIRC posts regular cyber security bulletins on the Public Safety Canada website. The CCIRC also operates an online reporting tool where individuals and organizations can report a cyber incident. Canada's Cyber Security Strategy is part of a larger, integrated approach to critical infrastructure protection, and functions as a counterpart document to the National Strategy and Action Plan for Critical Infrastructure.

On September 27, 2010, Public Safety Canada partnered with STOP.THINK.CONNECT, a coalition of non-profit, private sector, and government organizations dedicated to informing the general public on how to protect themselves online. On February 4, 2014, the Government of Canada launched the Cyber Security Cooperation Program. The program is a $1.5 million five-year initiative aimed at improving Canada's cyber systems through grants and contributions to projects in support of this objective. Public Safety Canada aims to begin an evaluation of Canada's Cyber Security Strategy in early 2015. Public Safety Canada administers and routinely updates the GetCyberSafe portal for Canadian citizens, and carries out Cyber Security Awareness Month during October.

China

China's network security and information technology leadership team was established February 27, 2014. The leadership team is tasked with national security and long-term development and co-ordination of major issues related to network security and information technology. Economic, political, cultural, social and military fields as related to network security and information technology strategy, planning and major macroeconomic policy are being researched. The promotion of national network security and information technology law are constantly under study for enhanced national security capabilities.

Germany

Berlin starts National Cyber Defense Initiative: On June 16, 2011, the German Minister for Home Affairs, officially opened the new German NCAZ (National Center for Cyber Defense) Nationales Cyber-Abwehrzentrum located in Bonn. The NCAZ closely cooperates with BSI (Federal Office for Information Security) Bundesamt für Sicherheit in der Informationstechnik, BKA (Federal Police Organisation) Bundeskriminalamt (Deutschland), BND (Federal Intelligence Service) Bundesnachrichtendienst, MAD (Military Intelligence Service) Amt für den Militärischen Abschirmdienst and other national organisations in Germany taking care of national security aspects. According to the Minister the primary task of the new organisation founded on February 23, 2011, is to detect and prevent attacks against the national infrastructure and mentioned incidents like Stuxnet.

India

Some provisions for cybersecurity have been incorporated into rules framed under the Information Technology Act 2000.

The National Cyber Security Policy 2013 is a policy framework by Department of Electronics and Information Technology (DeitY) which aims to protect the public and private infrastructure from cyber attacks, and safeguard "information, such as personal information (of web users), financial and banking information and sovereign data".

The Indian Companies Act 2013 has also introduced cyber law and cyber security obligations on the part of Indian directors.

Pakistan

Cyber-crime has risen rapidly in Pakistan. There are about 30 million Internet users with 15 million mobile subscribers in Pakistan. According to Cyber Crime Unit (CCU), a branch of Federal Investigation Agency, only 62 cases were reported to the unit in 2007, 287 cases in 2008, ratio dropped in 2009 but in 2010, more than 312 cases were registered. However, there are many unreported incidents of cyber-crime.

"Pakistan's Cyber Crime Bill 2007", the first pertinent law, focuses on electronic crimes, for example cyber-terrorism, criminal access, electronic system fraud, electronic forgery, and misuse of encryption.

National Response Centre for Cyber Crime (NR3C) – FIA is a law enforcement agency dedicated to fight cybercrime. Inception of this Hi-Tech crime fighting unit transpired in 2007 to identify and curb the phenomenon of technological abuse in society. However, certain private firms are also working in cohesion with the government to improve cyber security and curb cyberattacks.

South Korea

Following cyberattacks in the first half of 2013, when government, news-media, television station, and bank websites were compromised, the national government committed to the training of 5,000 new cybersecurity experts by 2017. The South Korean government blamed its northern counterpart for these attacks, as well as incidents that occurred in 2009, 2011, and 2012, but Pyongyang denies the accusations.

Other Countries

- CERT Brazil, member of FIRST (Forum for Incident Response and Security Teams)
- CARNet CERT, Croatia, member of FIRST
- AE CERT, United Arab Emirates
- SingCERT, Singapore
- CERT-LEXSI, France, Canada, Singapore

Modern Warfare

Cybersecurity is becoming increasingly important as more information and technology is being made available on cyberspace. There is growing concern among governments that cyberspace will become the next theatre of warfare. As Mark Clayton from the *Christian Science Monitor* described in an article titled "The New Cyber Arms Race":

In the future, wars will not just be fought by soldiers with guns or with planes that drop bombs. They will also be fought with the click of a mouse a half a world away that unleashes carefully weaponized computer programs that disrupt or destroy critical industries like utilities, transportation, communications, and energy. Such attacks could also disable military networks that control the movement of troops, the path of jet fighters, the command and control of warships.

This has led to new terms such as *cyberwarfare* and *cyberterrorism*. More and more critical infrastructure is being controlled via computer programs that, while increasing efficiency, exposes new vulnerabilities. The test will be to see if governments and corporations that control critical systems such as energy, communications and other information will be able to prevent attacks before they occur. As Jay Cross, the chief scientist of the Internet Time Group, remarked, "Connectedness begets vulnerability."

Job Market

Cybersecurity is a fast-growing field of IT concerned with reducing organizations' risk of hack or data breach. According to research from the Enterprise Strategy Group, 46% of organizations say that they have a "problematic shortage" of cybersecurity skills in 2016, up from 28% in 2015. Commercial, government and non-governmental organizations all employ cybersecurity professionals. The fastest increases in demand for cybersecurity workers are in industries managing increasing volumes of consumer data such as finance, health care, and retail. However, the use of the term "cybersecurity" is more prevalent in government job descriptions.

Typical cybersecurity job titles and descriptions include:

Security Analyst

Analyzes and assesses vulnerabilities in the infrastructure (software, hardware, networks), investigates available tools and countermeasures to remedy the detected vulnerabilities, and recommends solutions and best practices. Analyzes and assesses damage to the data/infrastructure as a result of security incidents, examines available recovery tools and processes, and recommends solutions. Tests for compliance with security policies and procedures. May assist in the creation, implementation, and/or management of security solutions.

Security Engineer

Performs security monitoring, security and data/logs analysis, and forensic analysis, to detect security incidents, and mounts incident response. Investigates and utilizes new technologies and processes to enhance security capabilities and implement improvements. May also review code or perform other security engineering methodologies.

Security Architect

Designs a security system or major components of a security system, and may head a security design team building a new security system.

Security Administrator

Installs and manages organization-wide security systems. May also take on some of the tasks of a security analyst in smaller organizations.

Chief Information Security Officer (Ciso)

A high-level management position responsible for the entire information security division/staff. The position may include hands-on technical work.

Chief Security Officer (CSO)

A high-level management position responsible for the entire security division/staff. A newer position now deemed needed as security risks grow.

Security Consultant/Specialist/Intelligence

Broad titles that encompass any one or all of the other roles/titles, tasked with protecting computers, networks, software, data, and/or information systems against viruses, worms, spyware, malware, intrusion detection, unauthorized access, denial-of-service attacks, and an ever increasing list of attacks by hackers acting as individuals or as part of organized crime or foreign governments.

Student programs are also available to people interested in beginning a career in cybersecurity. Meanwhile, a flexible and effective option for information security professionals of all experience levels to keep studying is online security training, including webcasts.

Terminology

The following terms used with regards to engineering secure systems are explained below.

- Access authorization restricts access to a computer to group of users through the use of authentication systems. These systems can protect either the whole computer – such as through an interactive login screen – or individual services, such as an FTP server. There are many methods for identifying and authenticating users, such as passwords, identification cards, and, more recently, smart cards and biometric systems.

- Anti-virus software consists of computer programs that attempt to identify, thwart and eliminate computer viruses and other malicious software (malware).

- Applications with known security flaws should not be run. Either leave it turned off until it can be patched or otherwise fixed, or delete it and replace it with some other application. Publicly known flaws are the main entry used by worms to automatically break into a system and then spread to other systems connected to it. The security website Secunia provides a search tool for unpatched known flaws in popular products.

- Authentication techniques can be used to ensure that communication end-points are who they say they are.

- Automated theorem proving and other verification tools can enable critical algorithms and code used in secure systems to be mathematically proven to meet their specifications.

- Backups are a way of securing information; they are another copy of all the important computer files kept in another location. These files are kept on hard disks, CD-Rs, CD-RWs, tapes and more recently on the cloud. Suggested locations for backups are a fireproof, waterproof, and heat proof safe, or in a separate, offsite location than that in which the original files are contained. Some individuals and companies also keep their backups in safe deposit boxes inside bank vaults. There is also a fourth option, which involves using one of the file hosting services that backs up files over the Internet for both business and individuals, known as the cloud.

  - Backups are also important for reasons other than security. Natural disasters, such as earthquakes, hurricanes, or tornadoes, may strike the building where the computer is located. The building can be on fire, or an explosion may occur. There needs to be a recent backup at an alternate secure location, in case of such kind of disaster. Further, it is recommended that the alternate location be placed where the same disaster would not affect both locations. Examples of alternate disaster recovery sites being compromised by the same disaster that affected the primary site include having had a primary site in World Trade Center I and the recovery site in 7 World Trade Center, both of which were destroyed in the 9/11 attack, and having one's primary site and recovery site in the same coastal region, which leads to both being vulnerable to hurricane damage (for example, primary site in New Orleans and recovery site in Jefferson Parish, both of which were hit by Hurricane Katrina in 2005). The backup media should be moved between the geographic sites in a secure manner, in order to prevent them from being stolen.

- Capability and access control list techniques can be used to ensure privilege separation and mandatory access control. This section discusses their use.

- Chain of trust techniques can be used to attempt to ensure that all software loaded has been certified as authentic by the system's designers.

- Confidentiality is the nondisclosure of information except to another authorized person.

- Cryptographic techniques can be used to defend data in transit between systems, reducing the probability that data exchanged between systems can be intercepted or modified.

- Cyberwarfare is an internet-based conflict that involves politically motivated attacks on information and information systems. Such attacks can, for example, disable official websites and networks, disrupt or disable essential services, steal or alter classified data, and cripple financial systems.

- Data integrity is the accuracy and consistency of stored data, indicated by an absence of any alteration in data between two updates of a data record.

This is secret stuff, PSE do not...

5a0 (k$hQ% ...

This is secret stuff, PSE do not...

Cryptographic techniques involve transforming information, scrambling it so it becomes unreadable during transmission. The intended recipient can unscramble the message; ideally, eavesdroppers cannot.

- Encryption is used to protect the message from the eyes of others. Cryptographically secure ciphers are designed to make any practical attempt of breaking infeasible. Symmetric-key ciphers are suitable for bulk encryption using shared keys, and public-key encryption using digital certificates can provide a practical solution for the problem of securely communicating when no key is shared in advance.

- Endpoint security software helps networks to prevent exfiltration (data theft) and virus infection at network entry points made vulnerable by the prevalence of potentially infected portable computing devices, such as laptops and mobile devices, and external storage devices, such as USB drives.

- Firewalls are an important method for control and security on the Internet and other networks. A network firewall can be a communications processor, typically a router, or a dedicated server, along with firewall software. A firewall serves as a gatekeeper system that protects a company's intranets and other computer networks from intrusion by providing a filter and safe transfer point for access to and from the Internet and other networks. It screens all network traffic for proper passwords or other security codes and only allows authorized transmission in and out of the network. Firewalls can deter, but not completely prevent, unauthorized access (hacking) into computer networks; they can also provide some protection from online intrusion.

- Honey pots are computers that are either intentionally or unintentionally left vulnerable to attack by crackers. They can be used to catch crackers or fix vulnerabilities.

- Intrusion-detection systems can scan a network for people that are on the network but who should not be there or are doing things that they should not be doing, for example trying a lot of passwords to gain access to the network.

- A microkernel is the near-minimum amount of software that can provide the mechanisms to implement an operating system. It is used solely to provide very low-level, very precisely defined machine code upon which an operating system can be developed. A simple example is the early '90s GEMSOS (Gemini Computers), which provided extremely low-level machine code, such as "segment" management, atop which an operating system could be built. The theory (in the case of "segments") was that—rather than have the operating system itself worry about mandatory access separation by means of military-style labeling—it is safer if a low-level, independently scrutinized module can be charged solely with the management of individually labeled segments, be they memory "segments" or file system "segments" or executable text "segments." If software below the visibility of the operating system is (as in this case) charged with labeling, there is no theoretically viable means for a clever hacker to subvert the labeling scheme, since the operating system *per se* does not provide mechanisms for interfering with labeling: the operating system is, essentially, a client (an "application," arguably) atop the microkernel and, as such, subject to its restrictions.

- Pinging The ping application can be used by potential crackers to find if an IP address is reachable. If a cracker finds a computer, they can try a port scan to detect and attack services on that computer.

- Social engineering awareness keeps employees aware of the dangers of social engineering and/or having a policy in place to prevent social engineering can reduce successful breaches of the network and servers.

Scholars

- Ross J. Anderson
- Annie Anton
- Adam Back
- Daniel J. Bernstein
- Matt Blaze
- Stefan Brands
- L. Jean Camp
- Lance Cottrell
- Lorrie Cranor
- Dorothy E. Denning
- Peter J. Denning
- Cynthia Dwork
- Deborah Estrin
- Joan Feigenbaum
- Ian Goldberg
- Shafi Goldwasser
- Lawrence A. Gordon
- Peter Gutmann
- Paul Kocher
- Monica S. Lam
- Butler Lampson
- Brian LaMacchia
- Carl Landwehr
- Kevin Mitnick
- Peter G. Neumann
- Susan Nycum
- Roger R. Schell

- Bruce Schneier

- Dawn Song

- Gene Spafford

- Joseph Steinberg

- Salvatore J. Stolfo

- Willis Ware

- Moti Yung

Mobile Security

Mobile security or mobile phone security has become increasingly important in mobile computing. Of particular concern is the security of personal and business information now stored on smartphones.

More and more users and businesses employ smartphones as communication tools, but also as a means of planning and organizing their work and private life. Within companies, these technologies are causing profound changes in the organization of information systems and therefore they have become the source of new risks. Indeed, smartphones collect and compile an increasing amount of sensitive information to which access must be controlled to protect the privacy of the user and the intellectual property of the company.

All smartphones, as computers, are preferred targets of attacks. These attacks exploit weaknesses related to smartphones that can come from means of communication like Short Message Service (SMS, aka text messaging), Multimedia Messaging Service (MMS), Wi-Fi networks, Bluetooth and GSM, the de facto global standard for mobile communications. There are also attacks that exploit software vulnerabilities from both the web browser and operating system. Finally, there are forms of malicious software that rely on the weak knowledge of average users.

Different security counter-measures are being developed and applied to smartphones, from security in different layers of software to the dissemination of information to end users. There are good practices to be observed at all levels, from design to use, through the development of operating systems, software layers, and downloadable apps.

Challenges of Mobile Security

Threats

A smartphone user is exposed to various threats when they use their phone. In just the last two quarters of 2012, the number of unique mobile threats grew by 261%, according to ABI Research. These threats can disrupt the operation of the smartphone, and transmit or modify user data. For these reasons, the applications deployed there must guarantee privacy and integrity of the information they handle. In addition, since some apps could themselves be malware, their functionality and activities should be limited (for example, restricting the apps from accessing location information via GPS, blocking access to the user's address book, preventing the transmission of data on the network, sending SMS messages that are billed to the user, etc.).

There are three prime targets for attackers:

- Data: smartphones are devices for data management, therefore they may contain sensitive data like credit card numbers, authentication information, private information, activity logs (calendar, call logs);

- Identity: smartphones are highly customizable, so the device or its contents are associated with a specific person. For example, every mobile device can transmit information related to the owner of the mobile phone contract, and an attacker may want to steal the identity of the owner of a smartphone to commit other offenses;

- Availability: by attacking a smartphone one can limit access to it and deprive the owner of the service.

- The source of these attacks are the same actors found in the non-mobile computing space:

- Professionals, whether commercial or military, who focus on the three targets mentioned above. They steal sensitive data from the general public, as well as undertake industrial espionage. They will also use the identity of those attacked to achieve other attacks;

- Thieves who want to gain income through data or identities they have stolen. The thieves will attack many people to increase their potential income;

- Black hat hackers who specifically attack availability. Their goal is to develop viruses, and cause damage to the device. In some cases, hackers have an interest in stealing data on devices.

- Grey hat hackers who reveal vulnerabilities. Their goal is to expose vulnerabilities of the device. Grey hat hackers do not intend on damaging the device or stealing data.

Consequences

When a smartphone is infected by an attacker, the attacker can attempt several things:

- The attacker can manipulate the smartphone as a zombie machine, that is to say, a machine with which the attacker can communicate and send commands which will be used to send unsolicited messages (spam) via sms or email;

- The attacker can easily force the smartphone to make phone calls. For example, one can use the API (library that contains the basic functions not present in the smartphone) PhoneMakeCall by Microsoft, which collects telephone numbers from any source such as yellow pages, and then call them. But the attacker can also use this method to call paid services, resulting in a charge to the owner of the smartphone. It is also very dangerous because the smartphone could call emergency services and thus disrupt those services;

- A compromised smartphone can record conversations between the user and others and send them to a third party. This can cause user privacy and industrial security problems;

- An attacker can also steal a user's identity, usurp their identity (with a copy of the user's sim card or even the telephone itself), and thus impersonate the owner. This raises security concerns in countries where smartphones can be used to place orders, view bank accounts or are used as an identity card;

- The attacker can reduce the utility of the smartphone, by discharging the battery. For example, they can launch an application that will run continuously on the smartphone processor, requiring a lot of energy and draining the battery. One factor that distinguishes mobile computing from traditional desktop PCs is their limited performance. Frank Stajano and Ross Anderson first described this form of attack, calling it an attack of "battery exhaustion" or "sleep deprivation torture";

- The attacker can prevent the operation and/or starting of the smartphone by making it unusable. This attack can either delete the boot scripts, resulting in a phone without a functioning OS, or modify certain files to make it unusable (e.g. a script that launches at startup that forces the smartphone to restart) or even embed a startup application that would empty the battery;

- The attacker can remove the personal (photos, music, videos, etc.) or professional data (contacts, calendars, notes) of the user.

Attacks Based on Communication

Attack Based on SMS and MMS

Some attacks derive from flaws in the management of SMS and MMS.

Some mobile phone models have problems in managing binary SMS messages. It is possible, by sending an ill-formed block, to cause the phone to restart, leading to denial of service attacks. If a user with a Siemens S55 received a text message containing a Chinese character, it would lead to a denial of service. In another case, while the standard requires that the maximum size of a Nokia Mail address is 32 characters, some Nokia phones did not verify this standard, so if a user enters an email address over 32 characters, that leads to complete dysfunction of the e-mail handler and puts it out of commission. This attack is called "curse of silence". A study on the safety of the SMS infrastructure revealed that SMS messages sent from the Internet can be used to perform a distributed denial of service (DDoS) attack against the mobile telecommunications infrastructure of a big city. The attack exploits the delays in the delivery of messages to overload the network.

Another potential attack could begin with a phone that sends an MMS to other phones, with an attachment. This attachment is infected with a virus. Upon receipt of the MMS, the user can choose to open the attachment. If it is opened, the phone is infected, and the virus sends an MMS with an infected attachment to all the contacts in the address book. There is a real-world example of this attack: the virus Commwarrior uses the address book and sends MMS messages including an infected file to recipients. A user installs the software, as received via MMS message. Then, the virus began to send messages to recipients taken from the address book.

Attacks Based on Communication Networks

Attacks Based on the GSM Networks

The attacker may try to break the encryption of the mobile network. The GSM network encryption algorithms belong to the family of algorithms called A5. Due to the policy of security through obscurity it has not been possible to openly test the robustness of these algorithms. There were originally two variants of the algorithm: A5/1 and A5/2 (stream ciphers), where the former was designed to be relatively strong, and the latter was designed to be weak on purpose to allow easy cryptanalysis and eavesdropping. ETSI forced some countries (typically outside Europe) to use A5/2. Since the encryption algorithm was made public, it was proved it was possible to break the encryption: A5/2 could be broken on the fly, and A5/1 in about 6 hours . In July 2007, the 3GPP approved a change request to prohibit the implementation of A5/2 in any new mobile phones, which means that is has been decommissioned and is no longer implemented in mobile phones. Stronger public algorithms have been added to the GSM standard, the A5/3 and A5/4 (Block ciphers), otherwise known as KASUMI or UEA1 published by the ETSI. If the network does not support A5/1, or any other A5 algorithm implemented by the phone, then the base station can specify A5/0 which is the null-algorithm, whereby the radio traffic is sent unencrypted. Even in case mobile phones are able to use 3G or 4G which have much stronger encryption than 2G GSM, the base station can downgrade the radio communication to 2G GSM and specify A5/0 (no encryption) . This is the basis for eavesdropping attacks on mobile radio networks using a fake base station commonly called an IMSI catcher.

In addition, tracing of mobile terminals is difficult since each time the mobile terminal is accessing or being accessed by the network, a new temporary identity (TMSI) is allocated to the mobile terminal. The TSMI is used as identity of the mobile terminal the next time it accesses the network. The TMSI is sent to the mobile terminal in encrypted messages.

Once the encryption algorithm of GSM is broken, the attacker can intercept all unencrypted communications made by the victim's smartphone.

Attacks Based on Wi-Fi

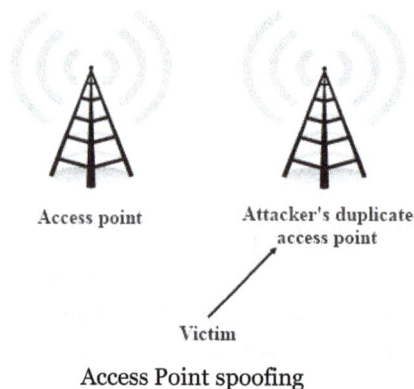

Access Point spoofing

An attacker can try to eavesdrop on Wi-Fi communications to derive information (e.g. username, password). This type of attack is not unique to smartphones, but they are very vulnerable to these attacks because very often the Wi-Fi is the only means of communication they have to access the internet. The security of wireless networks (WLAN) is thus an important subject. Initially wireless networks were secured by WEP keys. The weakness of WEP is a short encryption key which is the same for all connected clients. In addition, several reductions in the search space of the keys have been found by researchers. Now, most wireless networks are protected by the WPA security protocol. WPA is based on the "Temporal Key Integrity Protocol (TKIP)" which was designed to allow migration from WEP to WPA on the equipment already deployed. The major improvements in security are the dynamic encryption keys. For small networks, the WPA is a "pre-shared key" which is based on a shared key. Encryption can be vulnerable if the length of the shared key is short. With limited opportunities for input (i.e. only the numeric keypad) mobile phone users might define short encryption keys that contain only numbers. This increases the likelihood that an attacker succeeds with a brute-force attack. The successor to WPA, called WPA2, is supposed to be safe enough to withstand a brute force attack.

As with GSM, if the attacker succeeds in breaking the identification key, it will be possible to attack not only the phone but also the entire network it is connected to.

Many smartphones for wireless LANs remember they are already connected, and this mechanism prevents the user from having to re-identify with each connection. However, an attacker could create a WIFI access point twin with the same parameters and characteristics as the real network. Using the fact that some smartphones remember the networks, they could confuse the two networks and connect to the network of the attacker who can intercept data if it does not transmit its data in encrypted form.

Lasco is a worm that initially infects a remote device using the SIS file format. SIS file format (Software Installation Script) is a script file that can be executed by the system without user interaction. The smartphone thus believes the file to come from a trusted source and downloads it, infecting the machine.

Principle of Bluetooth-based Attacks

Security issues related to Bluetooth on mobile devices have been studied and have shown numerous problems on different phones. One easy to exploit vulnerability: unregistered services do not require authentication, and vulnerable applications have a virtual serial port used to control the phone. An attacker only needed to connect to the port to take full control of the device. Another example: a phone must be within reach and Bluetooth in discovery mode. The attacker sends a file via Bluetooth. If the recipient accepts, a virus is transmitted. For example: Cabir is a worm that spreads via Bluetooth connection. The worm searches for nearby phones with Bluetooth in discoverable mode and sends itself to the target device. The user must accept the incoming file and install the program. After installing, the worm infects the machine.

Attacks Based on Vulnerabilities in Software Applications

Other attacks are based on flaws in the OS or applications on the phone.

Web Browser

The mobile web browser is an emerging attack vector for mobile devices. Just as common Web browsers, mobile web browsers are extended from pure web navigation with widgets and plug-ins, or are completely native mobile browsers.

Jailbreaking the iPhone with firmware 1.1.1 was based entirely on vulnerabilities on the web browser. As a result, the exploitation of the vulnerability described here underlines the importance of the Web browser as an attack vector for mobile devices. In this case, there was a vulnerability based on a stack-based buffer overflow in a library used by the web browser (Libtiff).

A vulnerability in the web browser for Android was discovered in October 2008. As the iPhone vulnerability above, it was due to an obsolete and vulnerable library. A significant difference with the iPhone vulnerability was Android's sandboxing architecture which limited the effects of this vulnerability to the Web browser process.

Smartphones are also victims of classic piracy related to the web: phishing, malicious websites, etc. The big difference is that smartphones do not yet have strong antivirus software available.

Operating System

Sometimes it is possible to overcome the security safeguards by modifying the operating system itself. As real-world examples, this section covers the manipulation of firmware and malicious signature certificates. These attacks are difficult.

In 2004, vulnerabilities in virtual machines running on certain devices were revealed. It was possible to bypass the bytecode verifier and access the native underlying operating system. The results of this research were not published in detail. The firmware security of Nokia's Symbian Platform Security Architecture (PSA) is based on a central configuration file called SWIPolicy. In 2008 it was possible to manipulate the Nokia firmware before it is installed, and in fact in some downloadable versions of it, this file was human readable, so it was possible to modify and change the image of the firmware. This vulnerability has been solved by an update from Nokia.

In theory smartphones have an advantage over hard drives since the OS files are in ROM, and cannot be changed by malware. However, in some systems it was possible to circumvent this: in the Symbian OS it was possible to overwrite a file with a file of the same name. On the Windows OS, it was possible to change a pointer from a general configuration file to an editable file.

When an application is installed, the signing of this application is verified by a series of certificates. One can create a valid signature without using a valid certificate and add it to the list. In the Symbian OS all certificates are in the directory: c:\resource\swicertstore\dat. With firmware changes explained above it is very easy to insert a seemingly valid but malicious certificate.

Attacks Based on Hardware Vulnerabilities

Electromagnetic Waveforms

In 2015, researchers at the French government agency ANSSI demonstrated the capability to trigger the voice interface of certain smartphones remotely by using "specific electromagnetic waveforms". The exploit took advantage of antenna-properties of headphone wires while plugged into the audio-output jacks of the vulnerable smartphones and effectively spoofed audio input to inject commands via the audio interface.

Juice Jacking

Juice Jacking is a method of physical or a hardware vulnerability specific to mobile platforms. Utilizing the dual purpose of the USB charge port, many devices have been susceptible to having data ex-filtrated from, or malware installed on to a mobile device by utilizing malicious charging kiosks set up in public places, or hidden in normal charge adapters.

Password Cracking

In 2010, researcher from the University of Pennsylvania investigated the possibility of cracking a device's password through a smudge attack (literally imaging the finger smudges on the screen to discern the user's password). The researchers were able to discern the device password up to 68% of the time under certain conditions. Outsiders may perform over-the-shoulder on victims, such as watching specific keystrokes or pattern gestures, to unlock device password or passcode.

Malicious Software (Malware)

As smartphones are a permanent point of access to the internet (mostly on), they can be compromised as easily as computers with malware. A malware is a computer program that aims to harm the system in which it resides. Trojans, worms and viruses are all considered malware. A Trojan is a program that is on the smartphone and allows external users to connect discreetly. A worm is a program that reproduces on multiple computers across a network. A virus is malicious software designed to spread to other computers by inserting itself into legitimate programs and running programs in parallel. However, it must be said that the malware are far less numerous and important to smartphones as they are to computers.

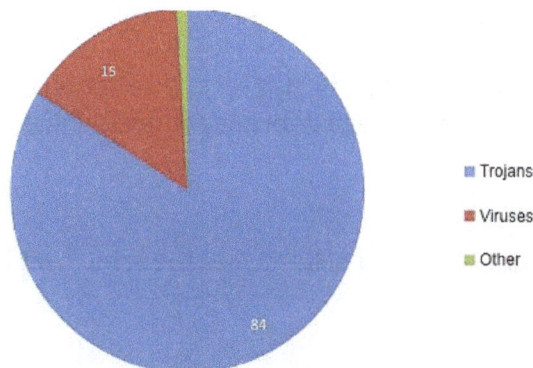

Nonetheless, recent studies show that the evolution of malware in smartphones have rocketed in the last few years posing a threat to analysis and detection.

The Three Phases of Malware Attacks

Typically an attack on a smartphone made by malware takes place in 3 phases: the infection of a host, the accomplishment of its goal, and the spread of the malware to other systems. Malware often use the resources offered by the infected smartphones. It will use the output devices such as Bluetooth or infrared, but it may also use the address book or email address of the person to infect the user's acquaintances. The malware exploits the trust that is given to data sent by an acquaintance.

Infection

Infection is the means used by the malware to get into the smartphone, it can either use one of the faults previously presented or may use the gullibility of the user. Infections are classified into four classes according to their degree of user interaction:

Explicit Permission

the most benign interaction is to ask the user if it is allowed to infect the machine, clearly indicating its potential malicious behavior. This is typical behavior of a proof of concept malware.

Implied Permission

this infection is based on the fact that the user has a habit of installing software. Most trojans try to seduce the user into installing attractive applications (games, useful applications etc.) that actually contain malware.

Common Interaction

this infection is related to a common behavior, such as opening an MMS or email.

No Interaction

the last class of infection is the most dangerous. Indeed, a worm that could infect a smartphone and could infect other smartphones without any interaction would be catastrophic.

Accomplishment of its Goal

Once the malware has infected a phone it will also seek to accomplish its goal, which is usually one of the following: monetary damage, damage data and/or device, and concealed damage:

Monetary Damages

the attacker can steal user data and either sell them to the same user, or sell to a third party.

Damage

malware can partially damage the device, or delete or modify data on the device.

Concealed Damage

the two aforementioned types of damage are detectable, but the malware can also leave a backdoor for future attacks or even conduct wiretaps.

Spread to Other Systems

Once the malware has infected a smartphone, it always aims to spread one way or another:

- It can spread through proximate devices using Wi-Fi, Bluetooth and infrared;

- It can also spread using remote networks such as telephone calls or SMS or emails.

Examples of Malware

Here are various malware that exist in the world of smartphones with a short description of each.

Viruses and Trojans

- Cabir (also known as Caribe, SybmOS/Cabir, Symbian/Cabir and EPOC.cabir) is the name of a computer worm developed in 2004 that is designed to infect mobile phones running Symbian OS. It is believed to be the first computer worm that can infect mobile phones

- Commwarrior, found March 7, 2005, is the first worm that can infect many machines from MMS. It is sent in the form of an archive file COMMWARRIOR.ZIP that contains a file COMMWARRIOR.SIS. When this file is executed, Commwarrior attempts to connect to nearby devices by Bluetooth or infrared under a random name. It then attempts to send MMS message to the contacts in the smartphone with different header messages for each person, who receive the MMS and often open them without further verification.

- Phage is the first Palm OS virus that was discovered. It transfers to the Palm from a PC via synchronization. It infects all applications that are in the smartphone and it embeds its own code to function without the user and the system detecting it. All that the system will detect is that its usual applications are functioning.

- RedBrowser is a Trojan which is based on java. The Trojan masquerades as a program called "RedBrowser" which allows the user to visit WAP sites without a WAP connection. During application installation, the user sees a request on their phone that the application needs permission to send messages. Therefore, if the user accepts, RedBrowser can send sms to paid call centers. This program uses the smartphone's connection to social networks (Facebook, Twitter, etc.) to get the contact information for the user's acquaintances (provided the required permissions have been given) and will send them messages.

- WinCE.PmCryptic.A is a malicious software on Windows Mobile which aims to earn money for its authors. It uses the infestation of memory cards that are inserted in the smartphone to spread more effectively.

- CardTrap is a virus that is available on different types of smartphone, which aims to deactivate the system and third party applications. It works by replacing the files used to start the smartphone and applications to prevent them from executing. There are different variants of this virus such as Cardtrap.A for SymbOS devices. It also infects the memory card with malware capable of infecting Windows.

- Ghost Push is a malicious software on Android OS which automatically root the android device and installs malicious applications directly to system partition then unroots the device to prevent users from removing the threat by master reset (The threat can be removed only by reflashing). It cripples the system resources, executes quickly, and harder to detect.

Ransomware

Mobile ransomware is a type of malware that locks users out of their mobile devices in a pay-to-unlock-your-device ploy, it has grown by leaps and bounds as a threat category since 2014. Specific to mobile computing platforms, users are often less security-conscious, particularly as it pertains to scrutinizing applications and web links trusting the native protection capability of the mobile device operating system. Mobile ransomware poses a significant threat to businesses reliant on instant access and availability of their proprietary information and contacts. The likelihood of a traveling businessman paying a ransom to unlock their device is significantly higher since they are at a disadvantage given inconveniences such as timeliness and less likely direct access to IT staff.

Spyware

- Flexispy is an application that can be considered as a trojan, based on Symbian. The program sends all information received and sent from the smartphone to a Flexispy server. It was originally created to protect children and spy on adulterous spouses.

Number of Malware

Below is a diagram which loads the different behaviors of smartphone malware in terms of their effects on smartphones:

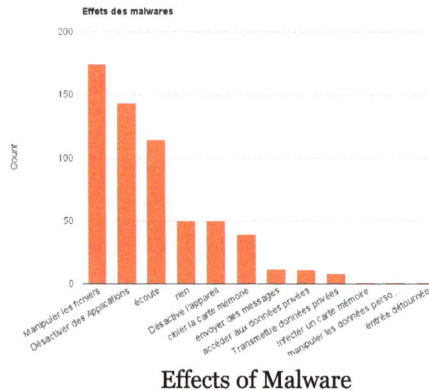

Effects of Malware

We can see from the graph that at least 50 malwares exhibit no negative behavior, except their ability to spread.

Portability of Malware Across Platforms

There is a multitude of malware. This is partly due to the variety of operating systems on smartphones. However attackers can also choose to make their malware target multiple platforms, and malware can be found which attacks an OS but is able to spread to different systems.

To begin with, malware can use runtime environments like Java virtual machine or the .NET Framework. They can also use other libraries present in many operating systems. Other malware carry several executable files in order to run in multiple environments and they utilize these during the propagation process. In practice, this type of malware requires a connection between the two operating systems to use as an attack vector. Memory cards can be used for this purpose, or synchronization software can be used to propagate the virus.

Countermeasures

The security mechanisms in place to counter the threats described above are presented in this section. They are divided into different categories, as all do not act at the same level, and they range from the management of security by the operating system to the behavioral education of the user. The threats prevented by the various measures are not the same depending on the case. Considering the two cases mentioned above, in the first case one would protect the system from corruption by an application, and in the second case the installation of a suspicious software would be prevented.

Security in Operating Systems

The first layer of security within a smartphone is at the level of the operating system (OS). Beyond the usual roles of an operating system (e.g. resource management, scheduling processes) on a smartphone, it must also establish the protocols for introducing external applications and data without introducing risk.

A central idea found in the mobile operating systems is the idea of a sandbox. Since smartphones are currently being designed to accommodate many applications, they must put in place mechanisms to ensure these facilities are safe for themselves, for other applications and data on the system, and the user. If a malicious program manages to reach a device, it is necessary that the vulnerable area presented by the system be as small as possible. Sandboxing extends this idea to compartmentalize different processes, preventing them from interacting and damaging each other. Based on the history of operating systems, sandboxing has different implementations. For example, where iOS will focus on limiting access to its public API for applications from the App Store by default, Managed Open In allows you to restrict which apps can access which types of data. Android bases its sandboxing on its legacy of Linux and TrustedBSD.

The following points highlight mechanisms implemented in operating systems, especially Android.

Rootkit Detectors

The intrusion of a rootkit in the system is a great danger in the same way as on a computer. It is important to prevent such intrusions, and to be able to detect them as often as possible. Indeed, there is concern that with this type of malicious program, the result could be a partial or complete bypass of the device security, and the acquisition of administrator rights by the attacker. If this happens, then nothing prevents the attacker from studying or disabling the safety features that were circumvented, deploying the applications they want, or disseminating a method of intrusion by a rootkit to a wider audience. We can cite, as a defense mechanism, the Chain of trust in iOS. This mechanism relies on the signature of the different applications required to start the operating system, and a certificate signed by Apple. In the event that the signature checks are inconclusive, the device detects this and stops the boot-up. If the Operating System is compromised due to Jailbreaking, root kit detection may not work if it is disabled by the Jailbreak method or software is loaded after Jailbreak disables Rootkit Detection.

Process Isolation

Android uses mechanisms of user process isolation inherited from Linux. Each application has a user associated with it, and a tuple (UID, GID). This approach serves as a sandbox: while applications can be malicious, they can not get out of the sandbox reserved for them by their identifiers, and thus cannot interfere with the proper functioning of the system. For example, since it is impossible for a process to end the process of another user, an application can thus not stop the execution of another.

File Permissions

From the legacy of Linux, there are also filesystem permissions mechanisms. They help with sandboxing: a process can not edit any files it wants. It is therefore not possible to freely corrupt files necessary for the operation of another application or system. Furthermore, in Android there is the method of locking memory permissions. It is not possible to change the permissions of files installed on the SD card from the phone, and consequently it is impossible to install applications.

Memory Protection

In the same way as on a computer, memory protection prevents privilege escalation. Indeed, if a process managed to reach the area allocated to other processes, it could write in the memory of a process with rights superior to their own, with root in the worst case, and perform actions which are beyond its permissions on the system. It would suffice to insert function calls are authorized by the privileges of the malicious application.

Development Through Runtime Environments

Software is often developed in high-level languages, which can control what is being done by a running program. For example, Java Virtual Machines continuously monitor the actions of the execution threads they manage, monitor and assign resources, and prevent malicious actions. Buffer overflows can be prevented by these controls.

Security Software

Above the operating system security, there is a layer of security software. This layer is composed of individual components to strengthen various vulnerabilities: prevent malware, intrusions, the identification of a user as a human, and user authentication. It contains software components that have learned from their experience with computer security; however, on smartphones, this software must deal with greater constraints.

Antivirus and Firewall

An antivirus software can be deployed on a device to verify that it is not infected by a known threat, usually by signature detection software that detects malicious executable files. A firewall, meanwhile, can watch over the existing traffic on the network and ensure that a malicious application does not seek to communicate through it. It may equally verify that an installed application does not seek to establish suspicious communication, which may prevent an intrusion attempt.

Visual Notifications

In order to make the user aware of any abnormal actions, such as a call they did not initiate, one can link some functions to a visual notification that is impossible to circumvent. For example, when a call is triggered, the called number should always be displayed. Thus, if a call is triggered by a malicious application, the user can see, and take appropriate action.

Turing Test

In the same vein as above, it is important to confirm certain actions by a user decision. The Turing test is used to distinguish between a human and a virtual user, and it often comes as a captcha.

Biometric Identification

Another method to use is biometrics. Biometrics is a technique of identifying a person by means of their morphology(by recognition of the eye or face, for example) or their behavior (their signature or way of writing for example). One advantage of using biometric security is that users can avoid having to remember a password or other secret combination to authenticate and prevent malicious users from accessing their device. In a system with strong biometric security, only the primary user can access the smartphone.

Resource Monitoring in the Smartphone

When an application passes the various security barriers, it can take the actions for which it was designed. When such actions are triggered, the activity of a malicious application can be sometimes detected if one monitors the various resources used on the phone. Depending on the goals of the malware, the consequences of infection are not always the same; all malicious applications are not intended to harm the devices on which they are deployed. The following sections describe different ways to detect suspicious activity.

Battery

Some malware is aimed at exhausting the energy resources of the phone. Monitoring the energy consumption of the phone can be a way to detect certain malware applications.

Memory Usage

Memory usage is inherent in any application. However, if one finds that a substantial proportion of memory is used by an application, it may be flagged as suspicious.

Network Traffic

On a smartphone, many applications are bound to connect via the network, as part of their normal operation. However, an application using a lot of bandwidth can be strongly suspected of attempting to communicate a lot of information, and disseminate data to many other devices. This observation only allows a suspicion, because some legitimate applications can be very resource-intensive in terms of network communications, the best example being streaming video.

Services

One can monitor the activity of various services of a smartphone. During certain moments, some services should not be active, and if one is detected, the application should be suspected. For example, the sending of an SMS when the user is filming video: this communication does not make sense and is suspicious; malware may attempt to send SMS while its activity is masked.

The various points mentioned above are only indications and do not provide certainty about the legitimacy of the activity of an application. However, these criteria can help target suspicious applications, especially if several criteria are combined.

Network Surveillance

Network traffic exchanged by phones can be monitored. One can place safeguards in network routing points in order to detect abnormal behavior. As the mobile's use of network protocols is much more constrained than that of a computer, expected network data streams can be predicted (e.g. the protocol for sending an SMS), which permits detection of anomalies in mobile networks.

Spam Filters

As is the case with email exchanges, we can detect a spam campaign through means of mobile communications (SMS, MMS). It is therefore possible to detect and minimize this kind of attempt by filters deployed on network infrastructure that is relaying these messages.

Encryption of Stored or Transmitted Information

Because it is always possible that data exchanged can be intercepted, communications, or even information storage, can rely on encryption to prevent a malicious entity from using any data obtained during communications. However, this poses the problem of key exchange for encryption algorithms, which requires a secure channel.

Telecom Network Monitoring

The networks for SMS and MMS exhibit predictable behavior, and there is not as much liberty compared with what one can do with protocols such as TCP or UDP. This implies that one cannot predict the use made of the common protocols of the web; one might generate very little traffic by consulting simple pages, rarely, or generate heavy traffic by using video streaming. On the other hand, messages exchanged via mobile phone have a framework and a specific model, and the user does not, in a normal case, have the freedom to intervene in the details of these communications. Therefore, if an abnormality is found in the flux of network data in the mobile networks, the potential threat can be quickly detected.

Manufacturer Surveillance

In the production and distribution chain for mobile devices, it is the responsibility of manufacturers to ensure that devices are delivered in a basic configuration without vulnerabilities. Most users are not experts and many of them are not aware of the existence of security vulnerabilities, so the device configuration as provided by manufacturers will be retained by many users. Below are listed several points which manufacturers should consider.

Remove Debug Mode

Phones are sometimes set in a debug mode during manufacturing, but this mode must be disabled before the phone is sold. This mode allows access to different features, not intended for routine use by a user. Due to the speed of development and production, distractions occur and some devices are sold in debug mode. This kind of deployment exposes mobile devices to exploits that utilize this oversight.

Default Settings

When a smartphone is sold, its default settings must be correct, and not leave security gaps. The default configuration is not always changed, so a good initial setup is essential for users. There are, for example, default configurations that are vulnerable to denial of service attacks.

Security Audit of Apps

Along with smart phones, appstores have emerged. A user finds themselves facing a huge range of applications. This is especially true for providers who manage appstores because they are tasked with examining the apps provided, from different points of view (e.g. security, content). The security audit should be particularly cautious, because if a fault is not detected, the application can spread very quickly within a few days, and infect a significant number of devices.

Detect Suspicious Applications Demanding Rights

When installing applications, it is good to warn the user against sets of permissions that, grouped together, seem potentially dangerous, or at least suspicious. Frameworks like such as Kirin, on Android, attempt to detect and prohibit certain sets of permissions.

Revocation Procedures

Along with appstores appeared a new feature for mobile apps: remote revocation. First developed by Android, this procedure can remotely and globally uninstall an application, on any device that has it. This means the spread of a malicious application that managed to evade security checks can be immediately stopped when the threat is discovered.

Avoid Heavily Customized Systems

Manufacturers are tempted to overlay custom layers on existing operating systems, with the dual purpose of offering customized options and disabling or charging for certain features. This has the dual effect of risking the introduction of new bugs in the system, coupled with an incentive for users to modify the systems to circumvent the manufacturer's restrictions. These systems are rarely as stable and reliable as the original, and may suffer from phishing attempts or other exploits.

Improve Software Patch Processes

New versions of various software components of a smartphone, including operating systems, are regularly published. They correct many flaws over time. Nevertheless, manufacturers often do not deploy these updates to their devices in a timely fashion, and sometimes not at all. Thus, vulnerabilities persist when they could be corrected, and if they are not, since they are known, they are easily exploitable.

User Awareness

Much malicious behavior is allowed by the carelessness of the user. From simply not leaving the device without a password, to precise control of permissions granted to applications added to the smartphone, the user has a large responsibility in the cycle of security: to not be the vector of intrusion. This precaution is especially important if the user is an employee of a company that stores business data on the device. Detailed below are some precautions that a user can take to manage security on a smartphone.

A recent survey by internet security experts BullGuard showed a lack of insight into the rising number of malicious threats affecting mobile phones, with 53% of users claiming that they are unaware of security software for Smartphones. A further 21% argued that such protection was unnecessary, and 42% admitted it hadn't crossed their mind ("Using APA," 2011). These statistics show consumers are not concerned about security risks because they believe it is not a serious problem. The key here is to always remember smartphones are effectively handheld computers and are just as vulnerable.

Being Skeptical

A user should not believe everything that may be presented, as some information may be phishing or attempting to distribute a malicious application. It is therefore advisable to check the reputation of the application that they want to buy before actually installing it.

Permissions given to Applications

The mass distribution of applications is accompanied by the establishment of different permissions mechanisms for each operating system. It is necessary to clarify these permissions mechanisms to users, as they differ from one system to another, and are not always easy to understand. In addition, it is rarely possible to modify a set of permissions requested by an application if the number of permissions is too great. But this last point is a source of risk because a user can grant rights to an application, far beyond the rights it needs. For example, a note taking application does not require access to the geolocation service. The user must ensure the privileges required by an application during installation and should not accept the installation if requested rights are inconsistent.

Be Careful

Protection of a user's phone through simple gestures and precautions, such as locking the smartphone when it is not in use, not leaving their device unattended, not trusting applications, not storing sensitive data, or encrypting sensitive data that cannot be separated from the device.

Ensure Data

Smartphones have a significant memory and can carry several gigabytes of data. The user must be careful about what data it carries and whether they should be protected. While it is usually not dramatic if a song is copied, a file containing bank information or business data can be more risky. The user must have the prudence to avoid the transmission of sensitive data on a smartphone, which can be easily stolen. Furthermore, when a user gets rid of a device, they must be sure to remove all personal data first.

These precautions are measures that leave no easy solution to the intrusion of people or malicious applications in a smartphone. If users are careful, many attacks can be defeated, especially phishing and applications seeking only to obtain rights on a device.

Centralized Storage of Text Messages

One form of mobile protection allows companies to control the delivery and storage of text messages, by hosting the messages on a company server, rather than on the sender or receiver's phone. When certain conditions are met, such as an expiration date, the messages are deleted.

Limitations of Certain Security Measures

The security mechanisms mentioned in this article are to a large extent inherited from knowledge and experience with computer security. The elements composing the two device types are similar, and there are common measures that can be used, such as antivirus and firewall. However, the implementation of these solutions is not necessarily possible or at least highly constrained within a mobile device. The reason for this difference is the technical resources offered by computers and mobile devices: even though the computing power of smartphones is becoming faster, they have other limitations than their computing power.

- Single-task system: Some operating systems, including some still commonly used, are single-tasking. Only the foreground task is executed. It is difficult to introduce applications such as antivirus and firewall on such systems, because they could not perform their monitoring while the user is operating the device, when there would be most need of such monitoring.

- Energy autonomy: A critical one for the use of a smartphone is energy autonomy. It is important that the security mechanisms not consume battery resources, without which the autonomy of devices will be affected dramatically, undermining the effective use of the smartphone.

- Network Directly related to battery life, network utilization should not be too high. It is indeed one of the most expensive resources, from the point of view of energy consumption. Nonetheless, some calculations may need to be relocated to remote servers in order to preserve the battery. This balance can make implementation of certain intensive computation mechanisms a delicate proposition.

Furthermore, it should be noted that it is common to find that updates exist, or can be developed or deployed, but this is not always done. One can, for example, find a user who does not know that there is a newer version of the operating system compatible with the smartphone, or a user may discover known vulnerabilities that are not corrected until the end of a long development cycle, which allows time to exploit the loopholes.

Next Generation of Mobile Security

There is expected to be four mobile environments that will make up the security framework:

Rich Operating System

In this category will fall traditional Mobile OS like Android, iOS, Symbian OS or Windows Phone. They will provide the traditional functionaity and security of an OS to the applications.

Secure Operating System (Secure OS)

A secure kernel which will run in parallel with a fully featured Rich OS, on the same processor core. It will include drivers for the Rich OS ("normal world") to communicate with the secure kernel ("secure world"). The trusted infrastructure could include interfaces like the display or keypad to regions of PCI-E address space and memories.

Trusted Execution Environment (TEE)

Made up of hardware and software. It helps in the control of access rights and houses sensitive applications, which need to be isolated from the Rich OS. It effectively acts as a firewall between the "normal world" and "secure world".

Secure Element (SE)

The SE consists of tamper resistant hardware and associated software. It can provide high levels of security and work in tandem with the TEE. The SE will be mandatory for hosting proximity payment applications or official electronic signatures.

References

- Harwood, Mike (29 June 2009). "Securing Wireless Networks". CompTIA Network+ N10-004 Exam Prep. Pearson IT Certification. p. 287. ISBN 978-0-7897-3795-3. Retrieved 9 July 2016.

- Gasser, Morrie (1988). Building a Secure Computer System (PDF). Van Nostrand Reinhold. p. 3. ISBN 0-442-23022-2. Retrieved 6 September 2015

- Umrigar, Zerksis D.; Pitchumani, Vijay (1983). "Formal verification of a real-time hardware design". Proceeding DAC '83 Proceedings of the 20th Design Automation Conference. IEEE Press. pp. 221–7. ISBN 0-8186-0026-8.

- Walker, Jesse. "A History of 802.11 Security" (PDF). Rutgers WINLAB. Intel Corporation. Archived from the original (PDF) on 9 July 2016. Retrieved 9 July 2016.

- Jim Finkle (23 April 2014). "Exclusive: FBI warns healthcare sector vulnerable to cyber attacks". Reuters. Retrieved 23 May 2016.

- Arachchilage, Nalin; Love, Steve; Scott, Michael (June 1, 2012). "Designing a Mobile Game to Teach Conceptual Knowledge of Avoiding 'Phishing Attacks'". International Journal for e-Learning Security. Infonomics Society. 2 (1): 127–132. Retrieved April 1, 2016.

- Scott, Michael; Ghinea, Gheorghita; Arachchilage, Nalin (7 July 2014). Assessing the Role of Conceptual Knowledge in an Anti-Phishing Educational Game (pdf). Proceedings of the 14th IEEE International Conference on Advanced Learning Technologies. IEEE. p. 218. doi:10.1109/ICALT.2014.70. Retrieved April 1, 2016.

- "U.S. GAO - Air Traffic Control: FAA Needs a More Comprehensive Approach to Address Cybersecurity As Agency Transitions to NextGen". Retrieved 23 May 2016.

- Pagliery, Jose. "Hackers attacked the U.S. energy grid 79 times this year". CNN Money. Cable News Network. Retrieved 16 April 2015.

- Sanders, Sam (4 June 2015). "Massive Data Breach Puts 4 Million Federal Employees' Records At Risk". NPR. Retrieved 5 June 2015.

- "Ensuring the Security of Federal Information Systems and Cyber Critical Infrastructure and Protecting the Privacy of Personally Identifiable Information". Government Accountability Office. Retrieved November 3, 2015.

- "Action Plan 2010–2015 for Canada's Cyber Security Strategy". Public Safety Canada. Government of Canada. Retrieved 3 November 2014.

- "Cyber Security Awareness Free Training and Webcasts". MS-ISAC (Multi-State Information Sharing & Analysis Center. Retrieved 9 January 2015.

- Gittleson, Kim (28 March 2014) Data-stealing Snoopy drone unveiled at Black Hat BBC News, Technology, Retrieved 29 March 2014

- Wilkinson, Glenn (25 September 2012) Snoopy: A distributed tracking and profiling framework Sensepost, Retrieved 29 March 2014

Applications of Telecommunications Engineering

Digital radio is the technology used to communicate across the radio spectrum and push buttons is a simple mechanism for controlling some aspect of a machine. The applications of telecommunication engineering such as photo phone, radio and free space optical communication are dealt within this chapter. This chapter discusses the methods of telecommunication in a critical manner providing key analysis to this subject matter.

Photophone

The photophone (later given the alternate name radiophone) is a telecommunications device which allowed for the transmission of speech on a beam of light. It was invented jointly by Alexander Graham Bell and his assistant Charles Sumner Tainter on February 19, 1880, at Bell's laboratory at 1325 L Street in Washington, D.C. Both were later to become full associates in the Volta Laboratory Association, created and financed by Bell.

A historical plaque on the side of the Franklin School in Washington, D.C. which marks one of the points from which the photophone was demonstrated

On June 3, 1880, Bell's assistant transmitted a wireless voice telephone message from the roof of the Franklin School to the window of Bell's laboratory, some 213 meters (about 700 ft.) away.

Bell believed the photophone was his most important invention. Of the 18 patents granted in Bell's name alone, and the 12 he shared with his collaborators, four were for the photophone, which Bell referred to as his "greatest achievement", telling a reporter shortly before his death that the photophone was "the greatest invention [I have] ever made, greater than the telephone".

The photophone was a precursor to the fiber-optic communication systems which achieved worldwide popular usage starting in the 1980s. The master patent for the photophone (U.S. Patent 235,199 *Apparatus for Signalling and Communicating, called Photophone*) was issued in December 1880, many decades before its principles came to have practical applications.

Design

A photophone receiver and headset, one half of Bell and Tainter's optical telecommunication system of 1880

The photophone was similar to a contemporary telephone, except that it used modulated light as a means of wireless transmission while the telephone relied on modulated electricity carried over a conductive wire circuit.

Bell's own description of the light modulator:

We have found that the simplest form of apparatus for producing the effect consists of a plane mirror of flexible material against the back of which the speaker's voice is directed. Under the action of the voice the mirror becomes alternately convex and concave and thus alternately scatters and condenses the light.

The brightness of a reflected beam of light, as observed from the location of the receiver, therefore varied in accordance with the audio-frequency variations in air pressure—the sound waves—which acted upon the mirror.

In its initial form, the photophone receiver was also non-electronic. Bell found that many substances could be used as direct light-to-sound transducers. Lampblack proved to be outstanding. Using a fully modulated beam of sunlight as a test signal, one experimental receiver design, employing only a deposit of lampblack, produced a tone that Bell described as "painfully loud" to an ear pressed close to the device.

In its ultimate electronic form, the photophone receiver used a simple selenium cell at the focus of a parabolic mirror. The cell's electrical resistance (between about 100 and 300 ohms) varied inversely with the light falling upon it, i.e., its resistance was higher when dimly lit, lower when brightly lit. The selenium cell took the place of a carbon microphone—also a variable-resistance device—in the circuit of what was otherwise essentially an ordinary telephone, consisting of a battery, an electromagnetic earphone, and the variable resistance, all connected in series. The selenium modulated the current flowing through the circuit, and the current was converted back into variations of air pressure—sound—by the earphone.

In his speech to the American Association for the Advancement of Science in August 1880, Bell gave credit to the first demonstration of speech transmission by light to Mr. A.C. Brown of London in the Fall of 1878.

The French scientist Ernest Mercadier suggested that the invention should not be named 'photophone', but 'radiophone', as its mirrors reflected the Sun's radiant energy in multiple bands including the invisible infrared band. For a period of time the invention also used the latter name.

First Successful Wireless Voice Communications

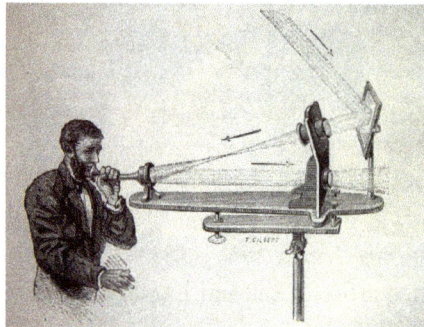

Illustration of a photophone transmitter, showing the path of reflected sunlight, before and after being modulated

Illustration of a photophone receiver, depicting the conversion of modulated light to sound, as well as its electrical power source (P)

While honeymooning in Europe with his bride Mabel Hubbard, Bell likely read of the newly discovered property of selenium having a variable resistance when acted upon by light, in a paper by Robert Sabine as published in *Nature* on 25 April 1878. In his experiments, Sabine used a meter to see the effects of light acting on selenium connected in a circuit to a battery. However Bell reasoned that by adding a telephone receiver to the same circuit he would be able to hear what Sabine could only see.

As Bell's former associate, Thomas Watson, was fully occupied as the superintendent of manufacturing for the nascent Bell Telephone Company back in Boston, Massachusetts, Bell hired Charles Sumner Tainter, an instrument maker who had previously been assigned to the U.S. 1874 Transit of Venus Commission, for his new 'L' Street laboratory in Washington, at the rate of $15 per week.

On February 19, 1880 the pair had managed to make a functional photophone in their new laboratory by attaching a set of metallic gratings to a diaphragm, with a beam of light being interrupted by the gratings movement in response to spoken sounds. When the modulated light beam fell upon their selenium receiver Bell, on his headphones, was able to clearly hear Tainter singing *Auld Lang Syne*.

In an April 1, 1880 Washington, D.C. experiment, Bell and Tainter communicated some 79 metres (259 ft) meters along an alleyway to the laboratory's rear window. Then a few months later on June 21 they succeeded in communicating clearly over a distance of some 213 meters (about 700 ft.), using plain sunlight as their light source, practical electrical lighting having only just been introduced to the U.S.A. by Edison. The transmitter in their latter experiments had sunlight reflected off the surface of a very thin mirror positioned at the end of a speaking tube; as words were spoken they cause the mirror to oscillate between convex and concave, altering the amount of light reflected from its surface to the receiver. Tainter, who was on the roof of the Franklin School, spoke to Bell, who was in his laboratory listening and who signaled back to Tainter by waving his hat vigorously from the window, as had been requested.

The receiver was a parabolic mirror with selenium cells at its focal point. Conducted from the roof of the Franklin School to Bell's laboratory at 1325 'L' Street, this was the world's first formal wireless telephone communication (away from their laboratory), thus making the photophone the world's earliest known radiophone and wireless telephone systems, at least 19 years ahead of the first spoken radio transmissions. Before Bell and Tainter had concluded their research in order to move on to the development of the Graphophone, they had devised some 50 different methods of modulating and demodulating light beams for optical telephony.

Reception and Adoption

The telephone itself was still something of a novelty, and radio was decades away from commercialization. The social resistance to the photophone's futuristic form of communications could be seen in an 1880 *New York Times* commentary:

The ordinary man ... will find a little difficulty in comprehending how sunbeams are to be used. Does Prof. Bell intend to connect Boston and Cambridge ... with a line of sunbeams hung on telegraph posts, and, if so, what diameter are the sunbeams to be ....[and] will it be necessary to insulate them against the weather ... until (the public) sees a man going through the streets with a coil of No. 12 sunbeams on his shoulder, and suspending them from pole to pole, there will be a general feeling that there is something about Professor Bell's photophone which places a tremendous strain on human credulity.

However at the time of their February 1880 breakthrough, Bell was immensely proud of the achievement, to the point that he wanted to name his new second daughter "Photophone", which was subtly discouraged by his wife Mabel Bell (they instead chose "Marian", with "Daisy" as her nickname). He wrote somewhat enthusiastically:

I have heard articulate speech by sunlight! I have heard a ray of the sun laugh and cough and sing! ...I have been able to hear a shadow and I have even perceived by ear the passage of a cloud across the sun's disk. You are the grandfather of the Photophone and I want to share my delight at my success.

—Alexander Graham Bell, in a letter to his father Alexander Melville Bell, dated February 26, 1880

Bell transferred the photophone's intellectual property rights to the American Bell Telephone Company in May 1880. While Bell had hoped his new photophone could be used by ships at sea and to also displace the plethora of telephone lines that were blooming along busy city boulevards, his design failed to protect its transmissions from outdoor interferences such as clouds, fog, rain, snow and such, that could easily disrupt the transmission of light. Factors such as the weather and the lack of light inhibited the use of Bell's invention. Not long after its invention laboratories within the Bell System continued to improve the photophone in the hope that it could supplement or replace expensive conventional telephone lines. Its earliest non-experimental use came with military communication systems during World War I and II, its key advantage being that its light-based transmissions could not be intercepted by the enemy.

Bell pondered the photophone's possible scientific use in the spectral analysis of artificial light sources, stars and sunspots. He later also speculated on its possible future applications, though he did not anticipate either the laser or fiber-optic telecommunications:

Can Imagination picture what the future of this invention is to be!.... We may talk by light to any visible distance without any conduction wire.... In general science, discoveries will be make by the Photophone that are undreamed of just now.

Further Development

Although Bell Telephone researchers made several modest incremental improvements on Bell and Tainter's design, Marconi's radio transmissions started to far surpass the maximum range of the photophone as early as 1897 and further development of the photophone was largely arrested until German-Austrian experiments began at the turn of the 20th century. The German Siemens & Halske Company boosted the photophone's range by utilizing current-modulated carbon arc lamps which provided a useful range of approximately 8 kilometres (5.0 mi). They produced units commercially for the German Navy, which were further adapted to increase their range to 11 kilometres (6.8 mi) using voice-modulated ship searchlights.

British Admiralty research during WWI resulted in the development of a vibrating mirror modulator in 1916. More sensitive molybdenite receiver cells, which also had greater sensitivity to infra-red radiation, replaced the older selenium cells in 1917. The United States and German governments also worked on technical improvements to Bell's system.

By 1935 the German Carl Zeiss Company had started producing infra-red photophones for the German Army's tank battalions, employing tungsten lamps with infra-red filters which were modulated by vibrating mirrors or prisms. These also used receivers which employed lead sulphide detector cells and amplifiers, boosting their range to 14 kilometres (8.7 mi) under optimal conditions. The Japanese and Italian armies also attempted similar development of lightwave telecommunications before 1945.

Several military laboratories, including those in the United States, continued R&D efforts on the photophone into the 1950s, experimenting with high-pressure vapour and mercury arc lamps of between 500 and 2,000 watts power.

Commemorations

1947 Franklin School Ceremony

On March 3, 1947, the centenary of Alexander Graham Bell's birth, the Telephone Pioneers of America dedicated a historical marker on the side of one of the buildings, the Franklin School, which Bell and Sumner Tainter used for their first formal trial involving a considerable distance. Tainter had originally stood on the roof of the school building and transmitted to Bell at the window of his laboratory. The plaque, which did not acknowledge Tainter's scientific and engineering contributions, read:

> FROM THE TOP FLOOR OF THIS BUILDING
> WAS SENT ON JUNE 3, 1880
> OVER A BEAM OF LIGHT TO 1325 'L' STREET
> THE FIRST WIRELESS TELEPHONE MESSAGE
> IN THE HISTORY OF THE WORLD.
> THE APPARATUS USED IN SENDING THE MESSAGE
> WAS THE PHOTOPHONE INVENTED BY
> ALEXANDER GRAHAM BELL
> INVENTOR OF THE TELEPHONE
> THIS PLAQUE WAS PLACED HERE BY
> ALEXANDER GRAHAM BELL CHAPTER
> TELEPHONE PIONEERS OF AMERICA
> MARCH 3, 1947
> THE CENTENNIAL OF DR. BELL'S BIRTH

1980 Centenary Commemoration

On February 19, 1980, exactly 100 years to the day after Bell and Tainter's first photophone transmission in their laboratory, staff from the Smithsonian Institution, the National Geographic Society and AT&T's Bell Labs gathered at the location of Bell's former 1325 'L' Street Volta Laboratory in Washington, D.C. for a commemoration of the event.

The Photophone Centenary commemoration had first been proposed by electronics researcher and writer Forrest M. Mims, who suggested it to Dr. Melville Bell Grosvenor, the inventor's grandson, during a visit to his office at the National Geographic Society. The historic grouping later observed the centennial of the photophone's first successful laboratory transmission by using Mims hand-made demonstration photophone, which functioned similar to Bell and Tainter's model.

Mims also built and provided a pair of modern hand-held battery-powered LED transceivers connected by 100 yards (91 m) of optical fiber. The Bell Labs' Richard Gundlach and the Smithsonian's Elliot Sivowitch used the device at the commemoration to demonstrate one of the photophone's modern-day descendants. The National Geographic Society also mounted a special educational exhibit in its Explorer's Hall, highlighting the photophone's invention with original items borrowed from the Smithsonian Institution.

Radio

The Alexandra Palace, here: mast of the broadcasting station

Classic radio receiver dial

Radio is the technology of using radio waves to carry information, such as sound, by systematically modulating some property of electromagnetic energy waves transmitted through space, such as their amplitude, frequency, phase, or pulse width. When radio waves strike an electrical conductor, the oscillating fields induce an alternating current in the conductor. The information in the waves can be extracted and transformed back into its original form.

Radio systems need a transmitter to modulate (change) some property of the energy produced to impress a signal on it, for example using amplitude modulation or angle modulation (which can be frequency modulation or phase modulation). Radio systems also need an antenna to convert electric currents into radio waves, and vice versa. An antenna can be used for both transmitting and receiving. The electrical resonance of tuned circuits in radios allow individual stations to be selected. The electromagnetic wave is intercepted by a tuned receiving antenna. A radio receiver receives its input from an antenna and converts it into a form usable for the consumer, such as sound, pictures, digital data, measurement values, navigational positions, etc. Radio frequencies occupy the range from a 3 kHz to 300 GHz, although commercially important uses of radio use only a small part of this spectrum.

A radio communication system sends signals by radio. The radio equipment involved in communication systems includes a transmitter and a receiver, each having an antenna and appropriate terminal equipment such as a microphone at the transmitter and a loudspeaker at the receiver in the case of a voice-communication system.

Etymology

The term "radio" is derived from the Latin word "radius", meaning "spoke of a wheel, beam of light, ray". It was first applied to communications in 1881 when, at the suggestion of French scientist Ernest Mercadier, Alexander Graham Bell adopted "radiophone" (meaning "radiated sound") as an alternate name for his photophone optical transmission system. However, this invention would not be widely adopted.

Following Heinrich Hertz's establishment of the existence of electromagnetic radiation in the late 1880s, a variety of terms were initially used for the phenomenon, with early descriptions of the radiation itself including "Hertzian waves", "electric waves", and "ether waves", while phrases describing its use in communications included "spark telegraphy", "space telegraphy", "aerography" and, eventually and most commonly, "wireless telegraphy". However, "wireless" included a broad variety of related electronic technologies, including electrostatic induction, electromagnetic induction and aquatic and earth conduction, so there was a need for a more precise term referring exclusively to electromagnetic radiation.

The first use of *radio-* in conjunction with electromagnetic radiation appears to have been by French physicist Édouard Branly, who in 1890 developed a version of a coherer receiver he called a *radio-conducteur*. The radio- prefix was later used to form additional descriptive compound and hyphenated words, especially in Europe, for example, in early 1898 the British publication *The Practical Engineer* included a reference to "the radiotelegraph" and "radiotelegraphy", while the French text of both the 1903 and 1906 Berlin Radiotelegraphic Conventions includes the phrases *radiotélégraphique* and *radiotélégrammes*.

The use of "radio" as a standalone word dates back to at least December 30, 1904, when instructions issued by the British Post Office for transmitting telegrams specified that "The word 'Radio'... is sent in the Service Instructions". This practice was universally adopted, and the word "radio" introduced internationally, by the 1906 Berlin Radiotelegraphic Convention, which included a Service Regulation specifying that "Radiotelegrams shall show in the preamble that the service is 'Radio'".

The switch to "radio" in place of "wireless" took place slowly and unevenly in the English-speaking world. Lee de Forest helped popularize the new word in the United States—in early 1907 he founded the DeForest Radio Telephone Company, and his letter in the June 22, 1907 *Electrical World* about the need for legal restrictions warned that "Radio chaos will certainly be the result until such stringent regulation is enforced". The United States Navy would also play a role. Although its translation of the 1906 Berlin Convention used the terms "wireless telegraph" and "wireless telegram", by 1912 it began to promote the use of "radio" instead. The term started to become preferred by the general public in the 1920s with the introduction of broadcasting. ("Broadcasting" is based upon an agricultural term meaning roughly "scattering seeds widely".) British Commonwealth countries continued to commonly use the term "wireless" until the mid-20th century, though the magazine of the British Broadcasting Corporation in the UK has been called Radio Times since its founding in the early 1920s.

In recent years the more general term "wireless" has gained renewed popularity, even for devices using electromagnetic radiation, through the rapid growth of short-range computer networking, e.g., Wireless Local Area Network (WLAN), Wi-Fi, and Bluetooth, as well as mobile telephony, e.g., GSM and UMTS cell phones. Today, the term "radio" specifies the transceiver device or chip, whereas "wireless" refers to the lack of physical connections; thus equipment employs embedded *radio* transceivers, but operates as *wireless* devices over *wireless* sensor networks.

Processes

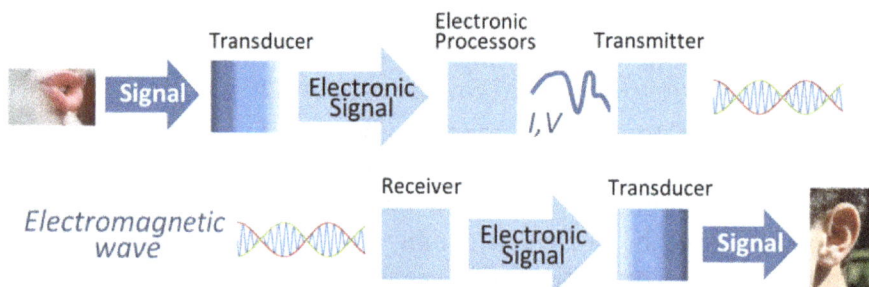

Radio communication. Information such as sound is converted by a transducer such as a microphone to an electrical signal, which modulates a radio wave sent from a transmitter. A receiver intercepts the radio wave and extracts the information-bearing electronic signal, which is converted back using another transducer such as a speaker.

Radio systems used for communication have the following elements. With more than 100 years of development, each process is implemented by a wide range of methods, specialized for different communications purposes.

Transmitter and Modulation

Each system contains a transmitter, This consists of a source of electrical energy, producing alternating current of a desired frequency of oscillation. The transmitter contains a system to modulate (change) some property of the energy produced to impress a signal on it. This modulation might be as simple as turning the energy on and off, or altering more subtle properties such as amplitude, frequency, phase, or combinations of these properties. The transmitter sends the modulated electrical energy to a tuned resonant antenna; this structure converts the rapidly changing alternating current into an electromagnetic wave that can move through free space (sometimes with a particular polarization).

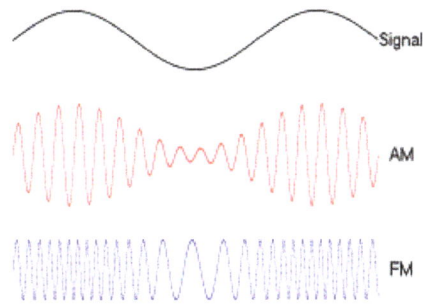

An audio signal (top) may be carried by an AM or FM radio wave.

Amplitude modulation of a carrier wave works by varying the strength of the transmitted signal in proportion to the information being sent. For example, changes in the signal strength can be used to reflect the sounds to be reproduced by a speaker, or to specify the light intensity of television pixels. It was the method used for the first audio radio transmissions, and remains in use today. "AM" is often used to refer to the medium wave broadcast band, but it is used in various radiotelephone services such as the Citizen Band, amateur radio and especially in aviation, due to its ability to be received under very weak signal conditions and its immunity to capture effect, allowing more than one signal to be heard simultaneously.

Frequency modulation varies the frequency of the carrier. The instantaneous frequency of the carrier is directly proportional to the instantaneous value of the input signal. FM has the "capture effect" whereby a receiver only receives the strongest signal, even when others are present. Digital data can be sent by shifting the carrier's frequency among a set of discrete values, a technique known as frequency-shift keying. FM is commonly used at Very high frequency (VHF) radio frequencies for high-fidelity broadcasts of music and speech. Analog TV sound is also broadcast using FM.

Angle modulation alters the instantaneous phase of the carrier wave to transmit a signal. It may be either FM or phase modulation (PM).

Antenna

Rooftop television antennas. Yagi-Uda antennas like these six are widely used at VHF and UHF frequencies.

An *antenna* (or *aerial*) is an electrical device which converts electric currents into radio waves, and vice versa. It is usually used with a radio transmitter or radio receiver. In transmission, a radio transmitter supplies an electric current oscillating at radio frequency (i.e. high frequency AC) to the antenna's terminals, and the antenna radiates the energy from the current as electromagnetic waves (radio waves). In reception, an antenna intercepts some of the power of an electromagnetic wave in order to produce a tiny voltage at its terminals, that is applied to a receiver to be amplified. Some antennas can be used for both transmitting and receiving, even simultaneously, depending on the connected equipment.

Propagation

Once generated, electromagnetic waves travel through space either directly, or have their path altered by reflection, refraction or diffraction. The intensity of the waves diminishes due to geometric dispersion (the inverse-square law); some energy may also be absorbed by the intervening medium in some cases. Noise will generally alter the desired signal; this electromagnetic interference comes from natural sources, as well as from artificial sources such as other transmitters and accidental radiators. Noise is also produced at every step due to the inherent properties of the devices used. If the magnitude of the noise is large enough, the desired signal will no longer be discernible; the signal-to-noise ratio is the fundamental limit to the range of radio communications.

Resonance

Electrical resonance of tuned circuits in radios allow individual stations to be selected. A resonant circuit will respond strongly to a particular frequency, and much less so to differing frequencies. This allows the radio receiver to discriminate between multiple signals differing in frequency.

Receiver and Demodulation

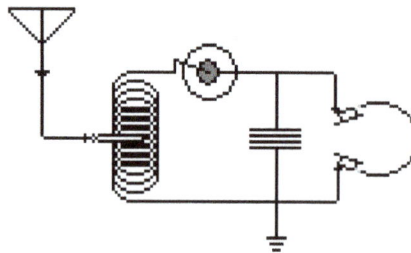

A crystal receiver, consisting of an antenna, adjustable electromagnetic coil, crystal rectifier, capacitor, headphones and ground connection.

The electromagnetic wave is intercepted by a tuned receiving antenna; this structure captures some of the energy of the wave and returns it to the form of oscillating electrical currents. At the receiver, these currents are demodulated, which is conversion to a usable signal form by a detector sub-system. The receiver is "tuned" to respond preferentially to the desired signals, and reject undesired signals.

Early radio systems relied entirely on the energy collected by an antenna to produce signals for the operator. Radio became more useful after the invention of electronic devices such as the vacuum tube and later the transistor, which made it possible to amplify weak signals. Today radio systems are used for applications from walkie-talkie children's toys to the control of space vehicles, as well as for broadcasting, and many other applications.

A *radio receiver* receives its input from an antenna, uses electronic filters to separate a wanted radio signal from all other signals picked up by this antenna, amplifies it to a level suitable for further processing, and finally converts through demodulation and decoding the signal into a form usable for the consumer, such as sound, pictures, digital data, measurement values, navigational positions, etc.

Radio Band

Light comparison		
Name	Frequency (Hz) (Wavelength)	Photon energy (eV)
Gamma ray	> 30 EHz (0.01 nm)	124 keV - 300+ GeV
X-Ray	30 EHz - 30 PHz (0.01 nm - 10 nm)	124 eV to 120 keV
Ultraviolet	30 PHz - 750 THz (10 nm - 400 nm)	3.1 eV to 124 eV
Visible	750 THz - 428.5 THz (400 nm - 700 nm)	1.7 eV - 3.1 eV
Infrared	428.5 THz - 300 GHz (700 nm - 1 mm)	1.24 meV - 1.7 eV
Microwave	300 GHz - 300 MHz (1 mm - 1 m)	1.24 μeV - 1.24 meV
Radio	300 MHz - 3 kHz (1 m - 100 km)	12.4 feV - 1.24 meV

Radio frequencies occupy the range from a 3 kHz to 300 GHz, although commercially important uses of radio use only a small part of this spectrum. Other types of electromagnetic radiation, with frequencies above the RF range, are infrared, visible light, ultraviolet, X-rays and gamma rays. Since the energy of an individual photon of radio frequency is too low to remove an electron from an atom, radio waves are classified as non-ionizing radiation.

Communication Systems

A *radio communication system* sends signals by radio. Types of radio communication systems deployed depend on technology, standards, regulations, radio spectrum allocation, user requirements, service positioning, and investment.

The radio equipment involved in communication systems includes a transmitter and a receiver, each having an antenna and appropriate terminal equipment such as a microphone at the transmitter and a loudspeaker at the receiver in the case of a voice-communication system.

The power consumed in a transmitting station varies depending on the distance of communication and the transmission conditions. The power received at the receiving station is usually only a tiny fraction of the transmitter's output, since communication depends on receiving the information, not the energy, that was transmitted.

Classical radio communications systems use frequency-division multiplexing (FDM) as a strategy to split up and share the available radio-frequency bandwidth for use by different parties communications concurrently. Modern radio communication systems include those that divide up a radio-frequency band by time-division multiplexing (TDM) and code-division multiplexing (CDM) as alternatives to the classical FDM strategy. These systems offer different tradeoffs in supporting multiple users, beyond the FDM strategy that was ideal for broadcast radio but less so for applications such as mobile telephony.

A radio communication system may send information only one way. For example, in broadcasting a single transmitter sends signals to many receivers. Two stations may take turns sending and receiving, using a single radio frequency; this is called "simplex." By using two radio frequencies, two stations may continuously and concurrently send and receive signals - this is called "duplex" operation.

History

In 1864 James Clerk Maxwell showed mathematically that electromagnetic waves could propagate through free space. The effects of electromagnetic waves (then-unexplained "action at a distance" sparking behavior) were actually observed before and after Maxwell's work by many inventors and experimenters including Luigi Galvani (1791), Peter Samuel Munk (1835), Joseph Henry (1842), Samuel Alfred Varley (1852), Edwin Houston, Elihu Thomson, Thomas Edison (1875) and David Edward Hughes (1878). Edison gave the effect the name "etheric force" and Hughes detected a spark impulse up to 500 yards (460 m) with a portable receiver, but none could identify what caused the phenomenon and it was usually written off as electromagnetic induction. In 1886 Heinrich Rudolf Hertz noticed the same sparking phenomenon and, in published experiments (1887-1888), was able to demonstrate the existence of electromagnetic waves in an experiment confirming Maxwell's theory of electromagnetism. The discovery of these "Hertzian waves" (radio waves) prompted many experiments by physicists. An August 1894 lecture by the British physicist Oliver Lodge, where he transmitted and received "Hertzian waves" at distances up to 50 meters, was followed up a year later with experiments by Indian physicist Jagadish Bose in radio microwave optics and construction of a radio based lightning detector by Russian physicist Alexander Stepanovich Popov. Starting in late 1894, Guglielmo Marconi began pursuing the idea of building a wireless telegraphy system based on Hertzian waves (radio). Marconi gained a patent on the system in 1896 and developed it into a commercial communication system over the next few years.

Early 20th century radio systems transmitted messages by continuous wave code only. Early attempts at developing a system of amplitude modulation for voice and music were demonstrated in 1900 and 1906, but had little success. World War I accelerated the development of radio for military communications, and in this era the first vacuum tubes were applied to radio transmitters and receivers. Electronic amplification was a key development in changing radio from an experimental practice by experts into a home appliance. After the war, commercial radio broadcasting began in the 1920s and became an important mass medium for entertainment and news.

World War II again accelerated development of radio for the wartime purposes of aircraft and land communication, radio navigation and radar. After the war, the experiments in television that had been interrupted were resumed, and it also became an important home entertainment medium.

Uses of Radio

Early uses were maritime, for sending telegraphic messages using Morse code between ships and land. The earliest users included the Japanese Navy scouting the Russian fleet during the Battle of

Tsushima in 1905. One of the most memorable uses of marine telegraphy was during the sinking of the RMS *Titanic* in 1912, including communications between operators on the sinking ship and nearby vessels, and communications to shore stations listing the survivors.

Radio was used to pass on orders and communications between armies and navies on both sides in World War I; Germany used radio communications for diplomatic messages once it discovered that its submarine cables had been tapped by the British. The United States passed on President Woodrow Wilson's Fourteen Points to Germany via radio during the war. Broadcasting began from San Jose, California in 1909, and became feasible in the 1920s, with the widespread introduction of radio receivers, particularly in Europe and the United States. Besides broadcasting, point-to-point broadcasting, including telephone messages and relays of radio programs, became widespread in the 1920s and 1930s. Another use of radio in the pre-war years was the development of detection and locating of aircraft and ships by the use of radar (*RAdio Detection And Ranging*).

Today, radio takes many forms, including wireless networks and mobile communications of all types, as well as radio broadcasting. Before the advent of television, commercial radio broadcasts included not only news and music, but dramas, comedies, variety shows, and many other forms of entertainment (the era from the late 1920s to the mid-1950s is commonly called radio's "Golden Age"). Radio was unique among methods of dramatic presentation in that it used only sound.

Audio

One-way

Bakelite radio at the Bakelite Museum, Orchard Mill, Williton, Somerset, UK.

A Fisher 500 AM/FM hi-fi receiver from 1959.

AM radio uses amplitude modulation, in which the amplitude of the transmitted signal is made proportional to the sound amplitude captured (transduced) by the microphone, while the transmitted frequency remains unchanged. Transmissions are affected by static and interference because lightning and other sources of radio emissions on the same frequency add their amplitudes to the original transmitted amplitude.

In the early part of the 20th century, American AM radio stations broadcast with powers as high as 500 kW, and some could be heard worldwide; these stations' transmitters were commandeered for military use by the US Government during World War II. Currently, the maximum broadcast power for a civilian AM radio station in the United States and Canada is 50 kW, and the majority of stations that emit signals this powerful were grandfathered in. In 1986 KTNN received the last granted 50,000-watt class A license. These 50 kW stations are generally called "clear channel" stations, because within North America each of these stations has exclusive use of its broadcast frequency throughout part or all of the broadcast day.

Bush House, old home of the BBC World Service.

FM broadcast radio sends music and voice with less noise than AM radio. It is often mistakenly thought that FM is higher fidelity than AM, but that is not true. AM is capable of the same audio bandwidth that FM employs. AM receivers typically use narrower filters in the receiver to recover the signal with less noise. AM stereo receivers can reproduce the same audio bandwidth that FM does due to the wider filter used in an AM stereo receiver, but today, AM radios limit the audio bandpass to 3–5 kHz. In frequency modulation, amplitude variation at the microphone causes the transmitter frequency to fluctuate. Because the audio signal modulates the frequency and not the amplitude, an FM signal is not subject to static and interference in the same way as AM signals. Due to its need for a wider bandwidth, FM is transmitted in the Very High Frequency (VHF, 30 MHz to 300 MHz) radio spectrum.

VHF radio waves act more like light, traveling in straight lines; hence the reception range is generally limited to about 50–200 miles (80–322 km). During unusual upper atmospheric conditions, FM signals are occasionally reflected back towards the Earth by the ionosphere, resulting in long distance FM reception. FM receivers are subject to the capture effect, which causes the radio to only receive the strongest signal when multiple signals appear on the same frequency. FM receivers are relatively immune to lightning and spark interference.

High power is useful in penetrating buildings, diffracting around hills, and refracting in the dense atmosphere near the horizon for some distance beyond the horizon. Consequently, 100,000-watt FM stations can regularly be heard up to 100 miles (160 km) away, and farther, 150 miles (240 km), if there are no competing signals.

A few old, "grandfathered" stations do not conform to these power rules. WBCT-FM (93.7) in Grand Rapids, Michigan, US, runs 320,000 watts ERP, and can increase to 500,000 watts ERP by the terms of its original license. Such a huge power level does not usually help to increase range as much as one might expect, because VHF frequencies travel in nearly straight lines over the horizon and off into space. Nevertheless, when there were fewer FM stations competing, this station could be heard near Bloomington, Illinois, US, almost 300 miles (480 km) away.

FM subcarrier services are secondary signals transmitted in a "piggyback" fashion along with the main program. Special receivers are required to utilize these services. Analog channels may contain alternative programming, such as reading services for the blind, background music or stereo sound signals. In some extremely crowded metropolitan areas, the sub-channel program might be an alternate foreign-language radio program for various ethnic groups. Sub-carriers can also transmit digital data, such as station identification, the current song's name, web addresses, or stock quotes. In some countries, FM radios automatically re-tune themselves to the same channel in a different district by using sub-bands.

Two-way

Aviation voice radios use Aircraft band VHF AM. AM is used so that multiple stations on the same channel can be received. (Use of FM would result in stronger stations blocking out reception of weaker stations due to FM's capture effect). Aircraft fly high enough that their transmitters can be received hundreds of miles away, even though they are using VHF.

Degen DE1103, an advanced world mini-receiver with single sideband modulation and dual conversion

Marine voice radios can use single sideband voice (SSB) in the shortwave High Frequency (HF—3 MHz to 30 MHz) radio spectrum for very long ranges or Marine VHF radio / *narrowband FM* in the VHF spectrum for much shorter ranges. Narrowband FM sacrifices fidelity to make more channels available within the radio spectrum, by using a smaller range of radio frequencies, usually with five kHz of deviation, versus the 75 kHz used by commercial FM broadcasts, and 25 kHz used for TV sound.

Government, police, fire and commercial voice services also use narrowband FM on special frequencies. Early police radios used AM receivers to receive one-way dispatches.

Civil and military HF (high frequency) voice services use shortwave radio to contact ships at sea, aircraft and isolated settlements. Most use single sideband voice (SSB), which uses less bandwidth than AM. On an AM radio SSB sounds like ducks quacking, or the adults in a Charlie Brown cartoon. Viewed as a graph of frequency versus power, an AM signal shows power where the frequencies of the voice add and subtract with the main radio frequency. SSB cuts the bandwidth in half by suppressing the carrier and one of the sidebands. This also makes the transmitter about three times more powerful, because it doesn't need to transmit the unused carrier and sideband.

TETRA, Terrestrial Trunked Radio is a digital cell phone system for military, police and ambulances. Commercial services such as XM, WorldSpace and Sirius offer encrypted digital satellite radio.

Telephony

Mobile phones transmit to a local cell site (transmitter/receiver) that ultimately connects to the public switched telephone network (PSTN) through an optic fiber or microwave radio and other network elements. When the mobile phone nears the edge of the cell site's radio coverage area, the central computer switches the phone to a new cell. Cell phones originally used FM, but now most use various digital modulation schemes. Recent developments in Sweden (such as DROPme) allow for the instant downloading of digital material from a radio broadcast (such as a song) to a mobile phone.

Satellite phones use satellites rather than cell towers to communicate.

Video

Analog television sends the picture as AM and the sound as AM or FM, with the sound carrier a fixed frequency (4.5 MHz in the NTSC system) away from the video carrier. Analog television also uses a vestigial sideband on the video carrier to reduce the bandwidth required.

Digital television uses 8VSB modulation in North America (under the ATSC digital television standard), and COFDM modulation elsewhere in the world (using the DVB-T standard). A Reed–Solomon error correction code adds redundant correction codes and allows reliable reception during moderate data loss. Although many current and future codecs can be sent in the MPEG transport stream container format, as of 2006 most systems use a standard-definition format almost identical to DVD: MPEG-2 video in Anamorphic widescreen and MPEG layer 2 (*MP2*) audio. High-definition television is possible simply by using a higher-resolution picture, but H.264/AVC is being considered as a replacement video codec in some regions for its improved compression. With the compression and improved modulation involved, a single "channel" can contain a high-definition program and several standard-definition programs.

Navigation

All satellite navigation systems use satellites with precision clocks. The satellite transmits its position, and the time of the transmission. The receiver listens to four satellites, and can figure its position as being on a line that is tangent to a spherical shell around each satellite, determined by the time-of-flight of the radio signals from the satellite. A computer in the receiver does the math.

Radio direction-finding is the oldest form of radio navigation. Before 1960 navigators used movable loop antennas to locate commercial AM stations near cities. In some cases they used marine radiolocation beacons, which share a range of frequencies just above AM radio with amateur radio operators. LORAN systems also used time-of-flight radio signals, but from radio stations on the ground.

Very High Frequency Omnidirectional Range (VOR), systems (used by aircraft), have an antenna array that transmits two signals simultaneously. A directional signal rotates like a lighthouse at a fixed rate. When the directional signal is facing north, an omnidirectional signal pulses. By measuring the difference in phase of these two signals, an aircraft can determine its bearing or radial from the station, thus establishing a line of position. An aircraft can get readings from two VORs and locate its position at the intersection of the two radials, known as a "fix."

When the VOR station is collocated with DME (Distance Measuring Equipment), the aircraft can determine its bearing and range from the station, thus providing a fix from only one ground station. Such stations are called VOR/DMEs. The military operates a similar system of navaids, called TACANs, which are often built into VOR stations. Such stations are called VORTACs. Because TACANs include distance measuring equipment, VOR/DME and VORTAC stations are identical in navigation potential to civil aircraft.

Radar

Radar (Radio Detection And Ranging) detects objects at a distance by bouncing radio waves off them. The delay caused by the echo measures the distance. The direction of the beam determines the direction of the reflection. The polarization and frequency of the return can sense the type of surface. Navigational radars scan a wide area two to four times per minute. They use very short waves that reflect from earth and stone. They are common on commercial ships and long-distance commercial aircraft.

General purpose radars generally use navigational radar frequencies, but modulate and polarize the pulse so the receiver can determine the type of surface of the reflector. The best general-purpose radars distinguish the rain of heavy storms, as well as land and vehicles. Some can superimpose sonar data and map data from GPS position.

Search radars scan a wide area with pulses of short radio waves. They usually scan the area two to four times a minute. Sometimes search radars use the Doppler effect to separate moving vehicles from clutter. Targeting radars use the same principle as search radar but scan a much smaller area far more often, usually several times a second or more. Weather radars resemble search radars, but use radio waves with circular polarization and a wavelength to reflect from water droplets. Some weather radar use the Doppler effect to measure wind speeds.

Data (Digital Radio)

Most new radio systems are digital, including Digital TV, satellite radio, and Digital Audio Broadcasting. The oldest form of digital broadcast was spark gap telegraphy, used by pioneers such as Marconi. By pressing the key, the operator could send messages in Morse code by energizing a rotating commutating spark gap. The rotating commutator produced a tone in the receiver, where a simple spark gap would produce a hiss, indistinguishable from static. Spark-gap transmitters are now illegal, because their transmissions span several hundred megahertz. This is very wasteful of both radio frequencies and power.

2008 Pure One Classic digital radio

The next advance was continuous wave telegraphy, or CW (Continuous Wave), in which a pure radio frequency, produced by a vacuum tube electronic oscillator was switched on and off by a key. A receiver with a local oscillator would "heterodyne" with the pure radio frequency, creating a whistle-like audio tone. CW uses less than 100 Hz of bandwidth. CW is still used, these days primarily by amateur radio operators (hams). Strictly, on-off keying of a carrier should be known as "Interrupted Continuous Wave" or ICW or on-off keying (OOK).

Radioteletype equipment usually operates on short-wave (HF) and is much loved by the military because they create written information without a skilled operator. They send a bit as one of two tones using frequency-shift keying. Groups of five or seven bits become a character printed by a teleprinter. From about 1925 to 1975, radioteletype was how most commercial messages were sent to less developed countries. These are still used by the military and weather services.

Aircraft use a 1200 Baud radioteletype service over VHF to send their ID, altitude and position, and get gate and connecting-flight data. Microwave dishes on satellites, telephone exchanges and TV stations usually use quadrature amplitude modulation (QAM). QAM sends data by changing both the phase and the amplitude of the radio signal. Engineers like QAM because it packs the most bits into a radio signal when given an exclusive (non-shared) fixed narrowband frequency range. Usually the bits are sent in "frames" that repeat. A special bit pattern is used to locate the beginning of a frame.

Modern GPS receivers.

Communication systems that limit themselves to a fixed narrowband frequency range are vulnerable to jamming. A variety of jamming-resistant spread spectrum techniques were initially developed for military use, most famously for Global Positioning System satellite transmissions. Commercial use of spread spectrum began in the 1980s. Bluetooth, most cell phones, and the 802.11b version of Wi-Fi each use various forms of spread spectrum.

Systems that need reliability, or that share their frequency with other services, may use "coded orthogonal frequency-division multiplexing" or COFDM. COFDM breaks a digital signal into as many as several hundred slower subchannels. The digital signal is often sent as QAM on the subchannels. Modern COFDM systems use a small computer to make and decode the signal with digital signal processing, which is more flexible and far less expensive than older systems that implemented separate electronic channels.

COFDM resists fading and ghosting because the narrow-channel QAM signals can be sent slowly. An adaptive system, or one that sends error-correction codes can also resist interference, because most interference can affect only a few of the QAM channels. COFDM is used for Wi-Fi, some cell phones, Digital Radio Mondiale, Eureka 147, and many other local area network, digital TV and radio standards.

Heating

Radio-frequency energy generated for heating of objects is generally not intended to radiate outside of the generating equipment, to prevent interference with other radio signals. Microwave ovens use intense radio waves to heat food. Diathermy equipment is used in surgery for sealing of blood vessels. Induction furnaces are used for melting metal for casting, and induction hobs for cooking.

Amateur Radio Service

Amateur radio station with multiple receivers and transceivers

Amateur radio, also known as "ham radio", is a hobby in which enthusiasts are licensed to communicate on a number of bands in the radio frequency spectrum non-commercially and for their own experiments. They may also provide emergency and service assistance in exceptional circumstances. This contribution has been very beneficial in saving lives in many instances.

Radio amateurs use a variety of modes, including efficient ones like Morse code and experimental ones like Low-Frequency Experimental Radio. Several forms of radio were pioneered by radio amateurs and later became commercially important, including FM, single-sideband (SSB), AM, digital packet radio and satellite repeaters. Some amateur frequencies may be disrupted illegally by power-line internet service.

Unlicensed Radio Services

Unlicensed, government-authorized personal radio services such as Citizens' band radio in Australia, most of the Americas, and Europe, and Family Radio Service and Multi-Use Radio Service in North America exist to provide simple, usually short range communication for individuals and small groups, without the overhead of licensing. Similar services exist in other parts of the world. These radio services involve the use of handheld units.

Wi-Fi also operates in unlicensed radio bands and is very widely used to network computers.

Free radio stations, sometimes called pirate radio or "clandestine" stations, are unauthorized, unlicensed, illegal broadcasting stations. These are often low power transmitters operated on sporadic schedules by hobbyists, community activists, or political and cultural dissidents. Some pirate stations operating offshore in parts of Europe and the United Kingdom more closely resembled legal stations, maintaining regular schedules, using high power, and selling commercial advertising time.

Radio Control (RC)

Radio remote controls use radio waves to transmit control data to a remote object as in some early forms of guided missile, some early TV remotes and a range of model boats, cars and airplanes. Large industrial remote-controlled equipment such as cranes and switching locomotives now usually use digital radio techniques to ensure safety and reliability.

In Madison Square Garden, at the Electrical Exhibition of 1898, Nikola Tesla successfully demonstrated a radio-controlled boat. He was awarded U.S. patent No. 613,809 for a "Method of and Apparatus for Controlling Mechanism of Moving Vessels or Vehicles."

Radio Resource Management

Radio resource management (RRM) is the system level management of co-channel interference, radio resources, and other radio transmission characteristics in wireless communication systems, for example cellular networks, wireless local area networks and wireless sensor systems. RRM involves strategies and algorithms for controlling parameters such as transmit power, user allocation, beamforming, data rates, handover criteria, modulation scheme, error coding scheme, etc. The objective is to utilize the limited radio-frequency spectrum resources and radio network infrastructure as efficiently as possible.

RRM concerns multi-user and multi-cell network capacity issues, rather than the point-to-point channel capacity. Traditional telecommunications research and education often dwell upon channel coding and source coding with a single user in mind, although it may not be possible to achieve the maximum channel capacity when several users and adjacent base stations share the same frequency channel. Efficient dynamic RRM schemes may increase the system spectral efficiency by an order of magnitude, which often is considerably more than what is possible by introducing advanced channel coding and source coding schemes. RRM is especially important in systems limited by co-channel interference rather than by noise, for example cellular systems and broadcast networks homogeneously covering large areas, and wireless networks consisting of many adjacent access points that may reuse the same channel frequencies.

The cost for deploying a wireless network is normally dominated by base station sites (real estate costs, planning, maintenance, distribution network, energy, etc.) and sometimes also by frequency license fees. The objective of radio resource management is therefore typically to maximize the system spectral efficiency in *bit/s/Hz/area unit* or *Erlang/MHz/site*, under some kind of user fairness constraint, for example, that the grade of service should be above a certain level. The latter involves covering a certain area and avoiding outage due to co-channel interference, noise, attenuation caused by path losses, fading caused by shadowing and multipath, Doppler shift and other forms of distortion. The grade of service is also affected by blocking due to admission control, scheduling starvation or inability to guarantee quality of service that is requested by the users.

While classical radio resource managements primarily considered the allocation of time and frequency resources (with fixed spatial reuse patterns), recent multi-user MIMO techniques enables adaptive resource management also in the spatial domain. In cellular networks, this means that the fractional frequency reuse in the GSM standard has been replaced by a universal frequency reuse in LTE standard.

Static Radio Resource Management

Static RRM involves manual as well as computer-aided fixed cell planning or radio network planning. Examples:

- Frequency allocation band plans decided by standardization bodies, by national frequency authorities and in frequency resource auctions.

- Deployment of base station sites (or broadcasting transmitter site)

- Antenna heights

- Channel frequency plans

- Sector antenna directions

- Selection of modulation and channel coding parameters

- Base station antenna space diversity, for example

    - Receiver micro diversity using antenna combining

    - Transmitter macro diversity such as OFDM single frequency networks (SFN)

Static RRM schemes are used in many traditional wireless systems, for example 1G and 2G cellular systems, in today's wireless local area networks and in non-cellular systems, for example broadcasting systems. Examples of static RRM schemes are:

- Circuit mode communication using FDMA and TDMA.

- Fixed channel allocation (FCA)

- Static handover criteria

Dynamic Radio Resource Management

Dynamic RRM schemes adaptively adjust the radio network parameters to the traffic load, user positions, user mobility, quality of service requirements, base station density, etc. Dynamic RRM schemes are considered in the design of wireless systems, in view to minimize expensive manual cell planning and achieve "tighter" frequency reuse patterns, resulting in improved system spectral efficiency.

Some schemes are centralized, where several base stations and access points are controlled by a Radio Network Controller (RNC). Others are distributed, either autonomous algorithms in mobile stations, base stations or wireless access points, or coordinated by exchanging information among these stations.

Examples of dynamic RRM schemes are:

- Power control algorithms

- Precoding algorithms

- Link adaptation algorithms

- Dynamic Channel Allocation (DCA) or Dynamic Frequency Selection (DFS) algorithms, allowing "cell breathing"

- Traffic adaptive handover criteria, allowing "cell breathing"

- Re-use partitioning

- Adaptive filtering

  - Single Antenna Interference Cancellation (SAIC)

- Dynamic diversity schemes, for example

  - Soft handover

  - Dynamic single-frequency networks (DSFN)

  - Phased array antenna with

- beamforming

- Multiple-input multiple-output communications (MIMO)

- Space-time coding

- Admission control

- Dynamic bandwidth allocation using resource reservation multiple access schemes or statistical multiplexing, for example Spread spectrum and/or packet radio

- Channel-dependent scheduling, for instance

  - Max-min fair scheduling using for example fair queuing

  - Proportionally fair scheduling using for example weighted fair queuing

- • Maximum throughput scheduling (gives low grade of service due to starvation)

- • Dynamic packet assignment (DPA)

- • Packet and Resource Plan Scheduling (PARPS) schemes

- Mobile ad hoc networks using multihop communication

- Cognitive radio

- Green communication

- QoS-aware RRM

- Femtocells

Inter-cell Radio Resource Management

Future networks like the LTE standard (defined by 3GPP) are designed for a frequency reuse of one. In such networks, neighboring cells use the same frequency spectrum. Such standards exploit Space Division Multiple Access (SDMA) and can thus be highly efficient in terms of spectrum, but required close coordination between cells to avoid excessive inter-cell interference. Like in most cellular system deployments, the overall system spectral efficiency is not range limited or noise limited, but interference limited. Inter-cell radio resource management coordinates resource allocation between different cell sites by using multi-user MIMO techniques. There are various means of Inter-Cell Interference Coordination (ICIC) already defined in the standard. Dynamic single-frequency networks, coordinated scheduling, multi-site MIMO or joint multi-cell precoding are other examples for inter-cell radio resource management.

Digital Radio

Digital radio is radio that uses digital technology to transmit and/or receive across the radio spectrum.

In many modern systems, the radio signal is a digital signal that has been produced using digital modulation.

Digital radio receivers also exist, that decode analog radio signals using digital electronics.

Types

1. Today the most common meaning is digital audio broadcasting systems. In these systems, the analog audio signal is digitized, compressed using formats such as mp2, and transmitted using a digital modulation scheme. The aim is to increase the number of radio programs in a given spectrum, to improve the audio quality, to eliminate fading problems in mobile environments, to allow additional datacasting services, and to decrease the transmission power or the number of transmitters required to cover a region. However, analog radio (AM and FM) is still more popular and listening to radio over IP (Internet Protocol) is growing in popularity.

In 2012 there are four digital wireless radio systems recognized by the International Telecommunication Union: the two European systems Digital Audio Broadcasting (DAB) and Digital Radio Mondiale (DRM), the Japanese ISDB-T and the in-band on-channel technique used in the US and Arab world and branded as HD Radio.

2. An older definition, still used in communication engineering literature, is wireless digital transmission technologies, i.e. microwave and radio frequency communication standards where analog information signals as well as digital data are carried by a digital signal, by means of a digital modulation method. This definition includes broadcasting systems such as digital TV and digital radio broadcasting, but also two-way digital radio standards such as the second generation (2G) cell-phones and later, short-range communication such as digital cordless phones, wireless computer networks, digital micro-wave radio links, deep space communication systems such as communications to and from the two Voyager space probes, etc.

3. A less common definition is radio receiver and transmitter implementations that are based on digital signal processing, but may transmit or receive analog radio transmission standards, for example FM radio. This may reduce noise and distortion induced in the electronics. It also allows software radio implementations, where the transmission technology is changed just by selecting another piece of software. In most cases, this would however increase the energy consumption of the receiver equipment.

One-way (Broadcasting) Systems

Broadcast Standards

Digital audio broadcasting standards may provide terrestrial or satellite radio service. Digital radio broadcasting systems are typically designed for handheld mobile devices, just like mobile-TV systems, but as opposed to other digital TV systems which typically require a fixed directional antenna. Some digital radio systems provide in-band on-channel (IBOC) solutions that may coexist with or simulcast with analog AM or FM transmissions, while others are designed for designated radio frequency bands. The latter allows one wideband radio signal to carry a multiplex consisting of several radio-channels of variable bitrate as well as data services and other forms of media. Some digital broadcasting systems allow single-frequency network (SFN), where all terrestrial transmitters in a region sending the same multiplex of radio programs may use the same frequency channel without self-interference problems, further improving the system spectral efficiency.

While digital broadcasting offers many potential benefits, its introduction has been hindered by a lack of global agreement on standards and many disadvantages. The DAB Eureka 147 standard for digital radio is coordinated by the World DMB Forum. This standard of digital radio technology was defined in the late 1980s, and is now being introduced in some European countries. Commercial DAB receivers began to be sold in 1999 and, by 2006, 500 million people were in the coverage area of DAB broadcasts, although by this time sales had only taken off in the UK and Denmark. In 2006 there are approximately 1,000 DAB stations in operation. There have been criticisms of the Eureka 147 standard and so a new 'DAB+' standard has been introduced.

The DRM standard has been used for several years to broadcast digitally on frequencies below 30 MHz (shortwave, mediumwave and longwave). Also there is now the extended standard DRM+

which make it possible to broadcast on frequencies above 30 MHz.This will make it possible to digitalize transmission on the FM-band. Successful tests of DRM+ has been made in several countries 2010-2012 as in Brazil, Germany, France, India, Sri Lanka, the UK, Slovakia and Italy (incl. the Vatican). DRM+ will be tested in Sweden 2012.

DRM+ is regarded as a more transparent and less costly standard than DAB+ and thus a better choice for local radio; commercial or community broadcasters. Although DAB+ has been introduced in Australia the government has concluded 2011 that a preference for DRM and DRM+ above HD Radio could be used to supplement DAB+ services in (some) local and regional areas.

All Digital Radio Broadcast system share many disadvantages which don't exist for Analogue to Digital TV changeover: About x20 more power consumption, Digital Cliff effect for Mobile use, very slow channel change, especially for a different DAB multiplex frequency, high transmission cost resulting poorer quality than FM and sometimes AM due to low bitrate (64K mono rather than 256K stereo), higher compression is more distorted for hearing aid users, usually poor user interfaces and Radio audio quality, not enough fill in stations for portable / mobile coverage (like 1950s UK FM). The Multiplex & SFN concepts are advantageous to State Broadcasters and Large Pan National Multi-channel companies and worse for all Local, Community and most Regional stations. In contrast almost all the aspects of Digital TV vs Analogue TV are positive with almost no negative effects. TVs could be used with a Set-box. Digital Radio requires replacement of all radios, though an awkward DAB receiver with FM output can be used with existing FM car Radios.

To date the following standards have been defined for one-way digital radio:

Digital Audio Broadcasting Systems

- Eureka 147 (branded as DAB)
- DAB+
- ISDB-TSB
- Internet radio (Technically not a true Broadcast system)
- FM band in-band on-channel (FM IBOC):
    - HD Radio (OFDM modulation over FM and AM band IBOC sidebands)
    - FMeXtra (FM band IBOC subcarriers)
    - Digital Radio Mondiale extension (DRM+) (OFDM modulation over AM band IBOC sidebands)
- AM band in-band on-channel (AM IBOC):
    - HD Radio (AM IBOC sideband)
    - Digital Radio Mondiale (branded as DRM) for the short, medium and long wavebands
- Satellite radio:
    - WorldSpace in Asia and Africa

- Sirius XM Radio in North America
- MobaHo! in Japan and the Republic of (South) Korea
- Systems also designed for digital TV:
    - DMB
    - DVB-H
    - ISDB-T
- Low-bandwidth digital data broadcasting over existing FM radio:
    - Radio Data System (branded as RDS)
- Radio pagers:
    - FLEX
    - ReFLEX
    - POCSAG
    - NTT

Digital Television (DTV) Broadcasting Systems

- Digital Video Broadcasting (DVB)
- Integrated Services Digital Broadcasting (ISDB)
- Digital Multimedia Broadcasting (DMB)
- Digital Terrestrial Television (DTTV or DTT) to fixed mainly roof-top antennas:
    - DVB-T (based on OFDM modulation)
    - ISDB-T (based on OFDM modulation)
    - ATSC (based on 8VSB modulation)
    - T-DMB
- Mobile TV reception in handheld devices:
    - DVB-H (based on OFDM modulation)
    - MediaFLO (based on OFDM modulation)
    - DMB (based on OFDM modulation)
    - Multimedia Broadcast Multicast Service (MBMS) via the GSM EDGE and UMTS cellular networks
    - DVB-SH (based on OFDM modulation)
- Satellite TV:

- DVB-S (for Satellite TV)
- ISDB-S
- 4DTV
- S-DMB
- MobaHo!

Status by Country

DAB Adopters

Digital Audio Broadcasting (DAB), also known as Eureka 147, has been under development since the early eighties, has been adopted by around 20 countries worldwide. It is based on the MPEG-1 Audio Layer II audio coding format and this has been co-ordinated by the WorldDMB.

WorldDMB announced in a press release in November 2006, that DAB would be adopting the HE-AACv2 audio coding format, which is also known as eAAC+. Also being adopted are the MPEG Surround format, and stronger error correction coding called Reed-Solomon coding. The update has been named DAB+. Receivers that support the new DAB standard began being released during 2007 with firmware updated available for some older receivers.

DAB and DAB+ cannot be used for mobile TV because they do not include any video codecs. DAB related standards Digital Multimedia Broadcasting (DMB) and DAB-IP are suitable for mobile radio and TV both because they have MPEG 4 AVC and WMV9 respectively as video coding formats. However a DMB video sub-channel can easily be added to any DAB transmission - as DMB was designed from the outset to be carried on a DAB subchannel. DMB broadcasts in Korea carry conventional MPEG 1 Layer II DAB audio services alongside their DMB video services.

United States

The United States has opted for a proprietary system called HD Radio technology, a type of in-band on-channel (IBOC) technology. Transmissions use orthogonal frequency-division multiplexing, a technique which is also used for European terrestrial digital TV broadcast (DVB-T). HD Radio technology was developed and is licensed by iBiquity Digital Corporation. It is widely believed that a major reason for HD radio technology is to offer some limited digital radio services while preserving the relative "stick values" of the stations involved and to insure that new programming services will be controlled by existing licensees.

The FM digital schemes in the U.S. provide audio at rates from 96 to 128 kilobits per second (kbit/s), with auxiliary "subcarrier" transmissions at up to 64 kbit/s. The AM digital schemes have data rates of about 48 kbit/s, with auxiliary services provided at a much lower data rate. Both the FM and AM schemes use lossy compression techniques to make the best use of the limited bandwidth.

Lucent Digital Radio, USA Digital Radio (USADR), and Digital Radio Express commenced tests in 1999 of their various schemes for digital broadcast, with the expectation that they would report their results to the National Radio Systems Committee (NRSC) in December 1999. Results of these tests remain unclear, which in general describes the status of the terrestrial digital radio broadcasting effort in North America. Some terrestrial analog broadcast stations are apprehensive about the impact of digital satellite radio on their business, while others plan to convert to digital broadcasting as soon as it is economically and technically feasible.

While traditional terrestrial radio broadcasters are trying to "go digital", most major US automobile manufacturers are promoting digital satellite radio. HD Radio technology has also made inroads in the automotive sector with factory-installed options announced by BMW, Ford, Hyundai, Jaguar, Lincoln, Mercedes, MINI, Mercury, Scion, and Volvo.

Satellite radio is distinguished by its freedom from FCC censorship in the United States, its relative lack of advertising, and its ability to allow people on the road to listen to the same stations at any location in the country. Listeners must currently pay an annual or monthly subscription fee in order to access the service, and must install a separate security card in each radio or receiver they use.

Ford and Daimler AG are working with Sirius Satellite Radio, previously CD Radio, of New York City, and General Motors and Honda are working with XM Satellite Radio of Washington, D.C. to build and promote satellite DAB radio systems for North America, each offering "CD quality" audio and about a hundred channels.

Sirius Satellite Radio launched a constellation of three Sirius satellites during the course of 2000. The satellites were built by Space Systems/Loral and were launched by Russian Proton boosters. As with XM Satellite Radio, Sirius implemented a series of terrestrial ground repeaters where satellite signal would otherwise be blocked by large structures including natural structures and high-rise buildings.

XM Satellite Radio has a constellation of three satellites, two of which were launched in the spring of 2001, with one following later in 2005. The satellites are Boeing (previously Hughes) 702 comsats, and were put into orbit by Sea Launch boosters. Back-up ground transmitters (repeaters) will be built in cities where satellite signals could be blocked by big buildings.

On February 19, 2007, Sirius Satellite Radio and XM Satellite Radio merged, to form Sirius XM Radio.

The FCC has auctioned bandwidth allocations for satellite broadcast in the S band range, around 2.3 GHz.

The perceived wisdom of the radio industry is that the terrestrial medium has two great strengths: it is free and it is local. Satellite radio is neither of these things; however, in recent years, it has grown to make a name for itself by providing uncensored content (most notably, the crossover of Howard Stern from terrestrial radio to satellite radio) and commercial-free, all-digital music channels that offer similar genres to local broadcast favorites.

- "Digital Radio" has a limited listening distance from the tower site. FCC laws currently show that 10% maximum digital signal of any US analog signal ratio. "There are still some concerns that HD Radio on FM will increase interference between different stations even though HD Radio at the 10% power level fits within the FCC spectral mask." HD Radio HD Radio#cite note-14. "HD Radio" is only 2 channels in the USA, side by side with analog stations. HD channel 1 may be on 93.2 FM, Analog station on 93.3, and HD channel 2 is on 93.4 FM. Differing stations are multicasting on different frequencies, respectively.

- Also note that according to iBiquity, "HD Radio" is the company's trade name for its proprietary digital radio system, but the name does not imply either high definition or "hybrid digital" as it is commonly incorrectly referenced.

Canada

Canada has begun allowing experimental HD Radio broadcasts and digital audio subchannels on a case-by-case basis, with the first stations in the country being CFRM-FM in Little Current, CING-FM in Hamilton, and CJSA-FM in Toronto (with a fourth, CFMS-FM in the Toronto suburb of Markham applying to operate HD Radio technology), all within the province of Ontario.

United Kingdom

In the United Kingdom, 44.3% of the population now has a DAB digital radio set and 34.4% of listening is to different digital platforms.

26 million people, or 50% of the population, now tune into digital radio each week, up 2.6 million year on year, according to RAJAR in Q1 2013. But FM listening has increased to 61% and DAB decreased to 21% DAB listeners may also use AM & FM too.

The UK currently has the world's biggest digital radio network, with 103 transmitters, two nationwide DAB ensembles and 48 local and regional DAB ensembles, broadcasting over 250 commercial and 34 BBC radio stations; 51 of these stations are broadcast in London. On DAB digital radio most listeners can receive around 20 additional stations, in addition to the analogue stations available on digital. The frequency band used is 217.5 to 230 MHz.

Some areas of the country are not yet covered by DAB but the BBC has announced plans to build out national coverage to 92% by the end of 2011 with 40 new transmitters being launched in 2011. The Government will make a decision on a radio switchover subject to listening and coverage criteria being met. A digital radio switchover would maintain FM as a platform, while moving some services to DAB-only distribution. Digital radio stations are also broadcast on digital television platforms such as Sky, Virgin Media and Freeview, as well as internet radio.

Germany

The first DAB station network was deployed in Bavaria since 1995 until full coverage in 1999. Other states had funded a station network but the lack of success led them to scrap the funding - the MDR switched off in 1998 already and Brandenburg declared a failure in 2004. Instead Berlin/Brandenburg began to switch to digital radio based on an audio-only DVB-T mode given the success of the DVB-T standard in the region when earlier analogue television was switched off in August 2003 (being the first region to switch in Germany). During that time the DVB-H variant of the DVB family was released for transmission to mobile receivers in 2004. During 2005 most radio stations left the DAB network with only one public service broadcaster ensemble to remain in the now fully state-funded station network. At last the KEF (*commission to determine the financial needs of broadcasters*) blocked federal funding on

15. July 2009 until economic viability of DAB broadcasting would be proven - and pointing to DVB-T as a viable alternative.

Digital radio deployment was rebooted during 2011 - a joint commission of public and private radio broadcasters decided upon "DAB+" as the new national standard in December 2010. The new station network started as planned on 1. August 2011 with 27 stations with 10 kW each giving a coverage of 70% across the nation. A single "Bundesmux" ("fed-mux" short of "federal multiplex") was created on band 5C as a single-frequency network with the band 5C to cross over to neighbouring countries. Berlin/Brandenburg joined the DAB+ broadcasting in January 2012 but other dark regions remain in Eastern Germany . With the initial market success of DAB+ the contractors decided on an expansion of the digital radio station network in November 2012. With DAB being available across Belgium, Netherlands, Switzerland and Northern Italy there is good coverage across the European Backbone area indicating a sufficient momentum on the market.

Australia

Australia commenced regular digital audio broadcasting using the DAB+ standard in May 2009, after many years of trialling alternative systems. Normal radio services operate on the AM and FM bands, as well as four stations (ABC and SBS) on digital TV channels. The services are currently operating in five state capital cities, namely, Adelaide, Brisbane, Melbourne, Perth and Sydney, and is being trialled in Canberra and Darwin.

Japan

Japan has started terrestrial sound broadcasting using ISDB-Tsb and MobaHO! 2.6 GHz Satellite Sound digital broadcasting

Korea

On 1 December 2005 South Korea launched its T-DMB service which includes both television and radio stations. T-DMB is a derivative of DAB with specifications published by ETSI. More than 110,000 receivers had been sold in one month only in 2005.

Developing Nations

Digital radio is now being provided to the developing world. A satellite communications company named WorldSpace was setting up a network of three satellites, including "AfriStar", "AsiaStar", and "AmeriStar", to provide digital audio information services to Africa, Asia, and Latin America. AfriStar and AsiaStar are in orbit. AmeriStar cannot be launched from the United States as Worldspace transmits on the L-band and would interfere with USA military as mentioned above.. in its heyday provided service to over 170,000 subscribers in eastern and southern Africa, the Middle East, and much of Asia with 96% coming from India. Timbre Media along with Saregama India plan to relaunch the company. As of 2013 Worldspace is defunct, but two satellites are in orbit which still have a few channels.

Each satellite provides three transmission beams that can support 50 channels each, carrying news, music, entertainment, and education, and including a computer multimedia service. Local, regional, and international broadcasters were working with WorldStar to provide services.

A consortium of broadcasters and equipment manufacturers are also working to bring the benefits of digital broadcasting to the radio spectrum currently used for terrestrial AM radio broadcasts, including international shortwave transmissions. Over seventy broadcasters are now transmitting programs using the new standard, known as Digital Radio Mondiale (DRM), and / commercial DRM receivers are available (though there are few models on the DRM website and some are discontinued). DRM's system uses the MPEG-4 based standard aacPlus to code the music and CELP or HVXC for speech programs. At present these are priced too high to be affordable by many in the third world, however. Take-up of DRM has been minuscule and many traditional Shortwave broadcasters now only stream on Internet, use fixed satellite (TV set-boxes) or Local Analogue FM relays to save on costs. Very few (expensive) DRM radio sets are available and some Broadcasters (RTE in Ireland on 252 kHz) have ceased trials without launching a service.

Low-cost DAB radio receivers are now available from various Japanese manufacturers, and WorldSpace has worked with Thomson Broadcast to introduce a village communications center known as a Telekiosk to bring communications services to rural areas. The Telekiosks are self-contained and are available as fixed or mobile units

Two-way Digital Radio Standards

The key breakthrough or key feature in digital radio transmission systems is that they allow lower transmission power, they can provide robustness to noise and cross-talk and other forms of interference, and thus allow the same radio frequency to be reused at shorter distance. Consequently, the spectral efficiency (the number of phonecalls per MHz and base station, or the number of bit/s per Hz and transmitter, etc.) may be sufficiently increased. Digital radio transmission can also carry any kind of information whatsoever — just as long at it has been expressed digitally. Earlier radio communication systems had to be made expressly for a given form of communications: telephone, telegraph, or television, for example. All kinds of digital communications can be multiplexed or encrypted at will.

- Digital cellular telephony (2G systems and later generations):
    - GSM
    - UMTS (sometimes called W-CDMA)
    - TETRA
    - IS-95 (cdmaOne)
    - IS-136 (D-AMPS, sometimes called TDMA)
    - IS-2000 (CDMA2000)
    - iDEN
- Digital Mobile Radio:
    - Project 25 a.k.a. "P25" or "APCO-25"

- TETRA
- TETRAPOL
- NXDN
- DMR
- Wireless networking:
  - Wi-Fi
  - HIPERLAN
  - Bluetooth
  - DASH7
  - ZigBee
  - 6LoWPAN
- Military radio systems for Network-centric warfare
  - JTRS (Joint Tactical Radio System- a flexible software-defined radio)
  - SINCGARS (Single channel ground to air radio system)
- Amateur packet radio:
  - AX.25
- Digital modems for HF:
  - PACTOR
- Satellite radio:
  - Satmodems
- Wireless local loop:
  - Basic Exchange Telephone Radio Service
- Broadband wireless access:
  - IEEE 802.16

Communications Satellite

A communications satellite is an artificial satellite that relays and amplifies radio telecommunications signals via a transponder; it creates a communication channel between a source transmitter and a receiver at different locations on Earth. Communications satellites are used for television, telephone, radio, internet, and military applications. There are over 2,000 communications satellites in Earth's orbit, used by both private and government organizations.

An Advanced Extremely High Frequency communications satellite relays secure communications for the United States and other allied countries.

Wireless communication uses electromagnetic waves to carry signals. These waves require line-of-sight, and are thus obstructed by the curvature of the Earth. The purpose of communications satellites is to relay the signal around the curve of the Earth allowing communication between widely separated points. Communications satellites use a wide range of radio and microwave frequencies. To avoid signal interference, international organizations have regulations for which frequency ranges or "bands" certain organizations are allowed to use. This allocation of bands minimizes the risk of signal interference.

History

The concept of the geostationary communications satellite was first proposed by Arthur C. Clarke, building on work by Konstantin Tsiolkovsky and on the 1929 work by Herman Potočnik (writing as Herman Noordung) *Das Problem der Befahrung des Weltraums — der Raketen-motor*. In October 1945 Clarke published an article titled "Extraterrestrial Relays" in the British magazine *Wireless World*. The article described the fundamentals behind the deployment of artificial satellites in geostationary orbits for the purpose of relaying radio signals. Thus, Arthur C. Clarke is often quoted as being the inventor of the communications satellite and the term 'Clarke Belt' employed as a description of the orbit.

Decades later a project named Communication Moon Relay was a telecommunication project carried out by the United States Navy. Its objective was to develop a secure and reliable method of wireless communication by using the Moon as a passive reflector and natural communications satellite.

The first artificial Earth satellite was Sputnik 1. Put into orbit by the Soviet Union on October 4, 1957, it was equipped with an on-board radio-transmitter that worked on two frequencies: 20.005 and 40.002 MHz. Sputnik 1 was launched as a step in the exploration of space and rocket development. While incredibly important it was not placed in orbit for the purpose of sending data from one point on earth to another. And it was the first artificial satellite in the steps leading to today's satellite communications.

The first artificial satellite used solely to further advances in global communications was a balloon named Echo 1. Echo 1 was the world's first artificial communications satellite capable of relaying signals to other points on Earth. It soared 1,600 kilometres (1,000 mi) above the planet after its Aug. 12, 1960 launch, yet relied on humanity's oldest flight technology — ballooning. Launched by NASA, Echo 1 was a 30-metre (100 ft) aluminised PET film balloon that served as a passive reflector for radio communications. The world's first inflatable satellite — or "satelloon", as they were informally known — helped lay the foundation of today's satellite communications. The idea behind a communications satellite is simple: Send data up into space and beam it back down to another spot on the globe. Echo 1 accomplished this by essentially serving as an enormous mirror, 10 stories tall, that could be used to reflect communications signals.

The first American satellite to relay communications was Project SCORE in 1958, which used a tape recorder to store and forward voice messages. It was used to send a Christmas greeting to the world from U.S. President Dwight D. Eisenhower.; Courier 1B, built by Philco, launched in 1960, was the world's first active repeater satellite.

There are two major classes of communications satellites, *passive* and *active*. Passive satellites only reflect the signal coming from the source, toward the direction of the receiver. With passive satellites, the reflected signal is not amplified at the satellite, and only a very small amount of the transmitted energy actually reaches the receiver. Since the satellite is so far above Earth, the radio signal is attenuated due to free-space path loss, so the signal received on Earth is very weak. Active satellites, on the other hand, amplify the received signal before re-transmitting it to the receiver on the ground. Passive satellites were the first communications satellites, but are little used now. Telstar was the second active, direct relay communications satellite. Belonging to AT&T as part of a multi-national agreement between AT&T, Bell Telephone Laboratories, NASA, the British General Post Office, and the French National PTT (Post Office) to develop satellite communications, it was launched by NASA from Cape Canaveral on July 10, 1962, the first privately sponsored space launch. Relay 1 was launched on December 13, 1962, and became the first satellite to broadcast across the Pacific on November 22, 1963.

An immediate antecedent of the geostationary satellites was Hughes' Syncom 2, launched on July 26, 1963. Syncom 2 was the first communications satellite in a geosynchronous orbit. It revolved around the earth once per day at constant speed, but because it still had north-south motion, special equipment was needed to track it. Its successor, Syncom 3 was the first geostationary communications satellite. Syncom 3 obtained a geosynchronous orbit, without a north-south motion, making it appear from the ground as a stationary object in the sky.

Beginning with the Mars Exploration Rovers, probes on the surface of Mars have used orbiting spacecraft as communications satellites for relaying their data to Earth. The orbiters were designed for this relay purpose to allow the landers to conserve power. The Orbiters with their solar power arrays, large antennas and more powerful transmitters enable them to transmit data to Earth with a much stronger, and as a result, clearer signal than a lander could manage on its own from the surface.

Satellite Orbits

Communications satellites usually have one of three primary types of orbit, while other orbital classifications are used to further specify orbital details.

- Geostationary satellites have a *geostationary orbit* (GEO), which is 35,786 kilometres (22,236 mi) from Earth's surface. This orbit has the special characteristic that the apparent position of the satellite in the sky when viewed by a ground observer does not change, the satellite appears to "stand still" in the sky. This is because the satellite's orbital period is the same as the rotation rate of the Earth. The advantage of this orbit is that ground antennas do not have to track the satellite across the sky, they can be fixed to point at the location in the sky the satellite appears.

- *Medium Earth orbit* (MEO) satellites are closer to Earth. Orbital altitudes range from 2,000 to 35,786 kilometres (1,243 to 22,236 mi) above Earth.

- The region below medium orbits is referred to as *low Earth orbit* (LEO), and is about 160 to 2,000 kilometres (99 to 1,243 mi) above Earth.

As satellites in MEO and LEO orbit the Earth faster, they do not remain visible in the sky to a fixed point on Earth continually like a geostationary satellite, but appear to a ground observer to cross the sky and "set" when they go behind the Earth. Therefore, to provide continuous communications capability with these lower orbits requires a larger number of satellites, so one will always be in the sky for transmission of communication signals. However, due to their relatively small distance to the Earth their signals are stronger.

Low Earth Orbiting (LEO) Satellites

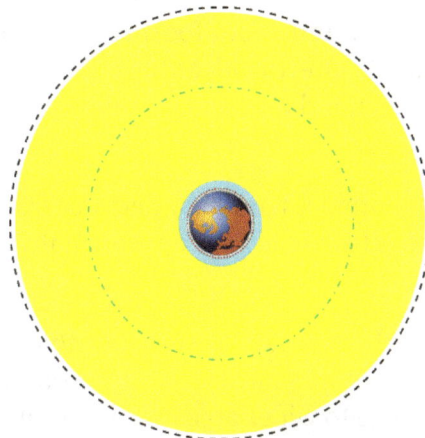

Low Earth orbit in Cyan

A low Earth orbit (LEO) typically is a circular orbit about 160 to 2,000 kilometres (99 to 1,243 mi) above the earth's surface and, correspondingly, a period (time to revolve around the earth) of about 90 minutes.

Because of their low altitude, these satellites are only visible from within a radius of roughly 1,000 kilometres (620 mi) from the sub-satellite point. In addition, satellites in low earth orbit change their position relative to the ground position quickly. So even for local applications, a large number of satellites are needed if the mission requires uninterrupted connectivity.

Low-Earth-orbiting satellites are less expensive to launch into orbit than geostationary satellites and, due to proximity to the ground, do not require as high signal strength (*Recall that signal strength falls off as the square of the distance from the source, so the effect is dramatic*). Thus there is a trade off between the number of satellites and their cost.

In addition, there are important differences in the onboard and ground equipment needed to support the two types of missions.

Satellite Constellation

A group of satellites working in concert is known as a satellite constellation. Two such constellations, intended to provide satellite phone services, primarily to remote areas, are the Iridium and Globalstar systems. The Iridium system has 66 satellites.

It is also possible to offer discontinuous coverage using a low-Earth-orbit satellite capable of storing data received while passing over one part of Earth and transmitting it later while passing over another part. This will be the case with the CASCADE system of Canada's CASSIOPE communications satellite. Another system using this store and forward method is Orbcomm.

Medium Earth Orbit (MEO)

A MEO is a satellite in orbit somewhere between 2,000 and 35,786 kilometres (1,243 and 22,236 mi) above the earth's surface. MEO satellites are similar to LEO satellites in functionality. MEO satellites are visible for much longer periods of time than LEO satellites, usually between 2 and 8 hours. MEO satellites have a larger coverage area than LEO satellites. A MEO satellite's longer duration of visibility and wider footprint means fewer satellites are needed in a MEO network than a LEO network. One disadvantage is that a MEO satellite's distance gives it a longer time delay and weaker signal than a LEO satellite, although these limitations are not as severe as those of a GEO satellite.

Like LEOs, these satellites don't maintain a stationary distance from the earth. This is in contrast to the geostationary orbit, where satellites are always approximately 35,786 kilometres (22,236 mi) from the earth.

Typically the orbit of a medium earth orbit satellite is about 16,000 kilometres (10,000 mi) above earth. In various patterns, these satellites make the trip around earth in anywhere from 2–12 hours, which provides better coverage to wider areas than that provided by LEOs.

Example

In 1962, the first communications satellite, Telstar, was launched. It was a medium earth orbit satellite designed to help facilitate high-speed telephone signals. Although it was the first practical way to transmit signals over the horizon, its major drawback was soon realized. Because its orbital period of about 2.5 hours did not match the Earth's rotational period of 24 hours, continuous coverage was impossible. It was apparent that multiple MEOs needed to be used in order to provide continuous coverage.

Geostationary Orbits (GEO)

To an observer on the earth, a satellite in a geostationary orbit appears motionless, in a fixed position in the sky. This is because it revolves around the earth at the earth's own angular velocity (360 degrees every 24 hours, in an equatorial orbit).

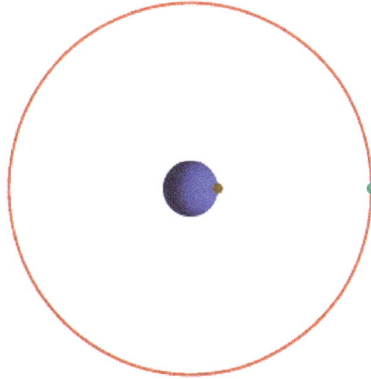

Geostationary orbit

A geostationary orbit is useful for communications because ground antennas can be aimed at the satellite without their having to track the satellite's motion. This is relatively inexpensive.

In applications that require a large number of ground antennas, such as DirecTV distribution, the savings in ground equipment can more than outweigh the cost and complexity of placing a satellite into orbit.

Examples

- The first geostationary satellite was Syncom 3, launched on August 19, 1964, and used for communication across the Pacific starting with television coverage of the 1964 Summer Olympics. Shortly after Syncom 3, Intelsat I, aka *Early Bird*, was launched on April 6, 1965 and placed in orbit at 28° west longitude. It was the first geostationary satellite for telecommunications over the Atlantic Ocean.

- On November 9, 1972, Canada's first geostationary satellite serving the continent, Anik A1, was launched by Telesat Canada, with the United States following suit with the launch of Westar 1 by Western Union on April 13, 1974.

- On May 30, 1974, the first geostationary communications satellite in the world to be three-axis stabilized was launched: the experimental satellite ATS-6 built for NASA.

- After the launches of the Telstar through Westar 1 satellites, RCA Americom (later GE Americom, now SES) launched Satcom 1 in 1975. It was Satcom 1 that was instrumental in helping early cable TV channels such as WTBS (now TBS Superstation), HBO, CBN (now ABC Family) and The Weather Channel become successful, because these channels distributed their programming to all of the local cable TV headends using the satellite. Additionally, it was the first satellite used by broadcast television networks in the United States, like ABC, NBC, and CBS, to distribute programming to their local affiliate stations. Satcom 1 was widely used because it had twice the communications capacity of the competing Westar 1 in America (24 transponders as opposed to the 12 of Westar 1), resulting in lower transponder-usage costs. Satellites in later decades tended to have even higher transponder numbers.

By 2000, Hughes Space and Communications (now Boeing Satellite Development Center) had built nearly 40 percent of the more than one hundred satellites in service worldwide. Other major satellite manufacturers include Space Systems/Loral, Orbital Sciences Corporation with the STAR Bus series, Indian Space Research Organisation, Lockheed Martin (owns the former RCA Astro Electronics/GE Astro Space business), Northrop Grumman, Alcatel Space, now Thales Alenia Space, with the Spacebus series, and Astrium.

Molniya Satellites

Geostationary satellites must operate above the equator and therefore appear lower on the horizon as the receiver gets the farther from the equator. This will cause problems for extreme northerly latitudes, affecting connectivity and causing multipath interference (caused by signals reflecting off the ground and into the ground antenna).

Thus, for areas close to the North (and South) Pole, a geostationary satellite may appear below the horizon. Therefore, Molniya orbit satellites have been launched, mainly in Russia, to alleviate this problem.

Molniya orbits can be an appealing alternative in such cases. The Molniya orbit is highly inclined, guaranteeing good elevation over selected positions during the northern portion of the orbit. (Elevation is the extent of the satellite's position above the horizon. Thus, a satellite at the horizon has zero elevation and a satellite directly overhead has elevation of 90 degrees.)

The Molniya orbit is designed so that the satellite spends the great majority of its time over the far northern latitudes, during which its ground footprint moves only slightly. Its period is one half day, so that the satellite is available for operation over the targeted region for six to nine hours every second revolution. In this way a constellation of three Molniya satellites (plus in-orbit spares) can provide uninterrupted coverage.

The first satellite of the Molniya series was launched on April 23, 1965 and was used for experimental transmission of TV signals from a Moscow uplink station to downlink stations located in Siberia and the Russian Far East, in Norilsk, Khabarovsk, Magadan and Vladivostok. In November 1967 Soviet engineers created a unique system of national TV network of satellite television, called Orbita, that was based on Molniya satellites.

Polar Orbit

In the United States, the National Polar-orbiting Operational Environmental Satellite System (NPOESS) was established in 1994 to consolidate the polar satellite operations of NASA (National Aeronautics and Space Administration) NOAA (National Oceanic and Atmospheric Administration). NPOESS manages a number of satellites for various purposes; for example, METSAT for meteorological satellite, EUMETSAT for the European branch of the program, and METOP for meteorological operations.

These orbits are sun synchronous, meaning that they cross the equator at the same local time each day. For example, the satellites in the NPOESS (civilian) orbit will cross the equator, going from south to north, at times 1:30 P.M., 5:30 P.M., and 9:30 P.M.

Structure

Communications Satellites are usually composed of the following subsystems:

- Communication Payload, normally composed of transponders, antennas, and switching systems

- Engines used to bring the satellite to its desired orbit

- Station Keeping Tracking and stabilization subsystem used to keep the satellite in the right orbit, with its antennas pointed in the right direction, and its power system pointed towards the sun

- Power subsystem, used to power the Satellite systems, normally composed of solar cells, and batteries that maintain power during solar eclipse

- Command and Control subsystem, which maintains communications with ground control stations. The ground control earth stations monitor the satellite performance and control its functionality during various phases of its life-cycle.

The bandwidth available from a satellite depends upon the number of transponders provided by the satellite. Each service (TV, Voice, Internet, radio) requires a different amount of bandwidth for transmission. This is typically known as link budgeting and a network simulator can be used to arrive at the exact value.

Frequency Allocation for Satellite Systems

Allocating frequencies to satellite services is a complicated process which requires international coordination and planning. This is carried out under the auspices of the International Telecommunication Union (ITU). To facilitate frequency planning, the world is divided into three regions: Region 1: Europe, Africa, what was formerly the Soviet Union, and Mongolia Region 2: North and South America and Greenland Region 3: Asia (excluding region 1 areas), Australia, and the southwest Pacific

Within these regions, frequency bands are allocated to various satellite services, although a given service may be allocated different frequency bands in different regions. Some of the services provided by satellites are:

- Fixed satellite service (FSS)

- Broadcasting satellite service (BSS)

- Mobile satellite service

- Radionavigation-satellite service

- Meteorological-satellite service

- Amateur-satellite service

Applications

Telephone

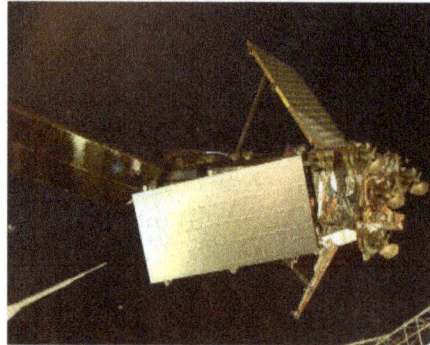

An Iridium satellite

The first and historically most important application for communication satellites was in intercontinental long distance telephony. The fixed Public Switched Telephone Network relays telephone calls from land line telephones to an earth station, where they are then transmitted to a geostationary satellite. The downlink follows an analogous path. Improvements in submarine communications cables through the use of fiber-optics caused some decline in the use of satellites for fixed telephony in the late 20th century.

Satellite communications are still used in many applications today. Remote islands such as Ascension Island, Saint Helena, Diego Garcia, and Easter Island, where no submarine cables are in service, need satellite telephones. There are also regions of some continents and countries where landline telecommunications are rare to nonexistent, for example large regions of South America, Africa, Canada, China, Russia, and Australia. Satellite communications also provide connection to the edges of Antarctica and Greenland. Other land use for satellite phones are rigs at sea, a back up for hospitals, military, and recreation. Ships at sea, as well as planes, often use satellite phones.

Satellite phone systems can be accomplished by a number of means. On a large scale, often there will be a local telephone system in an isolated area with a link to the telephone system in a main land area. There are also services that will patch a radio signal to a telephone system. In this example, almost any type of satellite can be used. Satellite phones connect directly to a constellation of either geostationary or low-earth-orbit satellites. Calls are then forwarded to a satellite teleport connected to the Public Switched Telephone Network .

Television

As television became the main market, its demand for simultaneous delivery of relatively few signals of large bandwidth to many receivers being a more precise match for the capabilities of geosynchronous comsats. Two satellite types are used for North American television and radio: Direct broadcast satellite (DBS), and Fixed Service Satellite (FSS).

The definitions of FSS and DBS satellites outside of North America, especially in Europe, are a bit more ambiguous. Most satellites used for direct-to-home television in Europe have the same high power output as DBS-class satellites in North America, but use the same linear polarization as

FSS-class satellites. Examples of these are the Astra, Eutelsat, and Hotbird spacecraft in orbit over the European continent. Because of this, the terms FSS and DBS are more so used throughout the North American continent, and are uncommon in Europe.

Fixed Service Satellites use the C band, and the lower portions of the K_u band. They are normally used for broadcast feeds to and from television networks and local affiliate stations (such as program feeds for network and syndicated programming, live shots, and backhauls), as well as being used for distance learning by schools and universities, business television (BTV), Videoconferencing, and general commercial telecommunications. FSS satellites are also used to distribute national cable channels to cable television headends.

Free-to-air satellite TV channels are also usually distributed on FSS satellites in the K_u band. The Intelsat Americas 5, Galaxy 10R and AMC 3 satellites over North America provide a quite large amount of FTA channels on their K_u band transponders.

The American Dish Network DBS service has also recently utilized FSS technology as well for their programming packages requiring their SuperDish antenna, due to Dish Network needing more capacity to carry local television stations per the FCC's "must-carry" regulations, and for more bandwidth to carry HDTV channels.

A direct broadcast satellite is a communications satellite that transmits to small DBS satellite dishes (usually 18 to 24 inches or 45 to 60 cm in diameter). Direct broadcast satellites generally operate in the upper portion of the microwave K_u band. DBS technology is used for DTH-oriented (Direct-To-Home) satellite TV services, such as DirecTV and DISH Network in the United States, Bell TV and Shaw Direct in Canada, Freesat and Sky in the UK, Ireland, and New Zealand and DSTV in South Africa.

Operating at lower frequency and lower power than DBS, FSS satellites require a much larger dish for reception (3 to 8 feet (1 to 2.5 m) in diameter for K_u band, and 12 feet (3.6 m) or larger for C band). They use linear polarization for each of the transponders' RF input and output (as opposed to circular polarization used by DBS satellites), but this is a minor technical difference that users do not notice. FSS satellite technology was also originally used for DTH satellite TV from the late 1970s to the early 1990s in the United States in the form of TVRO (TeleVision Receive Only) receivers and dishes. It was also used in its K_u band form for the now-defunct Primestar satellite TV service.

Some satellites have been launched that have transponders in the K_a band, such as DirecTV's SPACEWAY-1 satellite, and Anik F2. NASA and ISRO have also launched experimental satellites carrying K_a band beacons recently.

Some manufacturers have also introduced special antennas for mobile reception of DBS television. Using Global Positioning System (GPS) technology as a reference, these antennas automatically re-aim to the satellite no matter where or how the vehicle (on which the antenna is mounted) is situated. These mobile satellite antennas are popular with some recreational vehicle owners. Such mobile DBS antennas are also used by JetBlue Airways for DirecTV (supplied by LiveTV, a subsidiary of JetBlue), which passengers can view on-board on LCD screens mounted in the seats.

Radio Broadcasting

Satellite radio offers audio broadcast services in some countries, notably the United States.

Mobile services allow listeners to roam a continent, listening to the same audio programming anywhere.

A satellite radio or subscription radio (SR) is a digital radio signal that is broadcast by a communications satellite, which covers a much wider geographical range than terrestrial radio signals.

Satellite radio offers a meaningful alternative to ground-based radio services in some countries, notably the United States. Mobile services, such as SiriusXM, and Worldspace, allow listeners to roam across an entire continent, listening to the same audio programming anywhere they go. Other services, such as Music Choice or Muzak's satellite-delivered content, require a fixed-location receiver and a dish antenna. In all cases, the antenna must have a clear view to the satellites. In areas where tall buildings, bridges, or even parking garages obscure the signal, repeaters can be placed to make the signal available to listeners.

Initially available for broadcast to stationary TV receivers, by 2004 popular mobile direct broadcast applications made their appearance with the arrival of two satellite radio systems in the United States: Sirius and XM Satellite Radio Holdings. Later they merged to become the conglomerate SiriusXM.

Radio services are usually provided by commercial ventures and are subscription-based. The various services are proprietary signals, requiring specialized hardware for decoding and playback. Providers usually carry a variety of news, weather, sports, and music channels, with the music channels generally being commercial-free.

In areas with a relatively high population density, it is easier and less expensive to reach the bulk of the population with terrestrial broadcasts. Thus in the UK and some other countries, the contemporary evolution of radio services is focused on Digital Audio Broadcasting (DAB) services or HD Radio, rather than satellite radio.

Amateur Radio

Amateur radio operators have access to amateur satellites, which have been designed specifically to carry amateur radio traffic. Most such satellites operate as spaceborne repeaters, and are generally accessed by amateurs equipped with UHF or VHF radio equipment and highly directional antennas such as Yagis or dish antennas. Due to launch costs, most current amateur satellites are launched into fairly low Earth orbits, and are designed to deal with only a limited number of brief contacts at any given time. Some satellites also provide data-forwarding services using the X.25 or similar protocols.

Internet Access

After the 1990s, satellite communication technology has been used as a means to connect to the Internet via broadband data connections. This can be very useful for users who are located in remote areas, and cannot access a broadband connection, or require high availability of services.

Military

Communications satellites are used for military communications applications, such as Global Command and Control Systems. Examples of military systems that use communication satellites are the MILSTAR, the DSCS, and the FLTSATCOM of the United States, NATO satellites, United Kingdom satellites (for instance Skynet), and satellites of the former Soviet Union. India has launched its first Military Communication satellite GSAT-7, its transponders operate in UHF, F, C and K_u band bands. Typically military satellites operate in the UHF, SHF (also known as X-band) or EHF (also known as K_a band) frequency bands.

Fiber-optic Communication

An optical fiber junction box. The yellow cables are single mode fibers; the orange and blue cables are multi-mode fibers: 62.5/125 μm OM1 and 50/125 μm OM3 fibers, respectively.

Stealth installing a 432-count dark fibre cable underneath the streets of Midtown Manhattan, New York City

Fiber-optic communication is a method of transmitting information from one place to another by sending pulses of light through an optical fiber. The light forms an electromagnetic carrier wave that is modulated to carry information. First developed in the 1970s, fiber-optics have revolutionized the telecommunications industry and have played a major role in the advent of the Information Age. Because of its advantages over electrical transmission, optical fibers have largely replaced copper wire communications in core networks in the developed world. Optical fiber is used by many telecommunications companies to transmit telephone signals, Internet communication, and cable television signals. Researchers at Bell Labs have reached internet speeds of over 100 petabit×kilometer per second using fiber-optic communication.

The process of communicating using fiber-optics involves the following basic steps: Creating the optical signal involving the use of a transmitter, relaying the signal along the fiber, ensuring that the signal does not become too distorted or weak, receiving the optical signal, and converting it into an electrical signal.

Applications

Optical fiber is used by many telecommunications companies to transmit telephone signals, Internet communication, and cable television signals. Due to much lower attenuation and interference, optical fiber has large advantages over existing copper wire in long-distance and high-demand applications. However, infrastructure development within cities was relatively difficult and time-consuming, and fiber-optic systems were complex and expensive to install and operate. Due to these difficulties, fiber-optic communication systems have primarily been installed in long-distance applications, where they can be used to their full transmission capacity, offsetting the increased cost. Since 2000, the prices for fiber-optic communications have dropped considerably.

The price for rolling out fiber to the home has currently become more cost-effective than that of rolling out a copper based network. Prices have dropped to $850 per subscriber in the US and lower in countries like The Netherlands, where digging costs are low and housing density is high.

Since 1990, when optical-amplification systems became commercially available, the telecommunications industry has laid a vast network of intercity and transoceanic fiber communication lines. By 2002, an intercontinental network of 250,000 km of submarine communications cable with a capacity of 2.56 Tb/s was completed, and although specific network capacities are privileged information, telecommunications investment reports indicate that network capacity has increased dramatically since 2004.

History

In 1880 Alexander Graham Bell and his assistant Charles Sumner Tainter created a very early precursor to fiber-optic communications, the Photophone, at Bell's newly established Volta Laboratory in Washington, D.C. Bell considered it his most important invention. The device allowed for the transmission of sound on a beam of light. On June 3, 1880, Bell conducted the world's first wireless telephone transmission between two buildings, some 213 meters apart. Due to its use of an atmospheric transmission medium, the Photophone would not prove practical until advances in laser and optical fiber technologies permitted the secure transport of light. The Photophone's first practical use came in military communication systems many decades later.

In 1954 Harold Hopkins and Narinder Singh Kapany showed that rolled fiber glass allowed light to be transmitted. Initially it was considered that the light can traverse in only straight medium.

In 1966 Charles K. Kao and George Hockham proposed optical fibers at STC Laboratories (STL) at Harlow, England, when they showed that the losses of 1,000 dB/km in existing glass (compared to 5-10 dB/km in coaxial cable) was due to contaminants, which could potentially be removed.

Optical fiber was successfully developed in 1970 by Corning Glass Works, with attenuation low enough for communication purposes (about 20dB/km), and at the same time GaAs semiconductor lasers were developed that were compact and therefore suitable for transmitting light through fiber optic cables for long distances.

After a period of research starting from 1975, the first commercial fiber-optic communications system was developed, which operated at a wavelength around 0.8 µm and used GaAs semiconductor lasers. This first-generation system operated at a bit rate of 45 Mbps with repeater spacing of up to 10 km. Soon on 22 April 1977, General Telephone and Electronics sent the first live telephone traffic through fiber optics at a 6 Mbit/s throughput in Long Beach, California.

In October 1973, Corning Glass signed a development contract with CSELT and Pirelli aimed to test fiber optics in an urban environment: in September 1977, the second cable in this test series, named COS-2, was experimentally deployed in two lines (9 km) in Turin, for the first time in a big city, at a speed of 140 Mbit/s.

The second generation of fiber-optic communication was developed for commercial use in the early 1980s, operated at 1.3 µm, and used InGaAsP semiconductor lasers. These early systems were initially limited by multi mode fiber dispersion, and in 1981 the single-mode fiber was revealed to greatly improve system performance, however practical connectors capable of working with single mode fiber proved difficult to develop. In 1984, they had already developed a fiber optic cable that would help further their progress toward making fiber optic cables that would circle the globe. Canadian service provider SaskTel had completed construction of what was then the world's longest commercial fiberoptic network, which covered 3,268 km and linked 52 communities. By 1987, these systems were operating at bit rates of up to 1.7 Gb/s with repeater spacing up to 50 km.

The first transatlantic telephone cable to use optical fiber was TAT-8, based on Desurvire optimized laser amplification technology. It went into operation in 1988.

Third-generation fiber-optic systems operated at 1.55 µm and had losses of about 0.2 dB/km. This development was spurred by the discovery of Indium gallium arsenide and the development of the Indium Gallium Arsenide photodiode by Pearsall. Engineers overcame earlier difficulties with pulse-spreading at that wavelength using conventional InGaAsP semiconductor lasers. Scientists overcame this difficulty by using dispersion-shifted fibers designed to have minimal dispersion at 1.55 µm or by limiting the laser spectrum to a single longitudinal mode. These developments eventually allowed third-generation systems to operate commercially at 2.5 Gbit/s with repeater spacing in excess of 100 km.

The fourth generation of fiber-optic communication systems used optical amplification to reduce the need for repeaters and wavelength-division multiplexing to increase data capacity. These two improvements caused a revolution that resulted in the doubling of system capacity every six months starting in 1992 until a bit rate of 10 Tb/s was reached by 2001. In 2006 a bit-rate of 14 Tbit/s was reached over a single 160 km line using optical amplifiers.

The focus of development for the fifth generation of fiber-optic communications is on extending the wavelength range over which a WDM system can operate. The conventional wavelength window, known as the C band, covers the wavelength range 1.53-1.57 µm, and *dry fiber* has a low-loss window promising an extension of that range to 1.30-1.65 µm. Other developments include the concept of "optical solitons, " pulses that preserve their shape by counteracting the effects of dispersion with the nonlinear effects of the fiber by using pulses of a specific shape.

In the late 1990s through 2000, industry promoters, and research companies such as KMI, and RHK predicted massive increases in demand for communications bandwidth due to increased use of the Internet, and commercialization of various bandwidth-intensive consumer services, such as video on demand. Internet protocol data traffic was increasing exponentially, at a faster rate than integrated circuit complexity had increased under Moore's Law. From the bust of the dot-com bubble through 2006, however, the main trend in the industry has been consolidation of firms and offshoring of manufacturing to reduce costs. Companies such as Verizon and AT&T have taken advantage of fiber-optic communications to deliver a variety of high-throughput data and broadband services to consumers' homes.

Technology

Modern fiber-optic communication systems generally include an optical transmitter to convert an electrical signal into an optical signal to send into the optical fiber, a cable containing bundles of multiple optical fibers that is routed through underground conduits and buildings, multiple kinds of amplifiers, and an optical receiver to recover the signal as an electrical signal. The information transmitted is typically digital information generated by computers, telephone systems, and cable television companies.

Transmitters

A GBIC module (shown here with its cover removed), is an optical and electrical transceiver. The electrical connector is at top right, and the optical connectors are at bottom left

The most commonly used optical transmitters are semiconductor devices such as light-emitting diodes (LEDs) and laser diodes. The difference between LEDs and laser diodes is that LEDs produce incoherent light, while laser diodes produce coherent light. For use in optical communications, semiconductor optical transmitters must be designed to be compact, efficient, and reliable, while operating in an optimal wavelength range, and directly modulated at high frequencies.

In its simplest form, a LED is a forward-biased p-n junction, emitting light through spontaneous emission, a phenomenon referred to as electroluminescence. The emitted light is incoherent with a relatively wide spectral width of 30-60 nm. LED light transmission is also inefficient, with only about 1% of input power, or about 100 microwatts, eventually converted into launched power which has been coupled into the optical fiber. However, due to their relatively simple design, LEDs are very useful for low-cost applications.

Communications LEDs are most commonly made from Indium gallium arsenide phosphide (InGaAsP) or gallium arsenide (GaAs). Because InGaAsP LEDs operate at a longer wavelength than GaAs LEDs (1.3 micrometers vs. 0.81-0.87 micrometers), their output spectrum, while equivalent in energy is wider in wavelength terms by a factor of about 1.7. The large spectrum width of LEDs is subject to higher fiber dispersion, considerably limiting their bit rate-distance product (a common measure of usefulness). LEDs are suitable primarily for local-area-network applications with bit rates of 10-100 Mbit/s and transmission distances of a few kilometers. LEDs have also been developed that use several quantum wells to emit light at different wavelengths over a broad spectrum, and are currently in use for local-area WDM (Wavelength-Division Multiplexing) networks.

Today, LEDs have been largely superseded by VCSEL (Vertical Cavity Surface Emitting Laser) devices, which offer improved speed, power and spectral properties, at a similar cost. Common VCSEL devices couple well to multi mode fiber.

A semiconductor laser emits light through stimulated emission rather than spontaneous emission, which results in high output power (~100 mW) as well as other benefits related to the nature of coherent light. The output of a laser is relatively directional, allowing high coupling efficiency (~50 %) into single-mode fiber. The narrow spectral width also allows for high bit rates since it reduces the effect of chromatic dispersion. Furthermore, semiconductor lasers can be modulated directly at high frequencies because of short recombination time.

Commonly used classes of semiconductor laser transmitters used in fiber optics include VCSEL (Vertical-Cavity Surface-Emitting Laser), Fabry–Pérot and DFB (Distributed Feed Back).

Laser diodes are often directly modulated, that is the light output is controlled by a current applied directly to the device. For very high data rates or very long distance *links*, a laser source may be operated continuous wave, and the light modulated by an external device such as an electro-absorption modulator or Mach–Zehnder interferometer. External modulation increases the achievable link distance by eliminating laser chirp, which broadens the linewidth of directly modulated lasers, increasing the chromatic dispersion in the fiber.

A transceiver is a device combining a transmitter and a receiver in a single housing.

Receivers

The main component of an optical receiver is a photodetector, which converts light into electricity using the photoelectric effect. The primary photodetectors for telecommunications are made from Indium gallium arsenide The photodetector is typically a semiconductor-based photodiode. Several types of photodiodes include p-n photodiodes, p-i-n photodiodes, and avalanche photodiodes. Metal-semiconductor-metal (MSM) photodetectors are also used due to their suitability for circuit integration in regenerators and wavelength-division multiplexers.

Optical-electrical converters are typically coupled with a transimpedance amplifier and a limiting amplifier to produce a digital signal in the electrical domain from the incoming optical signal, which may be attenuated and distorted while passing through the channel. Further signal processing such as clock recovery from data (CDR) performed by a phase-locked loop may also be applied before the data is passed on.

Fiber Cable Types

An optical fiber cable consists of a core, cladding, and a buffer (a protective outer coating), in which the cladding guides the light along the core by using the method of total internal reflection. The core and the cladding (which has a lower-refractive-index) are usually made of high-quality silica glass, although they can both be made of plastic as well. Connecting two optical fibers is done by fusion splicing or mechanical splicing and requires special skills and interconnection technology due to the microscopic precision required to align the fiber cores.

A cable reel trailer with conduit that can carry optical fiber

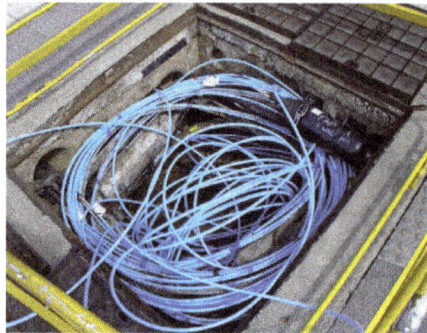

Multi-mode optical fiber in an underground service pit

Two main types of optical fiber used in optic communications include multi-mode optical fibers and single-mode optical fibers. A multi-mode optical fiber has a larger core (≥ 50 micrometers), allowing less precise, cheaper transmitters and receivers to connect to it as well as cheaper connectors. However, a multi-mode fiber introduces multimode distortion, which often limits the bandwidth and length of the link. Furthermore, because of its higher dopant content, multi-mode fibers are usually expensive and exhibit higher attenuation. The core of a single-mode fiber is smaller (<10 micrometers) and requires more expensive components and interconnection methods, but allows much longer, higher-performance links.

In order to package fiber into a commercially viable product, it typically is protectively coated by using ultraviolet (UV), light-cured acrylate polymers, then terminated with optical fiber connectors, and finally assembled into a cable. After that, it can be laid in the ground and then run through the walls of a building and deployed aerially in a manner similar to copper cables. These fibers require less maintenance than common twisted pair wires, once they are deployed.

Specialized cables are used for long distance subsea data transmission, e.g. transatlantic communications cable. New (2011–2013) cables operated by commercial enterprises (Emerald Atlantis, Hibernia Atlantic) typically have four strands of fiber and cross the Atlantic (NYC-London) in 60-70ms. Cost of each such cable was about $300M in 2011. *source: The Chronicle Herald.*

Another common practice is to bundle many fiber optic strands within long-distance power transmission cable. This exploits power transmission rights of way effectively, ensures a power company can own and control the fiber required to monitor its own devices and lines, is effectively immune to tampering, and simplifies the deployment of smart grid technology.

Amplifier

The transmission distance of a fiber-optic communication system has traditionally been limited by fiber attenuation and by fiber distortion. By using opto-electronic repeaters, these problems have been eliminated. These repeaters convert the signal into an electrical signal, and then use a transmitter to send the signal again at a higher intensity than was received, thus counteracting the loss incurred in the previous segment. Because of the high complexity with modern wavelength-division multiplexed signals (including the fact that they had to be installed about once every 20 km), the cost of these repeaters is very high.

An alternative approach is to use an optical amplifier, which amplifies the optical signal directly without having to convert the signal into the electrical domain. It is made by doping a length of fiber with the rare-earth mineral erbium, and *pumping* it with light from a laser with a shorter wavelength than the communications signal (typically 980 nm). Amplifiers have largely replaced repeaters in new installations.

Wavelength-division Multiplexing

Wavelength-division multiplexing (WDM) is the practice of multiplying the available capacity of optical fibers through use of parallel channels, each channel on a dedicated wavelength of light. This requires a wavelength division multiplexer in the transmitting equipment and a demultiplexer (essentially a spectrometer) in the receiving equipment. Arrayed waveguide gratings are commonly used for multiplexing and demultiplexing in WDM. Using WDM technology now commercially available, the bandwidth of a fiber can be divided into as many as 160 channels to support a combined bit rate in the range of 1.6 Tbit/s.

Parameters

Bandwidth–distance Product

Because the effect of dispersion increases with the length of the fiber, a fiber transmission system is often characterized by its *bandwidth–distance product*, usually expressed in units of MHz·km. This value is a product of bandwidth and distance because there is a trade off between the bandwidth of the signal and the distance it can be carried. For example, a common multi-mode fiber with bandwidth–distance product of 500 MHz·km could carry a 500 MHz signal for 1 km or a 1000 MHz signal for 0.5 km.

Engineers are always looking at current limitations in order to improve fiber-optic communication, and several of these restrictions are currently being researched.

Record Speeds

Each fiber can carry many independent channels, each using a different wavelength of light (wavelength-division multiplexing). The net data rate (data rate without overhead bytes) per fiber is the per-channel data rate reduced by the FEC overhead, multiplied by the number of channels (usually up to eighty in commercial dense WDM systems as of 2008).

Year	Organization	Effective speed	WDM channels	Per channel speed	Distance
2009	Alcatel-Lucent	15 Tbit/s	155	100 Gbit/s	90 km
2010	NTT	69.1 Tbit/s	432	171 Gbit/s	240 km
2011	KIT	26 Tbit/s	1	26 Tbit/s	50 km
2011	NEC	101 Tbit/s	370	273 Gbit/s	165 km
2012	NEC, Corning	1.05 Petabit/s	12 core fiber		52.4 km

While the physical limitations of electrical cable prevent speeds in excess of 10 Gigabits per second, the physical limitations of fiber optics have not yet been reached.

In 2013, *New Scientist* reported that a team at the University of Southampton had achieved a throughput of 73.7 Tbit per second, with the signal traveling at 99.7% the speed of light through a hollow-core photonic crystal fiber.

Dispersion

For modern glass optical fiber, the maximum transmission distance is limited not by direct material absorption but by several types of dispersion, or spreading of optical pulses as they travel along the fiber. Dispersion in optical fibers is caused by a variety of factors. Intermodal dispersion, caused by the different axial speeds of different transverse modes, limits the performance of multi-mode fiber. Because single-mode fiber supports only one transverse mode, intermodal dispersion is eliminated.

In single-mode fiber performance is primarily limited by chromatic dispersion (also called group velocity dispersion), which occurs because the index of the glass varies slightly depending on the wavelength of the light, and light from real optical transmitters necessarily has nonzero spectral width (due to modulation). Polarization mode dispersion, another source of limitation, occurs because although the single-mode fiber can sustain only one transverse mode, it can carry this mode with two different polarizations, and slight imperfections or distortions in a fiber can alter the propagation velocities for the two polarizations. This phenomenon is called fiber birefringence and can be counteracted by polarization-maintaining optical fiber. Dispersion limits the bandwidth of the fiber because the spreading optical pulse limits the rate that pulses can follow one another on the fiber and still be distinguishable at the receiver.

Some dispersion, notably chromatic dispersion, can be removed by a 'dispersion compensator'. This works by using a specially prepared length of fiber that has the opposite dispersion to that induced by the transmission fiber, and this sharpens the pulse so that it can be correctly decoded by the electronics.

Attenuation

Fiber attenuation, which necessitates the use of amplification systems, is caused by a combination of material absorption, Rayleigh scattering, Mie scattering, and connection losses. Although material absorption for pure silica is only around 0.03 dB/km (modern fiber has attenuation around 0.3 dB/km), impurities in the original optical fibers caused attenuation of about 1000 dB/km. Other forms of attenuation are caused by physical stresses to the fiber, microscopic fluctuations in density, and imperfect splicing techniques.

Transmission Windows

Each effect that contributes to attenuation and dispersion depends on the optical wavelength. There are wavelength bands (or windows) where these effects are weakest, and these are the most favorable for transmission. These windows have been standardized, and the currently defined bands are the following:

Band	Description	Wavelength Range
O band	Original	1260 to 1360 nm
E band	Extended	1360 to 1460 nm
S band	short wavelengths	1460 to 1530 nm
C band	conventional ("erbium window")	1530 to 1565 nm
L band	long wavelengths	1565 to 1625 nm
U band	ultralong wavelengths	1625 to 1675 nm

Note that this table shows that current technology has managed to bridge the second and third windows that were originally disjoint.

Historically, there was a window used below the O band, called the first window, at 800-900 nm; however, losses are high in this region so this window is used primarily for short-distance communications. The current lower windows (O and E) around 1300 nm have much lower losses. This region has zero dispersion. The middle windows (S and C) around 1500 nm are the most widely used. This region has the lowest attenuation losses and achieves the longest range. It does have some dispersion, so dispersion compensator devices are used to remove this.

Regeneration

When a communications link must span a larger distance than existing fiber-optic technology is capable of, the signal must be *regenerated* at intermediate points in the link by optical communications repeaters. Repeaters add substantial cost to a communication system, and so system designers attempt to minimize their use.

Recent advances in fiber and optical communications technology have reduced signal degradation so far that *regeneration* of the optical signal is only needed over distances of hundreds of kilometers. This has greatly reduced the cost of optical networking, particularly over undersea spans where the cost and reliability of repeaters is one of the key factors determining the performance of the whole cable system. The main advances contributing to these performance improvements are dispersion management, which seeks to balance the effects of dispersion against non-linearity; and solitons, which use nonlinear effects in the fiber to enable dispersion-free propagation over long distances.

Last Mile

Although fiber-optic systems excel in high-bandwidth applications, optical fiber has been slow to achieve its goal of fiber to the premises or to solve the last mile problem. However, as bandwidth demand increases, more and more progress towards this goal can be observed. In Japan, for instance EPON has largely replaced DSL as a broadband Internet source. South Korea's KT also provides a service called FTTH (Fiber To The Home), which provides fiber-optic connections to the subscriber's home. The largest FTTH deployments are in Japan, South Korea, and China. Singapore started implementation of their all-fiber Next Generation Nationwide Broadband Network (Next Gen NBN), which is slated for completion in 2012 and is being installed by OpenNet. Since they began rolling out services in September 2010, Network coverage in Singapore has reached 85% nationwide.

In the US, Verizon Communications provides a FTTH service called FiOS to select high-ARPU (Average Revenue Per User) markets within its existing territory. The other major surviving ILEC (or Incumbent Local Exchange Carrier), AT&T, uses a FTTN (Fiber To The Node) service called U-verse with twisted-pair to the home. Their MSO competitors employ FTTN with coax using HFC. All of the major access networks use fiber for the bulk of the distance from the service provider's network to the customer.

Also in the US, Wilson Utilities, located in Wilson, North Carolina, has implemented FTTH and has successfully achieved 1 gigabit fiber to the home. This was implemented in late 2013. Wilson Utilities first rolled out their FTTH (Fiber to the Home) in 2012 with speeds offerings of 20/40/60/100 megabits per second. Their service is referred to as GreenLight.

Some other small cities in the US, such as Morristown, TN, have had their local utility company, Morristown Utility Systems in this case, deploy FTTH, offering symmetric gigabit speeds to each subscriber (though most are 50/50 or 100/100 mbit). It's called MUS Fibernet. AT&T and others have aggressively sought legislation at the state level to prevent further competition from municipalities, despite their low investment in rural areas.

The globally dominant access network technology is EPON (Ethernet Passive Optical Network). In Europe, and among telcos in the United States, BPON (ATM-based Broadband PON) and GPON (Gigabit PON) had roots in the FSAN (Full Service Access Network) and ITU-T standards organizations under their control.

Comparison with Electrical Transmission

A mobile fiber optic splice lab used to access and splice underground cables

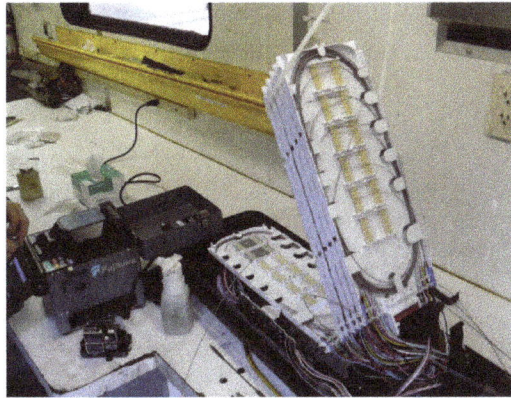

An underground fiber optic splice enclosure opened up

The choice between optical fiber and electrical (or copper) transmission for a particular system is made based on a number of trade-offs. Optical fiber is generally chosen for systems requiring higher bandwidth or spanning longer distances than electrical cabling can accommodate.

The main benefits of fiber are its exceptionally low loss (allowing long distances between amplifiers/repeaters), its absence of ground currents and other parasite signal and power issues common to long parallel electric conductor runs (due to its reliance on light rather than electricity for transmission, and the dielectric nature of fiber optic), and its inherently high data-carrying capacity. Thousands of electrical links would be required to replace a single high bandwidth fiber cable. Another benefit of fibers is that even when run alongside each other for long distances, fiber cables experience effectively no crosstalk, in contrast to some types of electrical transmission lines. Fiber can be installed in areas with high electromagnetic interference (EMI), such as alongside utility lines, power lines, and railroad tracks. Nonmetallic all-dielectric cables are also ideal for areas of high lightning-strike incidence.

For comparison, while single-line, voice-grade copper systems longer than a couple of kilometers require in-line signal repeaters for satisfactory performance; it is not unusual for optical systems to go over 100 kilometers (62 mi), with no active or passive processing. Single-mode fiber cables are commonly available in 12 km lengths, minimizing the number of splices required over a long cable run. Multi-mode fiber is available in lengths up to 4 km, although industrial standards only mandate 2 km unbroken runs.

In short distance and relatively low bandwidth applications, electrical transmission is often preferred because of its

- Lower material cost, where large quantities are not required

- Lower cost of transmitters and receivers

- Capability to carry electrical power as well as signals (in appropriately designed cables)

- Ease of operating transducers in linear mode.

- Crosstalk from nearby cables and other parasitical unwanted signals increase profits from replacement and mitigation devices.

Optical fibers are more difficult and expensive to splice than electrical conductors. And at higher powers, optical fibers are susceptible to fiber fuse, resulting in catastrophic destruction of the fiber core and damage to transmission components.

Because of these benefits of electrical transmission, optical communication is not common in short box-to-box, backplane, or chip-to-chip applications; however, optical systems on those scales have been demonstrated in the laboratory.

In certain situations fiber may be used even for short distance or low bandwidth applications, due to other important features:

- Immunity to electromagnetic interference, including nuclear electromagnetic pulses.

- High electrical resistance, making it safe to use near high-voltage equipment or between areas with different earth potentials.

- Lighter weight—important, for example, in aircraft.

- No sparks—important in flammable or explosive gas environments.

- Not electromagnetically radiating, and difficult to tap without disrupting the signal—important in high-security environments.

- Much smaller cable size—important where pathway is limited, such as networking an existing building, where smaller channels can be drilled and space can be saved in existing cable ducts and trays.

- Resistance to corrosion due to non-metallic transmission medium

Optical fiber cables can be installed in buildings with the same equipment that is used to install copper and coaxial cables, with some modifications due to the small size and limited pull tension and bend radius of optical cables. Optical cables can typically be installed in duct systems in spans of 6000 meters or more depending on the duct's condition, layout of the duct system, and installation technique. Longer cables can be coiled at an intermediate point and pulled farther into the duct system as necessary.

Governing Standards

In order for various manufacturers to be able to develop components that function compatibly in fiber optic communication systems, a number of standards have been developed. The International Telecommunications Union publishes several standards related to the characteristics and performance of fibers themselves, including

- ITU-T G.651, "Characteristics of a 50/125 μm multimode graded index optical fibre cable"

- ITU-T G.652, "Characteristics of a single-mode optical fibre cable"

Other standards specify performance criteria for fiber, transmitters, and receivers to be used together in conforming systems. Some of these standards are:

- 100 Gigabit Ethernet

- 10 Gigabit Ethernet

- Fibre Channel

- Gigabit Ethernet

- HIPPI

- Synchronous Digital Hierarchy

- Synchronous Optical Networking

- Optical Transport Network (OTN)

TOSLINK is the most common format for digital audio cable using plastic optical fiber to connect digital sources to digital receivers.

Push-button Telephone

Modern push-button telephone

The push-button telephone is a telephone that uses buttons or keys for dialing a telephone number to place a call to another telephone subscriber.

Western Electric experimented as early as 1941 with methods of using mechanically activated reeds to produce two tones for each of the ten digits and by the late 1940s such technology was field-tested in a No. 5 Crossbar switching system in Pennsylvania. But the technology proved unreliable and it was not until long after the invention of the transistor when push-button technology matured. On 18 November 1963, after approximately three years of customer testing, the Bell System in the United States officially introduced dual-tone multi-frequency (DTMF) technology under its registered Touch-Tone mark. Over the next few decades touch-tone service replaced traditional pulse dialing technology and it eventually became a world-wide standard for telecommunication signaling.

Although DTMF was the driving technology implemented in push-button telephones, some telephone manufacturers used push-button keypads to generate pulse dial signaling. Before the introduction of touch-tone telephone sets, the Bell System sometimes used the term *push-button telephone* to refer to key system telephones, which were rotary dial telephones that also had a set of push-buttons to select one of multiple telephone circuits, or to activate other features.

History

The concept of the use of push-buttons in telephony originated around 1887 with a device called the micro-telephone push-button, but it was not an automatic dialing system as understood later. This use even predated the invention of the rotary dial by Almon Brown Strowger in 1891. The Bell System in the United States relied on manual switched service until 1919, when it reversed its decisions and embraced dialed, automatic switching. The 1951 introduction of direct distance dialing required automatic transmission of dialed numbers between distant exchanges, leading to use of inband multi-frequency signaling within the Long Lines network while individual local subscribers continued to dial using standard pulses.

As direct distance dialing expanded to a growing number of communities, local numbers (often four, five or six digits) were extended to standardized seven-digit named exchanges. A toll call to another area code was eleven digits, including the leading 1. In the 1950s, AT&T conducted extensive studies of product engineering and efficiency and concluded that push-button dialing was preferable to rotary dialing.

After initial customer trials in Connecticut and Illinois, approximately one fourth of the central office in Findlay, Ohio, was equipped in 1960 with touch-tone digit registers for the first commercial deployment of push-button dialing, starting on 1 November 1960.

On 22 April 1963 President John F. Kennedy started the countdown for the opening of the 1964 World's Fair by keying "1964" on a touch-tone telephone in the Oval Office, starting "a contraption which will count off the seconds until the opening". On November 18, 1963, the first electronic push-button system with touch-tone dialing was commercially offered by Bell Telephone to customers in the Pittsburgh area towns of Carnegie and Greensburg, Pennsylvania, after the DTMF system had been tested for several years in multiple locations, including Greensburg.

Typical push-button phone in the 1970s and early 80s, with 12 keys

This phone, the Western Electric 1500, had only ten buttons. In 1968 it was replaced by the twelve-button model 2500, adding the asterisk or star (*) and pound or hash (#) keys. The use of tones instead of dial pulses relied heavily on technology already developed for the long line network, although the 1963 touch-tone deployment adopted a different frequency set for its dual-tone multi-frequency signaling.

Although push-button touch-tone telephones made their debut to the general public in 1963, the rotary dial telephone still was common for many years. In the 1970s the majority of telephone subscribers still had rotary phones, which in the Bell System of that era were leased from telephone companies instead of being owned outright. Adoption of the push-button phone was steady, but it took a long time for them to appear in some areas. At first it was primarily businesses that adopted push-button phones. By 1979, the touch-tone phone was gaining popularity, but it wasn't until the 1980s that the majority of customers owned push-button telephones in their homes; by the 1990s, it was the overwhelming majority.

Some exchanges no longer support pulse-dialing or charge their few remaining pulse-dial users the higher tone-dial monthly rate as rotary telephones become increasingly rare. Dial telephones are not compatible with some modern telephone features, including interactive voice response systems, though enthusiasts may adapt pulse-dialing telephones using a pulse-to-tone converter.

Touch-tone

The international standard for telephone signaling utilizes dual-tone multi-frequency (DTMF) signaling, more commonly known as touch-tone dialing. It replaced the older and slower pulse dial system. The push-button format is also used for all cell phones, but with out-of-band signaling of the dialed number.

The touch-tone system uses audible tones for each of the digits zero through nine. Later this was expanded by two keys labeled with an asterisk (*) and the pound or hash sign (#) to represent the 11th and 12th DTMF signals. These signals accommodate various additional services and customer-controlled calling features.

The DTMF standard assigns specific frequencies to each column and row of push-buttons in the telephone keypad; the columns in the push-button pad have higher-frequency tones, and rows have lower-frequency tones in the audible range. When a button is pressed the dial generates a combination signal of the two frequencies for the selected row and column, a dual-tone signal, which is transmitted over the phone line to the telephone exchange.

When announced, the DTMF technology was not immediately available on all switching systems. The circuits of subscribers requesting the feature often had to be moved from older switches that supported only pulse dialing to a newer crossbar, or later an electronic switching system, requiring the assignment of a new telephone number which was billed at a higher monthly rate. Community dial office subscribers would often find the service initially unavailable as these villages were served by a single unattended exchange, often step by step, with service from a foreign exchange impractically expensive. Rural party line service was typically based on mechanical switching equipment which could not be upgraded.

While a tone-to-pulse converter could be deployed to any existing mechanical office line using 1970s technology, its speed would be limited to pulse dialing rates. The new central office switches were backward-compatible with rotary dialing.

DTMF Keypad Layout

The standard layout of the keys on the touch-tone telephone was the result of research of the human-engineering department at Bell Laboratories in the 1950s under the leadership of South African-born psychologist John Elias Karlin (1918–2013), who was previously a leading proponent in the introduction of all-number-dialing in the Bell System. This research resulted in the design of the DTMF keypad that arranged the push-buttons into 12 positions in a 3-by-4 position rectangular array, and placed the 1, 2, and 3 keys in the top row for most accurate dialing. The remaining digits occupied the lower rows in sequence from left to right, however, placing the 0 into the center of the fourth row, while omitting the lower left, and lower right positions. These two positions were later assigned to the asterisk and pound key when the keypad was expanded for twelve buttons in 1969.

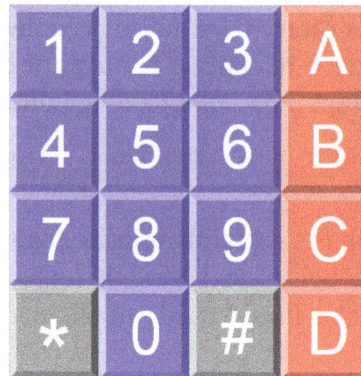

DTMF keypad layout

The engineers had envisioned telephones being used to access computers, and surveyed a business customers for possible uses. This led to the addition of the number sign (#, *pound* or *diamond* in this context, *hash*, *square* or *gate* in the UK, and *octothorpe* by the original engineers) and asterisk or *star* (*) keys. In military telephone systems four additional signals (A, B, C, D) were defined for signaling call priority. Later, the hash and asterisk keys were used in vertical service codes, such as *67 to suppress caller ID in the Bell System.

The DTMF keyboard layout broke with the tradition established in cash registers (and later adopted in calculators and computers) of having the 0 key at the bottom and the remaining keys increasing in rows above. This was due to research conducted by Bell Labs using test subjects unfamiliar with keypads. Comparing various layouts including two-row, two-column, and circular configurations, the study concluded that while there was little difference in speed or accuracy between any of the layouts, the now familiar arrangement with 1 at the top was the most favourably rated.

Pulse Dialing

Historically, not all push-button telephones used DTMF dialing technology. Some manufacturers implemented pulse dialing with push-button keypads and even Western Electric produced several telephone models with a push-button keypad that could also emit traditional dial pulses. Sometimes the mode was user-selectable with a switch on the telephone. Pulse-mode push-button keypads typically stored the dialed number sequence in a digit collector register to permit rapid dialing for the user.

As telephone companies continued to levy surcharges for touch-tone service long after any technical justification ceased to exist, a push-button telephone with pulse dialing capability represented a means for a user to obtain the convenience of push-button dialing without incurring the touch-tone surcharge.

Features

Telia *Mox* and *Fido*

Electronics within push-button telephones may provide several usability features, such as last-number re-dial and storage of commonly called numbers. Some telephone models support additional features, such as retrieval of information and data or code and PIN entry.

Most analog telephone adapters for Internet-based telecommunications (VoIP) recognize and translate DTMF tones but ignore dial pulses, an issue which also exists for some PBX systems. Like cellular handsets, telephones designed for voice-over-IP use out-of-band signaling to send the dialed number.

Dual-tone Multi-frequency Signalling

The keypads on telephones for the Autovon systems used all 16 DTMF signals. The red keys in the fourth column produce the A, B, C, and D DTMF events.

Dual-tone multi-frequency signaling (DTMF) is an in-band telecommunication signaling system using the voice-frequency band over telephone lines between telephone equipment and other communications devices and switching centers. DTMF was first developed in the Bell System in the United States, and became known under the trademark Touch-Tone for use in push-button telephones supplied to telephone customers, starting in 1963. DTMF is standardized by ITU-T Recommendation Q.23. It is also known in the UK as *MF4*.

The Touch-Tone system using a telephone keypad gradually replaced the use of rotary dial and has become the industry standard for landline and mobile service. Other multi-frequency systems are used for internal signaling within the telephone network.

Multifrequency Signaling

Prior to the development of DTMF, telephone numbers were dialed by users with a loop-disconnect (LD) signaling, more commonly known as pulse dialing (dial pulse, DP) in the U.S. It functions by interrupting the current in the local loop between the telephone exchange and the calling party's telephone at a precise rate with a switch in the telephone that is operated by the rotary dial as it spins back to its rest position after having been rotated to each desired number. The exchange equipment responds to the dial pulses either directly by operating relays, or by storing the number in a digit register recording the dialed number. The physical distance for which this type of dialing was possible was restricted by electrical distortions and was only possible on direct metallic links between end points of a line. Placing calls over longer distances required either operator assistance or provision of special subscriber trunk dialing equipment. Operators used an earlier type of multi-frequency signaling.

Multi-frequency signaling is a group of signaling methods that use a mixture of two pure tone (pure sine wave) sounds. Various MF signaling protocols were devised by the Bell System and CCITT. The earliest of these were for in-band signaling between switching centers, where long-distance telephone operators used a 16-digit keypad to input the next portion of the destination telephone number in order to contact the next downstream long-distance telephone operator. This semi-automated signaling and switching proved successful in both speed and cost effectiveness. Based on this prior success with using MF by specialists to establish long-distance telephone calls, dual-tone multi-frequency signaling was developed for end-user signaling without the assistance of operators.

The DTMF system uses a set of eight audio frequencies transmitted in pairs to represent 16 signals, represented by the ten digits, the letters A to D, and the symbols # and *. As the signals are audible tones in the voice frequency range, they can be transmitted through electrical repeaters and amplifiers, and over radio and microwave links, thus eliminating the need for intermediate operators on long-distance circuits.

AT&T described the product as "a method for pushbutton signaling from customer stations using the voice transmission path." In order to prevent consumer telephones from interfering with the MF-based routing and switching between telephone switching centers, DTMF frequencies differ from all of the pre-existing MF signaling protocols between switching centers: MF/R1, R2, CCS4, CCS5, and others that were later replaced by SS7 digital signaling. DTMF was known throughout the Bell System by the trademark *Touch-Tone*. The term was first used by AT&T in commerce on July 5, 1960 and was introduced to the public on November 18, 1963, when the first push-button telephone was made available to the public. It was a registered trademark by AT&T from September 4, 1962 to March 13, 1984. It is standardized by ITU-T Recommendation Q.23. In the UK, it is also known as MF4.

Other vendors of compatible telephone equipment called the Touch-Tone feature *tone dialing* or *DTMF*, or used their other trade names such as *Digitone* by Northern Electric Company in Canada.

As a method of in-band signaling, DTMF signals were also used by cable television broadcasters to indicate the start and stop times of local commercial insertion points during station breaks for the benefit of cable companies. Until out-of-band signaling equipment was developed in the 1990s, fast, unacknowledged DTMF tone sequences could be heard during the commercial breaks of cable channels in the United States and elsewhere. Previously, terrestrial television stations used DTMF tones to control remote transmitters.

#, *, A, B, C, and D

The engineers had envisioned telephones being used to access computers, and automated response systems. They consulted with companies to determine the requirements. This led to the addition of the number sign (#, ''pound'' or "diamond" in this context, "hash", "square" or "gate" in the UK, and "octothorpe" by the original engineers) and asterisk or "star" (*) keys as well as a group of keys for menu selection: A, B, C and D. In the end, the lettered keys were dropped from most phones, and it was many years before the two symbol keys became widely used for vertical service codes such as *67 in the United States of America and Canada to suppress caller ID.

Public payphones that accept credit cards use these additional codes to send the information from the magnetic strip.

The AUTOVON telephone system of the United States Armed Forces used these signals to assert certain privilege and priority levels when placing telephone calls. Precedence is still a feature of military telephone networks, but using number combinations. For example, entering 93 before a number is a priority call.

Present-day uses of the A, B, C and D signals on telephone networks are few, and are exclusive to network control. For example, the A key is used on some networks to cycle through different carriers at will. The A, B, C and D tones are used in radio phone patch and repeater operations to allow, among other uses, control of the repeater while connected to an active phone line.

The *, #, A, B, C and D keys are still widely used worldwide by amateur radio operators and commercial two-way radio systems for equipment control, repeater control, remote-base operations and some telephone communications systems.

DTMF signaling tones can also be heard at the start or end of some VHS (Video Home System) cassette tapes. Information on the master version of the video tape is encoded in the DTMF tone. The encoded tone provides information to automatic duplication machines, such as format, duration and volume levels, in order to replicate the original video as closely as possible.

DTMF tones are used in some caller ID systems to transfer the caller ID information, but in the United States only Bell 202 modulated FSK signaling is used to transfer the data.

Keypad

The DTMF telephone keypad is laid out in a 4×4 matrix of push buttons in which each row represents the *low* frequency component and each column represents the *high* frequency component of the DTMF signal. Pressing a key sends a combination of the row and column frequencies. For example, the key *1* produces a superimposition of tones of 697 and 1209 hertz (Hz). Initial push-button designs employed levers, so that each button activated two contacts. The tones are decoded by the switching center to determine the keys pressed by the user.

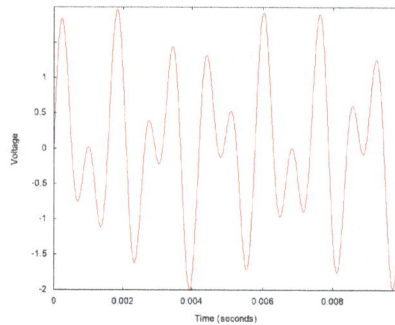

1209 Hz on 697 Hz to make the 1 tone

DTMF keypad frequencies (with sound clips)			
	1209 Hz	1336 Hz	1477 Hz	1633 Hz
697 Hz	1	2	3	A
770 Hz	4	5	6	B
852 Hz	7	8	9	C
941 Hz	*	0	#	D

Decoding

DTMF was originally decoded by tuned filter banks. By the end of the 20th century, digital signal processing became the predominant technology for decoding. DTMF decoding algorithms often use the Goertzel algorithm to detect tones.

Other Multiple Frequency Signals

National telephone systems define other tones that indicate the status of lines, equipment, or the result of calls. Such call-progress tones are often also composed of multiple frequencies and are standardized in each country. The Bell System defines them in the Precise Tone Plan. However, such signaling systems are not considered to belong to the DTMF system.

Free-space Optical Communication

An 8-beam free space optics laser link, rated for 1 Gbit/s. The receptor is the large disc in the middle, the transmitters the smaller ones. At the top right corner is a monocular for assisting the alignment of the two heads.

Free-space optical communication (FSO) is an optical communication technology that uses light propagating in free space to wirelessly transmit data for telecommunications or computer networking. "Free space" means air, outer space, vacuum, or something similar. This contrasts with using solids such as optical fiber cable or an optical transmission line. The technology is useful where the physical connections are impractical due to high costs or other considerations.

History

A photophone receiver and headset, one half of Bell and Tainter's optical telecommunication system of 1880

Optical communications, in various forms, have been used for thousands of years. The Ancient Greeks used a coded alphabetic system of signalling with torches developed by Cleoxenus, Democleitus and Polybius. In the modern era, semaphores and wireless solar telegraphs called heliographs were developed, using coded signals to communicate with their recipients.

In 1880 Alexander Graham Bell and his assistant Charles Sumner Tainter created the Photophone, at Bell's newly established Volta Laboratory in Washington, DC. Bell considered it his most important invention. The device allowed for the transmission of sound on a beam of light. On June 3, 1880, Bell conducted the world's first wireless telephone transmission between two buildings, some 213 meters (700 feet) apart.

Its first practical use came in military communication systems many decades later, first for optical telegraphy. German colonial troops used Heliograph telegraphy transmitters during the 1904/05 Herero Genocide in German South-West Africa (today's Namibia) as did British, French, US or Ottoman signals.

WW I German Blinkgerät

During the trench warfare of World War I when wire communications were often cut, German signals used three types of optical Morse transmitters called *Blinkgerät*, the intermediate type for distances of up to 4 km (2.5 miles) at daylight and of up to 8 km (5 miles) at night, using red filters for undetected communications. Optical telephone communications were tested at the end of the war, but not introduced at troop level. In addition, special blinkgeräts were used for communication with airplanes, balloons, and tanks, with varying success.

A major technological step was to replace the Morse code by modulating optical waves in speech transmission. Carl Zeiss Jena developed the *Lichtsprechgerät 80/80* (literal translation: optical speaking device) that the German army used in their World War II anti-aircraft defense units, or in bunkers at the Atlantic Wall.

The invention of lasers in the 1960s revolutionized free space optics. Military organizations were particularly interested and boosted their development. However the technology lost market momentum when the installation of optical fiber networks for civilian uses was at its peak.

Many simple and inexpensive consumer remote controls use low-speed communication using infrared (IR) light. This is known as consumer IR technologies.

A recently declassified 1987 Pentagon report reveals free-space lasers have been mounted on Israeli F-15 fighter jets for the purposes of surveillance, missile-tracking, and targeted weaponry.

Usage and Technologies

Free-space point-to-point optical links can be implemented using infrared laser light, although low-data-rate communication over short distances is possible using LEDs. Infrared Data Association (IrDA) technology is a very simple form of free-space optical communications. On the communications side the FSO technology is considered as a part of the Optical Wireless Communications applications. Free-space optics can be used for communications between spacecraft.

Current Market Demands

The demand for a high-speed (10 GBps+) and long range (3 – 5 km) FSO system is apparent in the market place.

- In 2008, MRV Communications introduced a free-space optics (FSO)-based system with a data rate of 10GB/s initially claiming a distance of 2 km at high availability. This equipment is no longer available; before end-of-life, the product's useful distance was changed down to 350m.

- In 2013, the company MOSTCOM started to serially produce a new wireless communication system that also had a data rate of 10Gb/s as well as an improved range of up to 2.5 km, but to get to 99.99% up-time the designers used an RF hybrid solution, meaning the data rate drops to extremely low levels during atmospheric disturbances (typically down to 10MB/s).

- LightPointe offers many similar hybrid solutions to MOSTCOM's offering.

Useful Distances

The reliability of FSO units has always been a problem for commercial telecommunications. Consistently, studies find too many dropped packets and signal errors over small ranges (400 to 500 meters). This is from both independent studies, such as in the Czech republic, as well as formal internal nationwide studies, such as one conducted by MRV FSO staff. Military based studies consistently produce longer estimates for reliability, projecting the maximum range for terrestrial links is of the order of 2 to 3 km (1.2 to 1.9 mi). All studies agree the stability and quality of the link is highly dependent on atmospheric factors such as rain, fog, dust and heat.

Extending the Useful Distance

DARPA ORCA Official Concept Art created circa 2008

The main reason terrestrial communications have been limited to non-commercial telecommunications functions is fog. Fog consistently keeps FSO laser links over 500 meters from achieving a year-round bit error rate of 99.999%. Several entities are continually attempting to overcome these key disadvantages to FSO communications and field a system with a better quality of service. DARPA has sponsored over $130 million USD in research towards this effort, with the ORCA and ORCLE programs.

Other non-government groups are fielding tests to evaluate different technologies that some claim have the ability to address key FSO adoption challenges. As of October 2014, none have fielded a working system that addresses the most common atmospheric events.

FSO research from 1998-2006 in the private sector totaled $407.1 million, divided primarily among 4 start-up companies. All four failed to deliver products that would meet telecommunications quality and distance standards:

- Terabeam received approximately $226 million in funding. AT&T and Lucent backed this attempt. The work ultimately failed, and the company reorganized in 2004.

- AirFiber received $96.1 million in funding, and never solved the weather issue. They sold out to MRV communications in 2003, and MRV sold their FSO units until 2012 when the end-of-life was abruptly announced for the Terescope series.

- LightPointe Communications received $76 million in start-up funds, and eventually reorganized to sell hybrid FSO-RF units to overcome the weather-based challenges.

- The Maxima Corporation published its operating theory in Science (magazine), and received $9 million in funding before permanently shutting down. No known spin-off or purchase followed this effort.

One private company published a paper on Nov 20,2014, claiming they had achieved commercial reliability (99.999% availability) in extreme fog. There is no indication this product is currently commercially available.

Extraterrestrial

The massive advantages of laser communication in space have multiple space agencies racing to develop a stable space communication platform, with many significant demonstrations and achievements. To date (18 December 2014), *no laser communication system is in use in space.*

Demonstrations in Space:

The first gigabit laser-based communication was achieved by the European Space Agency and called the European Data Relay System (EDRS) on November 28, 2014. The initial images have just been demonstrated, and a working system is expected to be in place in the 2015-2016 time frame.

NASA's OPALS announced a breakthrough in space-to-ground communication December 9, 2014, uploading 175 megabytes in 3.5 seconds. Their system is also able to re-acquire tracking after the signal was lost due to cloud cover.

In January 2013, NASA used lasers to beam an image of the Mona Lisa to the Lunar Reconnaissance Orbiter roughly 390,000 km (240,000 mi) away. To compensate for atmospheric interference, an error correction code algorithm similar to that used in CDs was implemented.

A two-way distance record for communication was set by the Mercury laser altimeter instrument aboard the MESSENGER spacecraft, and was able to communicate across a distance of 24 million km (15 million miles), as the craft neared Earth on a fly-by in May, 2005. The previous record had been set with a one-way detection of laser light from Earth, by the Galileo probe, of 6 million km in 1992. Quote from Laser Communication in Space Demonstrations (EDRS)

LEDs

RONJA is a free implementation of FSO using high-intensity LEDs.

In 2001, Twibright Labs released Ronja Metropolis, an open source DIY 10Mbit/s full duplex LED FSO over 1.4 km In 2004, a Visible Light Communication Consortium was formed in Japan. This was based on work from researchers that used a white LED-based space lighting system for indoor local area network (LAN) communications. These systems present advantages over traditional UHF RF-based systems from improved isolation between systems, the size and cost of receivers/ transmitters, RF licensing laws and by combining space lighting and communication into the same system. In January 2009 a task force for visible light communication was formed by the Institute of Electrical and Electronics Engineers working group for wireless personal area network standards known as IEEE 802.15.7. A trial was announced in 2010 in St. Cloud, Minnesota.

Amateur radio operators have achieved significantly farther distances using incoherent sources of light from high-intensity LEDs. One reported 173 miles (278 km) in 2007. However, physical limitations of the equipment used limited bandwidths to about 4 kHz. The high sensitivities required of the detector to cover such distances made the internal capacitance of the photodiode used a dominant factor in the high-impedance amplifier which followed it, thus naturally forming a low-pass filter with a cut-off frequency in the 4 kHz range. From the other side use of lasers radiation source allows to reach very high data rates which are comparable to fiber communications.

Projected data rates and future data rate claims vary. A low-cost white LED (GaN-phosphor) which could be used for space lighting can typically be modulated up to 20 MHz. Data rates of over 100 Mbit/s can be easily achieved using efficient modulation schemes and Siemens claimed to have achieved over 500 Mbit/s in 2010. Research published in 2009 used a similar system for traffic control of automated vehicles with LED traffic lights.

In September 2013, pureLiFi, the Edinburgh start-up working on Li-Fi, also demonstrated high speed point-to-point connectivity using any off-the-shelf LED light bulb. In previous work, high bandwidth specialist LEDs have been used to achieve the high data rates. The new system, the Li-1st, maximizes the available optical bandwidth for any LED device, thereby reducing the cost and improving the performance of deploying indoor FSO systems.

Engineering Details

Typically, best use scenarios for this technology are:

- LAN-to-LAN connections on campuses at Fast Ethernet or Gigabit Ethernet speeds

- LAN-to-LAN connections in a city, a metropolitan area network

- To cross a public road or other barriers which the sender and receiver do not own

- Speedy service delivery of high-bandwidth access to optical fiber networks

- Converged Voice-Data-Connection

- Temporary network installation (for events or other purposes)

- Reestablish high-speed connection quickly (disaster recovery)

- As an alternative or upgrade add-on to existing wireless technologies

    - Especially powerful in combination with auto aiming systems, this way you could power moving cars or you can power your laptop while you move or use auto-aiming nodes to create a network with other nodes.

- As a safety add-on for important fiber connections (redundancy)

- For communications between spacecraft, including elements of a satellite constellation

- For inter- and intra -chip communication.

The light beam can be very narrow, which makes FSO hard to intercept, improving security. In any case, it is comparatively easy to encrypt any data traveling across the FSO connection for additional security. FSO provides vastly improved electromagnetic interference (EMI) behavior compared to using microwaves.

Technical Advantages

- Ease of deployment

- Can be used to power devices

- License-free long-range operation (in contrast with radio communication)

- High bit rates

- Low bit error rates

- Immunity to electromagnetic interference

- Full duplex operation

- Protocol transparency

- Increased security when working with narrow beam(s)

- No Fresnel zone necessary

- Reference open source implementation

Range Limiting Factors

For terrestrial applications, the principal limiting factors are:

- Fog (10..~100 dB/km attenuation)

- Beam dispersion

- Atmospheric absorption

- Rain

- Snow

- Terrestrial scintillation

- Interference from background light sources (including the Sun)

- Shadowing

- Pointing stability in wind

- Pollution / smog

These factors cause an attenuated receiver signal and lead to higher bit error ratio (BER). To overcome these issues, vendors found some solutions, like multi-beam or multi-path architectures, which use more than one sender and more than one receiver. Some state-of-the-art devices also have larger fade margin (extra power, reserved for rain, smog, fog). To keep an eye-safe environment, good FSO systems have a limited laser power density and support laser classes 1 or 1M. Atmospheric and fog attenuation, which are exponential in nature, limit practical range of FSO devices to several kilometres.

References

- Clint Smith, Curt Gervelis (2003). Wireless Network Performance Handbook. McGraw-Hill Professional. ISBN 0-07-140655-7.

- Free radio: electronic civil disobedience by Lawrence C. Soley. Published by Westview Press, 1998. ISBN 0-8133-9064-8, ISBN 978-0-8133-9064-2

- Rebel Radio: The Full Story of British Pirate Radio by John Hind, Stephen Mosco. Published by Pluto Press, 1985. ISBN 0-7453-0055-3, ISBN 978-0-7453-0055-9

- G. Miao, J. Zander, K-W Sung, and B. Slimane, Fundamentals of Mobile Data Networks, Cambridge University Press, ISBN 1107143217, 2016.

- N. D. Tripathi, J. H. Reed, H. F. Vanlandingham, Radio Resource Management in Cellular Systems, Springer, ISBN 0-7923-7374-X, 2001.

- Martin, Donald; Anderson, Paul; Bartamian, Lucy (March 16, 2007). "Communications Satellites" (5th ed.). AIAA. ISBN 978-1884989193.

- Mary Kay Carson (2007). Alexander Graham Bell: Giving Voice To The World. Sterling Biographies. New York: Sterling Publishing. pp. 76–78. ISBN 978-1-4027-3230-0.

- Mary Kay Carson (2007). Alexander Graham Bell: Giving Voice To The World. Sterling Biographies. New York: Sterling Publishing. pp. 76–78. ISBN 978-1-4027-3230-0.

- "Communication: How the rover can communicate through Mars-orbiting spacecraft". Jet Propulsion Laboratory. Retrieved 21 January 2016.

- Eric Korevaar, Isaac I. Kim and Bruce McArthur (2001). "Atmospheric Propagation Characteristics of Highest Importance to Commercial Free Space Optics" (PDF). Optical Wireless Communications IV, SPIE Vol. 4530 p. 84. Retrieved October 27, 2014.

- Bruce V. Bigelow (June 16, 2006). "Zapped of its potential, Rooftop laser startups falter, but debate on high-speed data technology remains". Retrieved October 26, 2014.

- Fred Dawson (May 1, 2000). "TeraBeam, Lucent Extend Bandwidth Limits, Multichannel News, Vol 21 Issue 18 Pg 160". Retrieved October 27, 2014.

- "DIRECTV's Spaceway F1 Satellite Launches New Era in High-Definition Programming; Next Generation Satellite Will Initiate Historic Expansion of DIRECTV". SpaceRef. Retrieved 2012-05-11.

- Tom Garlington, Joel Babbitt and George Long (March 2005). "Analysis of Free Space Optics as a Transmission Technology" (PDF). WP No. AMSEL-IE-TS-05001. U.S. Army Information Systems Engineering Command. p. 3. Archived from the original (PDF) on June 13, 2007. Retrieved June 28, 2011.

- Clint Turner (October 3, 2007). "A 173-mile 2-way all-electronic optical contact". Modulated light web site. Retrieved June 28, 2011.

- "14 Tbit/s over a single optical fiber: successful demonstration of world's largest capacity". News release (NTT). September 29, 2006. Retrieved June 17, 2011.

- "IEEE 802.15 WPAN Task Group 7 (TG7) Visible Light Communication". IEEE 802 local and metro area network standards committee. 2009. Retrieved June 28, 2011.

- s"World Record 69-Terabit Capacity for Optical Transmission over a Single Optical Fiber" (Press release). NTT. 2010-03-25. Retrieved 2010-04-03.

- Lee, M. M.; J. M. Roth; T. G. Ulmer; C. V. Cryan (2006). "The Fiber Fuse Phenomenon in Polarization-Maintaining Fibers at 1.55 μm" (PDF). Conference on Lasers and Electro-Optics/Quantum Electronics and Laser Science Conference and Photonic Applications Systems Technologies. paper JWB66 (Optical Society of America). Retrieved March 14, 2010.

Allied Fields of Telecommunications Engineering

Telecommunications engineering is an interdisciplinary subject. This section will provide a glimpse of the allied fields of telecommunications engineering. Some of these allied fields are electrical engineering, computer engineering, telegraphy and informatics. The topics discussed in the chapter are of great importance to broaden the existing knowledge on this field.

Electrical Engineering

Electrical engineering is a field of engineering that generally deals with the study and application of electricity, electronics, and electromagnetism. This field first became an identifiable occupation in the latter half of the 19th century after commercialization of the electric telegraph, the telephone, and electric power distribution and use. Subsequently, broadcasting and recording media made electronics part of daily life. The invention of the transistor, and later the integrated circuit, brought down the cost of electronics to the point they can be used in almost any household object.

Electrical engineers design complex power systems ...

Electrical engineering has now subdivided into a wide range of subfields including electronics, digital computers, power engineering, telecommunications, control systems, radio-frequency engineering, signal processing, instrumentation, and microelectronics. The subject of electronic engineering is often treated as its own subfield but it intersects with all the other subfields, including the power electronics of power engineering.

Electrical engineers typically hold a degree in electrical engineering or electronic engineering. Practicing engineers may have professional certification and be members of a professional body. Such bodies include the Institute of Electrical and Electronics Engineers (IEEE) and the Institution of Engineering and Technology (professional society) (IET).

... and electronic circuits.

Electrical engineers work in a very wide range of industries and the skills required are likewise variable. These range from basic circuit theory to the management skills required of a project manager. The tools and equipment that an individual engineer may need are similarly variable, ranging from a simple voltmeter to a top end analyzer to sophisticated design and manufacturing software.

History

Electricity has been a subject of scientific interest since at least the early 17th century. A prominent early electrical scientist was William Gilbert who was the first to draw a clear distinction between magnetism and static electricity and is credited with establishing the term electricity. He also designed the versorium: a device that detected the presence of statically charged objects. Then in 1762 Swedish professor Johan Carl Wilcke invented, and in 1775 Alessandro Volta improved, a device (for which Volta coined the name electrophorus) that produced a static electric charge, and by 1800 Volta had developed the voltaic pile, a forerunner of the electric battery.

19th Century

The discoveries of Michael Faraday formed the foundation of electric motor technology

In the 19th century, research into the subject started to intensify. Notable developments in this century include the work of Georg Ohm, who in 1827 quantified the relationship between the electric current and potential difference in a conductor, of Michael Faraday, the discoverer of electromagnetic induction in 1831, and of James Clerk Maxwell, who in 1873 published a unified theory of electricity and magnetism in his treatise *Electricity and Magnetism*.

Electrical engineering became a profession in the later 19th century. Practitioners had created a global electric telegraph network and the first professional electrical engineering institutions were founded in the UK and USA to support the new discipline. Although it is impossible to precisely pinpoint a first electrical engineer, Francis Ronalds stands ahead of the field, who created the first working electric telegraph system in 1816 and documented his vision of how the world could be transformed by electricity. Over 50 years later, he joined the new Society of Telegraph Engineers (soon to be renamed the Institution of Electrical Engineers) where he was regarded by other members as the first of their cohort. By the end of the 19th century, the world had been forever changed by the rapid communication made possible by the engineering development of land-lines, submarine cables, and, from about 1890, wireless telegraphy.

Practical applications and advances in such fields created an increasing need for standardised units of measure. They led to the international standardization of the units volt, ampere, coulomb, ohm, farad, and henry. This was achieved at an international conference in Chicago in 1893. The publication of these standards formed the basis of future advances in standardisation in various industries, and in many countries the definitions were immediately recognised in relevant legislation.

During these years, the study of electricity was largely considered to be a subfield of physics. That's because early electrical technology was electromechanical in nature. The Technische Universität Darmstadt founded the world's first department of electrical engineering in 1882. The first electrical engineering degree program was started at Massachusetts Institute of Technology (MIT) in the physics department under Professor Charles Cross, though it was Cornell University to produce the world's first electrical engineering graduates in 1885. The first course in electrical engineering was taught in 1883 in Cornell's Sibley College of Mechanical Engineering and Mechanic Arts. It was not until about 1885 that Cornell President Andrew Dickson White established the first Department of Electrical Engineering in the United States. In the same year, University College London founded the first chair of electrical engineering in Great Britain. Professor Mendell P. Weinbach at University of Missouri soon followed suit by establishing the electrical engineering department in 1886. Afterwards, universities and institutes of technology gradually started to offer electrical engineering programs to their students all over the world.

Thomas Edison, electric light and (DC) power supply networks

Károly Zipernowsky, Ottó Bláthy, Miksa Déri, the ZDB transformer

William Stanley, Jr., transformers

During these decades use of electrical engineering increased dramatically. In 1882, Thomas Edison switched on the world's first large-scale electric power network that provided 110 volts — direct current (DC) — to 59 customers on Manhattan Island in New York City. In 1884, Sir Charles Parsons invented the steam turbine allowing for more efficient electric power generation. Alternating current, with its ability to transmit power more efficiently over long distances via the use of transformers, developed rapidly in the 1880s and 1890s with transformer designs by Károly Zipernowsky, Ottó Bláthy and Miksa Déri (later called ZBD transformers), Lucien Gaulard, John Dixon Gibbs and William Stanley, Jr.. Practical AC motor designs including induction motors were independently invented by Galileo Ferraris and Nikola Tesla and further developed into a practical three-phase form by Mikhail Dolivo-Dobrovolsky and Charles Eugene Lancelot Brown. Charles Steinmetz and Oliver Heaviside contributed to the theoretical basis of alternating current engineering. The spread in the use of AC set off in the United States what has been called the *War of Currents* between a George Westinghouse backed AC system and a Thomas Edison backed DC power system, with AC being adopted as the overall standard.

More Modern Developments

During the development of radio, many scientists and inventors contributed to radio technology and electronics. The mathematical work of James Clerk Maxwell during the 1850s had shown the relationship of different forms of electromagnetic radiation including possibility of invisible airborne waves (later called "radio waves"). In his classic physics experiments of 1888, Heinrich Hertz proved Maxwell's theory by transmitting radio waves with a spark-gap transmitter, and detected them by using simple electrical devices. Other physicists experimented with these new waves and in the process developed devices for transmitting and detecting them. In 1895, Guglielmo Marconi began work on a way to adapt the known methods of transmitting and detecting these "Hertzian waves" into a purpose built commercial wireless telegraphic system. Early on, he sent wireless signals over a distance of one and a half miles. In December 1901, he sent wireless waves that were not affected by the curvature of the Earth. Marconi later transmitted the wireless signals across the Atlantic between Poldhu, Cornwall, and St. John's, Newfoundland, a distance of 2,100 miles (3,400 km).

Guglielmo Marconi known for his pioneering work on long distance radio transmission

In 1897, Karl Ferdinand Braun introduced the cathode ray tube as part of an oscilloscope, a crucial enabling technology for electronic television. John Fleming invented the first radio tube, the diode, in 1904. Two years later, Robert von Lieben and Lee De Forest independently developed the amplifier tube, called the triode.

In 1920, Albert Hull developed the magnetron which would eventually lead to the development of the microwave oven in 1946 by Percy Spencer. In 1934, the British military began to make strides toward radar (which also uses the magnetron) under the direction of Dr Wimperis, culminating in the operation of the first radar station at Bawdsey in August 1936.

A replica of the first working transistor.

In 1941, Konrad Zuse presented the Z3, the world's first fully functional and programmable computer using electromechanical parts. In 1943, Tommy Flowers designed and built the Colossus, the world's first fully functional, electronic, digital and programmable computer. In 1946, the ENIAC

(Electronic Numerical Integrator and Computer) of John Presper Eckert and John Mauchly followed, beginning the computing era. The arithmetic performance of these machines allowed engineers to develop completely new technologies and achieve new objectives, including the Apollo program which culminated in landing astronauts on the Moon.

Solid-state Transistors

The invention of the transistor in late 1947 by William B. Shockley, John Bardeen, and Walter Brattain of the Bell Telephone Laboratories opened the door for more compact devices and led to the development of the integrated circuit in 1958 by Jack Kilby and independently in 1959 by Robert Noyce. Starting in 1968, Ted Hoff and a team at the Intel Corporation invented the first commercial microprocessor, which foreshadowed the personal computer. The Intel 4004 was a four-bit processor released in 1971, but in 1973 the Intel 8080, an eight-bit processor, made the first personal computer, the Altair 8800, possible.

Subdisciplines

Electrical engineering has many subdisciplines, the most common of which are listed below. Although there are electrical engineers who focus exclusively on one of these subdisciplines, many deal with a combination of them. Sometimes certain fields, such as electronic engineering and computer engineering, are considered separate disciplines in their own right.

Power

Power pole

Power engineering deals with the generation, transmission, and distribution of electricity as well as the design of a range of related devices. These include transformers, electric generators, electric motors, high voltage engineering, and power electronics. In many regions of the world, governments maintain an electrical network called a power grid that connects a variety of generators together with users of their energy. Users purchase electrical energy from the grid, avoiding the costly exercise of having to generate their own. Power engineers may work on the design and maintenance of the power grid as well as the power systems that connect to it. Such systems are called *on-grid* power systems and may supply the grid with additional power, draw power from the grid, or do both. Power engineers may also work on systems that do not connect to the grid, called *off-grid* power systems, which in some cases are preferable to on-grid systems. The future includes Satellite controlled power systems, with feedback in real time to prevent power surges and prevent blackouts.

Control

Control systems play a critical role in space flight.

Control engineering focuses on the modeling of a diverse range of dynamic systems and the design of controllers that will cause these systems to behave in the desired manner. To implement such controllers, electrical engineers may use electronic circuits, digital signal processors, microcontrollers, and programmable logic controls (PLCs). Control engineering has a wide range of applications from the flight and propulsion systems of commercial airliners to the cruise control present in many modern automobiles. It also plays an important role in industrial automation.

Control engineers often utilize feedback when designing control systems. For example, in an automobile with cruise control the vehicle's speed is continuously monitored and fed back to the system which adjusts the motor's power output accordingly. Where there is regular feedback, control theory can be used to determine how the system responds to such feedback.

Electronics

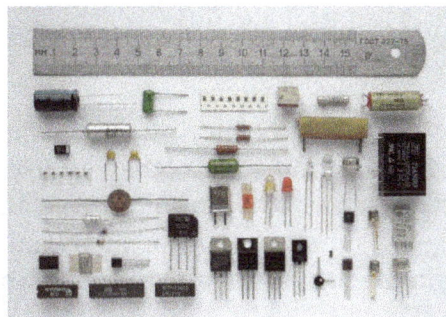

Electronic components

Electronic engineering involves the design and testing of electronic circuits that use the properties of components such as resistors, capacitors, inductors, diodes, and transistors to achieve a particular functionality. The tuned circuit, which allows the user of a radio to filter out all but a single station, is just one example of such a circuit. Another example (of a pneumatic signal conditioner) is shown in the adjacent photograph.

Prior to the Second World War, the subject was commonly known as *radio engineering* and basically was restricted to aspects of communications and radar, commercial radio, and early television. Later, in post war years, as consumer devices began to be developed, the field grew to include modern television, audio systems, computers, and microprocessors. In the mid-to-late 1950s, the term *radio engineering* gradually gave way to the name *electronic engineering*.

Before the invention of the integrated circuit in 1959, electronic circuits were constructed from discrete components that could be manipulated by humans. These discrete circuits consumed much space and power and were limited in speed, although they are still common in some applications. By contrast, integrated circuits packed a large number—often millions—of tiny electrical components, mainly transistors, into a small chip around the size of a coin. This allowed for the powerful computers and other electronic devices we see today.

Microelectronics

Microprocessor

Microelectronics engineering deals with the design and microfabrication of very small electronic circuit components for use in an integrated circuit or sometimes for use on their own as a general electronic component. The most common microelectronic components are semiconductor transistors, although all main electronic components (resistors, capacitors etc.) can be created at a microscopic level. Nanoelectronics is the further scaling of devices down to nanometer levels. Modern devices are already in the nanometer regime, with below 100 nm processing having been standard since about 2002.

Microelectronic components are created by chemically fabricating wafers of semiconductors such as silicon (at higher frequencies, compound semiconductors like gallium arsenide and indium phosphide) to obtain the desired transport of electronic charge and control of current. The field of microelectronics involves a significant amount of chemistry and material science and requires the electronic engineer working in the field to have a very good working knowledge of the effects of quantum mechanics.

Signal Processing

Signal processing deals with the analysis and manipulation of signals. Signals can be either analog, in which case the signal varies continuously according to the information, or digital, in which case the signal varies according to a series of discrete values representing the information. For analog signals, signal processing may involve the amplification and filtering of audio signals for audio equipment or the modulation and demodulation of signals for telecommunications. For digital signals, signal processing may involve the compression, error detection and error correction of digitally sampled signals.

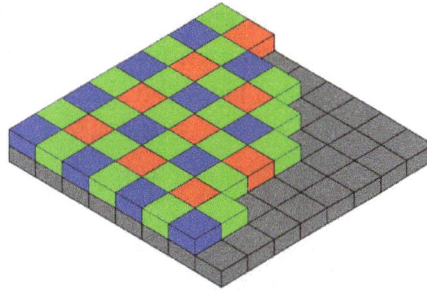

A Bayer filter on a CCD requires signal processing to get a red, green, and blue value at each pixel.

Signal Processing is a very mathematically oriented and intensive area forming the core of digital signal processing and it is rapidly expanding with new applications in every field of electrical engineering such as communications, control, radar, audio engineering, broadcast engineering, power electronics, and biomedical engineering as many already existing analog systems are replaced with their digital counterparts. Analog signal processing is still important in the design of many control systems.

DSP processor ICs are found in every type of modern electronic systems and products including, SDTV | HDTV sets, radios and mobile communication devices, Hi-Fi audio equipment, Dolby noise reduction algorithms, GSM mobile phones, mp3 multimedia players, camcorders and digital cameras, automobile control systems, noise cancelling headphones, digital spectrum analyzers, intelligent missile guidance, radar, GPS based cruise control systems, and all kinds of image processing, video processing, audio processing, and speech processing systems.

Telecommunications

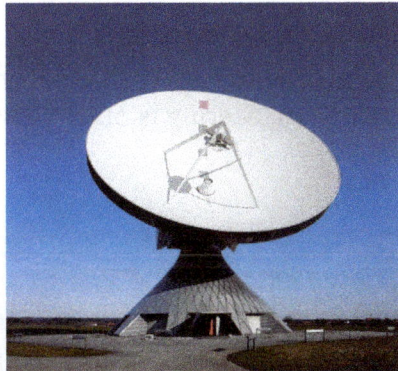

Satellite dishes are a crucial component in the analysis of satellite information.

Telecommunications engineering focuses on the transmission of information across a channel such as a coax cable, optical fiber or free space. Transmissions across free space require information to be encoded in a carrier signal to shift the information to a carrier frequency suitable for transmission; this is known as modulation. Popular analog modulation techniques include amplitude modulation and frequency modulation. The choice of modulation affects the cost and performance of a system and these two factors must be balanced carefully by the engineer.

Once the transmission characteristics of a system are determined, telecommunication engineers design the transmitters and receivers needed for such systems. These two are sometimes combined to form a two-way communication device known as a transceiver. A key consideration in the design of transmitters is their power consumption as this is closely related to their signal strength. If the signal strength of a transmitter is insufficient the signal's information will be corrupted by noise.

Instrumentation

Flight instruments provide pilots with the tools to control aircraft analytically.

Instrumentation engineering deals with the design of devices to measure physical quantities such as pressure, flow, and temperature. The design of such instrumentation requires a good understanding of physics that often extends beyond electromagnetic theory. For example, flight instruments measure variables such as wind speed and altitude to enable pilots the control of aircraft analytically. Similarly, thermocouples use the Peltier-Seebeck effect to measure the temperature difference between two points.

Often instrumentation is not used by itself, but instead as the sensors of larger electrical systems. For example, a thermocouple might be used to help ensure a furnace's temperature remains constant. For this reason, instrumentation engineering is often viewed as the counterpart of control engineering.

Computers

Supercomputers are used in fields as diverse as computational biology and geographic information systems.

Computer engineering deals with the design of computers and computer systems. This may involve the design of new hardware, the design of PDAs, tablets, and supercomputers, or the use of computers to control an industrial plant. Computer engineers may also work on a system's software. However, the design of complex software systems is often the domain of software engineering, which is usually considered a separate discipline. Desktop computers represent a tiny fraction of the devices a computer engineer might work on, as computer-like architectures are now found in a range of devices including video game consoles and DVD players.

Related Disciplines

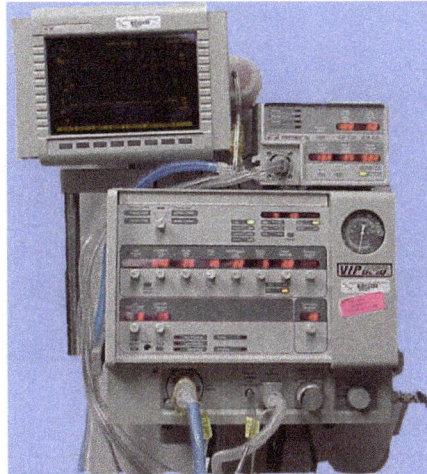

The Bird VIP Infant ventilator

Mechatronics is an engineering discipline which deals with the convergence of electrical and mechanical systems. Such combined systems are known as electromechanical systems and have widespread adoption. Examples include automated manufacturing systems, heating, ventilation and air-conditioning systems, and various subsystems of aircraft and automobiles.

The term *mechatronics* is typically used to refer to macroscopic systems but futurists have predicted the emergence of very small electromechanical devices. Already, such small devices, known as Microelectromechanical systems (MEMS), are used in automobiles to tell airbags when to deploy, in digital projectors to create sharper images, and in inkjet printers to create nozzles for high definition printing. In the future it is hoped the devices will help build tiny implantable medical devices and improve optical communication.

Biomedical engineering is another related discipline, concerned with the design of medical equipment. This includes fixed equipment such as ventilators, MRI scanners, and electrocardiograph monitors as well as mobile equipment such as cochlear implants, artificial pacemakers, and artificial hearts.

Aerospace engineering and robotics an example is the most recent electric propulsion and ion propulsion.

Education

Electrical engineers typically possess an academic degree with a major in electrical engineering, electronics engineering, electrical engineering technology, or electrical and electronic engineering. The same fundamental principles are taught in all programs, though emphasis may vary according to title. The length of study for such a degree is usually four or five years and the completed degree may be designated as a Bachelor of Science in Electrical/Electronics Engineering Technology,

Bachelor of Engineering, Bachelor of Science, Bachelor of Technology, or Bachelor of Applied Science depending on the university. The bachelor's degree generally includes units covering physics, mathematics, computer science, project management, and a variety of topics in electrical engineering. Initially such topics cover most, if not all, of the subdisciplines of electrical engineering. At some schools, the students can then choose to emphasize one or more subdisciplines towards the end of their courses of study.

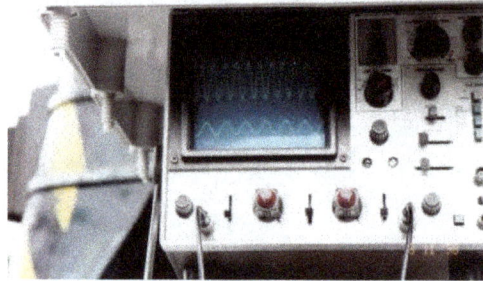

Oscilloscope

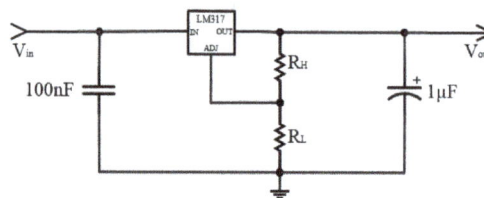

Typical electrical engineering diagram used as a troubleshooting tool

At many schools, electronic engineering is included as part of an electrical award, sometimes explicitly, such as a Bachelor of Engineering (Electrical and Electronic), but in others electrical and electronic engineering are both considered to be sufficiently broad and complex that separate degrees are offered.

Some electrical engineers choose to study for a postgraduate degree such as a Master of Engineering/Master of Science (M.Eng./M.Sc.), a Master of Engineering Management, a Doctor of Philosophy (Ph.D.) in Engineering, an Engineering Doctorate (Eng.D.), or an Engineer's degree. The master's and engineer's degrees may consist of either research, coursework or a mixture of the two. The Doctor of Philosophy and Engineering Doctorate degrees consist of a significant research component and are often viewed as the entry point to academia. In the United Kingdom and some other European countries, Master of Engineering is often considered to be an undergraduate degree of slightly longer duration than the Bachelor of Engineering rather than postgraduate.

Practicing Engineers

Belgian electrical engineers inspecting the rotor of a 40,000 kilowatt turbine of the General Electric Company in New York City

In most countries, a bachelor's degree in engineering represents the first step towards professional certification and the degree program itself is certified by a professional body. After completing a certified degree program the engineer must satisfy a range of requirements (including work experience requirements) before being certified. Once certified the engineer is designated the title of Professional Engineer (in the United States, Canada and South Africa), Chartered Engineer or Incorporated Engineer (in India, Pakistan, the United Kingdom, Ireland and Zimbabwe), Chartered Professional Engineer (in Australia and New Zealand) or European Engineer (in much of the European Union).

The IEEE corporate office is on the 17th floor of 3 Park Avenue in New York City

The advantages of certification vary depending upon location. For example, in the United States and Canada "only a licensed engineer may seal engineering work for public and private clients". This requirement is enforced by state and provincial legislation such as Quebec's Engineers Act. In other countries, no such legislation exists. Practically all certifying bodies maintain a code of ethics that they expect all members to abide by or risk expulsion. In this way these organizations play an important role in maintaining ethical standards for the profession. Even in jurisdictions where certification has little or no legal bearing on work, engineers are subject to contract law. In cases where an engineer's work fails he or she may be subject to the tort of negligence and, in extreme cases, the charge of criminal negligence. An engineer's work must also comply with numerous other rules and regulations such as building codes and legislation pertaining to environmental law.

Professional bodies of note for electrical engineers include the Institute of Electrical and Electronics Engineers (IEEE) and the Institution of Engineering and Technology (IET). The IEEE claims to produce 30% of the world's literature in electrical engineering, has over 360,000 members worldwide and holds over 3,000 conferences annually. The IET publishes 21 journals, has a worldwide membership of over 150,000, and claims to be the largest professional engineering society in Europe. Obsolescence of technical skills is a serious concern for electrical engineers. Membership and participation in technical societies, regular reviews of periodicals in the field and a habit of continued learning are therefore essential to maintaining proficiency. An MIET(Member of the Institution of Engineering and Technology) is recognised in Europe as an Electrical and computer (technology) engineer.

In Australia, Canada, and the United States electrical engineers make up around 0.25% of the labor force.

Tools and Work

From the Global Positioning System to electric power generation, electrical engineers have contributed to the development of a wide range of technologies. They design, develop, test, and supervise the deployment of electrical systems and electronic devices. For example, they may work on the design of telecommunication systems, the operation of electric power stations, the lighting and wiring of buildings, the design of household appliances, or the electrical control of industrial machinery.

Satellite communications is typical of what electrical engineers work on.

Fundamental to the discipline are the sciences of physics and mathematics as these help to obtain both a qualitative and quantitative description of how such systems will work. Today most engineering work involves the use of computers and it is commonplace to use computer-aided design programs when designing electrical systems. Nevertheless, the ability to sketch ideas is still invaluable for quickly communicating with others.

The Shadow robot hand system

Although most electrical engineers will understand basic circuit theory (that is the interactions of elements such as resistors, capacitors, diodes, transistors, and inductors in a circuit), the theories employed by engineers generally depend upon the work they do. For example, quantum mechanics and solid state physics might be relevant to an engineer working on VLSI (the design of integrated circuits), but are largely irrelevant to engineers working with macroscopic electrical systems.

Even circuit theory may not be relevant to a person designing telecommunication systems that use off-the-shelf components. Perhaps the most important technical skills for electrical engineers are reflected in university programs, which emphasize strong numerical skills, computer literacy, and the ability to understand the technical language and concepts that relate to electrical engineering.

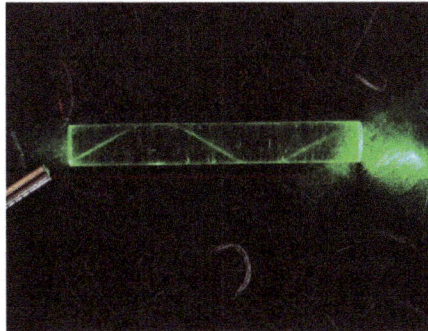

A laser bouncing down an acrylic rod, illustrating the total internal reflection of light in a multi-mode optical fiber.

A wide range of instrumentation is used by electrical engineers. For simple control circuits and alarms, a basic multimeter measuring voltage, current, and resistance may suffice. Where time-varying signals need to be studied, the oscilloscope is also an ubiquitous instrument. In RF engineering and high frequency telecommunications, spectrum analyzers and network analyzers are used. In some disciplines, safety can be a particular concern with instrumentation. For instance, medical electronics designers must take into account that much lower voltages than normal can be dangerous when electrodes are directly in contact with internal body fluids. Power transmission engineering also has great safety concerns due to the high voltages used; although voltmeters may in principle be similar to their low voltage equivalents, safety and calibration issues make them very different. Many disciplines of electrical engineering use tests specific to their discipline. Audio electronics engineers use audio test sets consisting of a signal generator and a meter, principally to measure level but also other parameters such as harmonic distortion and noise. Likewise, information technology have their own test sets, often specific to a particular data format, and the same is true of television broadcasting.

Radome at the Misawa Air Base Misawa Security Operations Center, Misawa, Japan

For many engineers, technical work accounts for only a fraction of the work they do. A lot of time may also be spent on tasks such as discussing proposals with clients, preparing budgets and determining project schedules. Many senior engineers manage a team of technicians or other engineers and for this reason project management skills are important. Most engineering projects involve some form of documentation and strong written communication skills are therefore very important.

The workplaces of engineers are just as varied as the types of work they do. Electrical engineers may be found in the pristine lab environment of a fabrication plant, the offices of a consulting firm or on site at a mine. During their working life, electrical engineers may find themselves supervising a wide range of individuals including scientists, electricians, computer programmers, and other engineers.

Electrical engineering has an intimate relationship with the physical sciences. For instance, the physicist Lord Kelvin played a major role in the engineering of the first transatlantic telegraph cable. Conversely, the engineer Oliver Heaviside produced major work on the mathematics of transmission on telegraph cables. Electrical engineers are often required on major science projects. For instance, large particle accelerators such as CERN need electrical engineers to deal with many aspects of the project: from the power distribution, to the instrumentation, to the manufacture and installation of the superconducting electromagnets.

Computer Engineering

The motherboard used in a HD DVD player, the result of computer engineering efforts.

Computer engineering is a discipline that integrates several fields of electrical engineering and computer science required to develop computer hardware and software. Computer engineers usually have training in electronic engineering (or electrical engineering), software design, and hardware-software integration instead of only software engineering or electronic engineering. Computer engineers are involved in many hardware and software aspects of computing, from the design of individual microcontrollers, microprocessors, personal computers, and supercomputers, to circuit design. This field of engineering not only focuses on how computer systems themselves work, but also how they integrate into the larger picture.

Usual tasks involving computer engineers include writing software and firmware for embedded microcontrollers, designing VLSI chips, designing analog sensors, designing mixed signal circuit boards, and designing operating systems. Computer engineers are also suited for robotics research, which relies heavily on using digital systems to control and monitor electrical systems like motors, communications, and sensors.

In many institutions, computer engineering students are allowed to choose areas of in-depth study in their junior and senior year, because the full breadth of knowledge used in the design and application of computers is beyond the scope of an undergraduate degree. Other institutions may require engineering students to complete one or two years of General Engineering before declaring computer engineering as their primary focus.

History

The first computer engineering degree program in the United States was established at Case Western Reserve University in 1972. As of 2015, there were 238 ABET-accredited computer engineering programs in the US. In Europe, accreditation of computer engineering schools is done by a variety of agencies part of the EQANIE network. Due to increasing job requirements for engineers who can concurrently design hardware, software, firmware, and manage all forms of computer systems used in industry, some tertiary institutions around the world offer a bachelor's degree generally called computer engineering. Both computer engineering and electronic engineering programs include analog and digital circuit design in their curriculum. As with most engineering disciplines, having a sound knowledge of mathematics and science is necessary for computer engineers.

Work

There are two major specialties in computer engineering: software and hardware.

Computer Software Engineering

Computer software engineers develop, design, and test software. Some software engineers design, construct, and maintain computer programs for companies. Some set up networks such as "intranets" for companies. Others make or install new software or upgrade computer systems. Computer software engineers can also work in application design. This involves designing or coding new programs and applications to meet the needs of a business or individual. Computer software engineers can also work as freelancers and sell their software products/applications to an enterprise/individual.

Computer Hardware Engineering

Most computer hardware engineers research, develop, design, and test various computer equipment. This can range from circuit boards and microprocessors to routers. Some update existing computer equipment to be more efficient and work with newer software. Most computer hardware engineers work in research laboratories and high-tech manufacturing firms. Some also work for the federal government. According to BLS, 95% of computer hardware engineers work in metropolitan areas. They generally work full-time. Approximately 33% of their work requires more than 40 hours a week. The median salary for employed qualified computer hardware engineers (2012) was $100,920 per year or $48.52 per hour. Computer hardware engineers held 83,300 jobs in 2012 in the USA.

Specialty Areas

There are many specialty areas in the field of computer engineering.

Coding, Cryptography, and Information Protection

Computer engineers work in coding, cryptography, and information protection to develop new methods for protecting various information, such as digital images and music, fragmentation, copyright infringement and other forms of tampering. Examples include work on wireless communications, multi-antenna systems, optical transmission, and digital watermarking.

Communications and Wireless Networks

Those focusing on communications and wireless networks, work advancements in telecommunications systems and networks (especially wireless networks), modulation and error-control coding, and information theory. High-speed network design, interference suppression and modulation, design and analysis of fault-tolerant system, and storage and transmission schemes are all a part of this specialty.

Compilers and Operating Systems

This specialty focuses on compilers and operating systems design and development. Engineers in this field develop new operating system architecture, program analysis techniques, and new techniques to assure quality. Examples of work in this field includes post-link-time code transformation algorithm development and new operating system development.

Computational Science and Engineering

Computational Science and Engineering is a relatively new discipline. According to the Sloan Career Cornerstone Center, individuals working in this area, "computational methods are applied to formulate and solve complex mathematical problems in engineering and the physical and the social sciences. Examples include aircraft design, the plasma processing of nanometer features on semiconductor wafers, VLSI circuit design, radar detection systems, ion transport through biological channels, and much more".

Computer Networks, Mobile Computing, and Distributed Systems

In this specialty, engineers build integrated environments for computing, communications, and information access. Examples include shared-channel wireless networks, adaptive resource management in various systems, and improving the quality of service in mobile and ATM environments. Some other examples include work on wireless network systems and fast Ethernet cluster wired systems.

Computer Systems: Architecture, Parallel Processing, and Dependability

Engineers working in computer systems work on research projects that allow for reliable, secure, and high-performance computer systems. Projects such as designing processors for multi-threading and parallel processing are included in this field. Other examples of work in this field include development of new theories, algorithms, and other tools that add performance to computer systems.

Computer Vision and Robotics

In this specialty, computer engineers focus on developing visual sensing technology to sense an environment, representation of an environment, and manipulation of the environment. The gathered three-dimensional information is then implemented to perform a variety of tasks. These include, improved human modeling, image communication, and human-computer interfaces, as well as devices such as special-purpose cameras with versatile vision sensors.

Embedded Systems

Examples of devices that use embedded systems.

Individuals working in this area design technology for enhancing the speed, reliability, and performance of systems. Embedded systems are found in many devices from a small FM radio to the space shuttle. According to the Sloan Cornerstone Career Center, ongoing developments in embedded systems include "automated vehicles and equipment to conduct search and rescue, automated transportation systems, and human-robot coordination to repair equipment in space."

Integrated Circuits, VLSI Design, Testing and CAD

This specialty of computer engineering requires adequate knowledge of electronics and electrical systems. Engineers working in this area work on enhancing the speed, reliability, and energy efficiency of next-generation very-large-scale integrated (VLSI) circuits and microsystems. An example of this specialty is work done on reducing the power consumption of VLSI algorithms and architecture.

Signal, Image and Speech Processing

Computer engineers in this area develop improvements in human–computer interaction, including speech recognition and synthesis, medical and scientific imaging, or communications systems. Other work in this area includes computer vision development such as recognition of human facial features.

Education

Most entry-level computer engineering jobs require at least a bachelor's degree in computer engineering. Sometimes a degree in electronic engineering is accepted, due to the similarity of the two fields. Because hardware engineers commonly work with computer software systems, a background in computer programming usually is needed. According to BLS, "a computer engineering major is similar to electrical engineering but with some computer science courses added to the curriculum". Some large firms or specialized jobs require a master's degree.

It is also important for computer engineers to keep up with rapid advances in technology. Therefore, many continue learning throughout their careers. This can be helpful, especially when it comes to learning new skills or improving existing ones. For example, as the relative cost of fixing a bug increases the further along it is in the software development cycle, there can be greater cost savings attributed to developing and testing for quality code as soon as possible in the process, and particularly before release.

Job Outlook in the United States

Computer Software Engineering

According to the U.S. Bureau of Labor Statistics (BLS), "computer applications software engineers and computer systems software engineers are projected to be among the faster than average growing occupations" from 2014-24, with a projected growth rate of 17%. This is down from the 2012 to 2022 BLS estimate of 22% for software developers. And, further down from the 30% 2010 to 2020 BLS estimate. In addition, growing concerns over cyber security add up to put computer software engineering high above the average rate of increase for all fields. However, some of the work will be outsourced in foreign countries. Due to this, job growth will not be as fast as during the last decade, as jobs that would have gone to computer software engineers in the United States would instead go to computer software engineers in countries such as India. In addition the BLS Job Outlook for Computer Programmers, 2014-24 has an -8% (a decline in their words) for those who program computers (i.e. embedded systems) who are not computer application developers.

Computer Hardware Engineering

According to the BLS, Job Outlook employment for computer hardware engineers, 2014-24 is 3% ("Slower than average" in their own words when compared to other occupations)" and is down from 7% for 2012 to 2022 BLS estimate and is further down from 9% in the BLS 2010 to 2020 estimate." Today, computer hardware is somehow equal to Electronic and Computer Engineering (ECE) and has divided to many subcategories, the most significant of them is Embedded system design.

Similar Occupations and Fields

- Computer programming
- Electrical engineering
- Software development
- Systems analyst

Telegraphy

Telegraphy is the long-distance transmission of textual or symbolic (as opposed to verbal or audio) messages without the physical exchange of an object bearing the message. Thus semaphore is a method of telegraphy, whereas pigeon post is not.

Replica of Claude Chappe's optical telegraph on the Litermont near Nalbach, Germany

Telegraphy requires that the method used for encoding the message be known to both sender and receiver. Such methods are designed according to the limits of the signalling medium used. The use of smoke signals, beacons, reflected light signals, and flag semaphore signals are early examples. In the 19th century, the harnessing of electricity led to the invention of electrical telegraphy. The advent of radio in the early 1900s brought about radiotelegraphy and other forms of wireless telegraphy. In the Internet age, telegraphic means developed greatly in sophistication and ease of use, with natural language interfaces that hide the underlying code, allowing such technologies as electronic mail and instant messaging.

Terminology

The word "telegraph" was first coined by the French inventor of the Semaphore line, Claude Chappe, who also coined the word "semaphore".

A "telegraph" is a device for transmitting and receiving messages over long distances, i.e., for telegraphy. The word "telegraph" alone now generally refers to an electrical telegraph.

Wireless telegraphy is also known as "CW", for continuous wave (a carrier modulated by on-off keying), as opposed to the earlier radio technique of using a spark gap.

Contrary to the extensive definition used by Chappe, Morse argued that the term *telegraph* can strictly be applied only to systems that transmit *and* record messages at a distance. This is to be distinguished from *semaphore*, which merely transmits messages. Smoke signals, for instance, are to be considered semaphore, not telegraph. According to Morse, telegraph dates only from 1832, when Pavel Schilling invented one of the earliest electrical telegraphs.

A telegraph message sent by an electrical telegraph operator or telegrapher using Morse code (or a printing telegraph operator using plain text) was known as a *telegram*. A *cablegram* was a message sent by a submarine telegraph cable, often shortened to a *cable* or a *wire*. Later, a *Telex* was a message sent by a Telex network, a switched network of teleprinters similar to a telephone network.

A *wire picture* or *wire photo* was a newspaper picture that was sent from a remote location by a facsimile telegraph. A *diplomatic telegram*, also known as a diplomatic cable, is the term given to a confidential communication between a diplomatic mission and the foreign ministry of its parent country. These continue to be called telegrams or cables regardless of the method used for transmission.

History

Even though early telegraphic precedents, such as signalling through the lighting of pyres, have existed since ancient times, long-distance telegraphy (transmission of complex messages) started in 1792 in the form of semaphore lines, or optical telegraphs, that sent messages to a distant observer through line-of-sight signals. Commercial electrical telegraphs were introduced from 1837.

Optical Telegraph

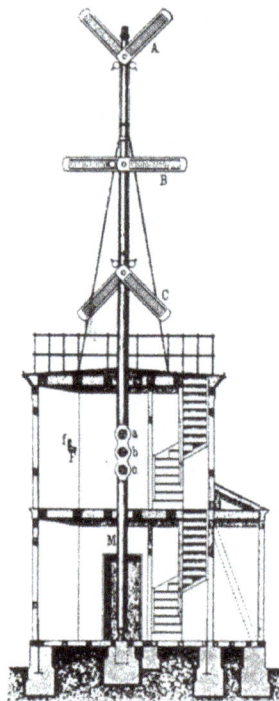

Construction schematic of a Prussian optical telegraph (or semaphore) tower, C. 1835

The first telegraphs came in the form of optical telegraph, including the use of smoke signals, beacons or reflected light, which have existed since ancient times. Early proposals for an optical telegraph system were made to the Royal Society by Robert Hooke in 1684 and were first implemented on an experimental level by Sir Richard Lovell Edgeworth in 1767.

The first successful semaphore network was invented by Claude Chappe and operated in France from 1793 through 1846.

During 1790–1795, at the height of the French Revolution, France needed a swift and reliable communication system to thwart the war efforts of its enemies. In 1790, the Chappe brothers set about devising a system of communication that would allow the central government to receive intelligence and to transmit orders in the shortest possible time. On 2 March 1791 at 11 am, they sent the message "si vous réussissez, vous serez bientôt couverts de gloire" (If you succeed, you will soon bask in glory) between Brulon and Parce, a distance of 16 kilometres (9.9 mi). The first means used a combination of black and white panels, clocks, telescopes, and codebooks to send their message.

Demonstration of the semaphore

In 1792 Claude was appointed *Ingénieur-Télégraphiste* and charged with establishing a line of stations between Paris and Lille, a distance of 230 kilometres (about 143 miles). It was used to carry dispatches for the war between France and Austria. In 1794, it brought news of a French capture of Condé-sur-l'Escaut from the Austrians less than an hour after it occurred.

The Prussian system was put into effect in the 1830s. However they were highly dependent on good weather and daylight to work and even then could accommodate only about two words per minute. . The last commercial semaphore link ceased operation in Sweden in 1880. As of 1895, France still operated coastal commercial semaphore telegraph stations, for ship-to-shore communication.

Electrical Telegraphs

Early Developments

The first suggestion for using electricity as a means of communication appeared in the "Scots Magazine" in 1753. Using one wire for each letter of the alphabet, a message could be transmitted by connecting the wire terminals in turn to an electrostatic machine, and observing the deflection of pith balls at the far end. Telegraphs employing electrostatic attraction were the basis of early experiments in electrical telegraphy in Europe, but were abandoned as being impractical and were never developed into a useful communication system.

One very early experiment in electrical telegraphy was an *electrochemical telegraph* created by the German physician, anatomist, and inventor Samuel Thomas von Sömmering in 1809, based on an earlier, less robust design of 1804 by Spanish polymath and scientist Francisco Salva Campillo. Both their designs employed multiple wires (up to 35) in order to visually represent most Latin letters and numerals. Thus, messages could be conveyed electrically up to a few kilometers (in von Sömmering's design), with each of the telegraph receiver's wires immersed in a separate glass tube of acid. As an electric current was applied by the sender representing each digit of a message, it would at the recipient's end electrolyse the acid in its corresponding tube, releasing a stream of hydrogen bubbles next to its associated letter or numeral. The telegraph receiver's operator would visually observe the bubbles and could then record the transmitted message, albeit at a very low baud rate.

The first working telegraph was built by the English inventor Francis Ronalds in 1816 and used static electricity. At the family home on Hammersmith Mall, he set up a complete subterranean system in a 175 yard long trench as well as an eight mile long overhead telegraph. The lines were connected at both ends to clocks marked with the letters of the alphabet and electrical impulses sent along the wire were used to transmit messages. Offering his invention to the Admiralty in July 1816, it was rejected as "wholly unnecessary". His account of the scheme and the possibilities of rapid global communication in *Descriptions of an Electrical Telegraph and of some other Electrical Apparatus* was the first published work on electric telegraphy and even described the risk of signal retardation due to induction. Elements of Ronalds' design were utilised in the subsequent commercialisation of the telegraph over 20 years later.

Pavel Schilling, an early pioneer of electrical telegraphy

An early electromagnetic telegraph design was created by Russian diplomat Pavel Schilling in 1832. He set it up in his apartment in St. Petersburg and demonstrated the long-distance transmission of signals by positioning two telegraphs of his invention in two different rooms of his apartment. Schilling was the first to put into practice the idea of a binary system of signal transmissions.

Carl Friedrich Gauss and Wilhelm Weber built the first electromagnetic telegraph used for *regular* communication in 1833 in Göttingen, connecting Göttingen Observatory and the Institute of Physics, covering a distance of about 1 km. The setup consisted of a coil that could be moved up and down over the end of two magnetic steel bars. The resulting induction current was transmitted through two wires to the receiver, consisting of a galvanometer. The direction of the current could be reversed by commuting the two wires in a special switch. Therefore, Gauss and Weber chose to encode the alphabet in a binary code, using positive and negative currents as the two states.

Commercial Telegraphy

Telegraph networks were expensive to build, but financing was readily available, especially from London bankers. By 1852, National systems were in operation major countries as follows:

- United States, 20 companies with 23,000 miles of wire.

- Great Britain, Cooke-Wheatstone company and minor companies, with 2200 miles of wire.

- Prussia, 1400 miles of wire, Siemens system.

- Austria, 1000 miles of wire, Siemens system.

- Canada, 900 miles of wire

- France, 700 miles of wire; optical systems dominant.

- Australia, 0. First line opened in 1854.

Cooke and Wheatstone System

The first commercial electrical telegraph was co-developed by Sir William Fothergill Cooke and Charles Wheatstone. In May 1837 they patented the Cooke and Wheatstone system, which used a number of needles on a board that could be moved to point to letters of the alphabet. The patent recommended a five-needle system, but any number of needles could be used depending on the number of characters it was required to code. A four-needle system was installed between Euston and Camden Town in London on a rail line being constructed by Robert Stephenson between London and Birmingham. It was successfully demonstrated on 25 July 1837. Euston needed to signal to an engine house at Camden Town to start hauling the locomotive up the incline. As at Liverpool, the electric telegraph was in the end rejected in favour of a pneumatic system with whistles.

Cooke and Wheatstone's five-needle, six-wire telegraph

Cooke and Wheatstone had their first commercial success with a system installed on the Great Western Railway over the 13 miles (21 km) from Paddington station to West Drayton in 1838, the first commercial telegraph in the world. This was a five-needle, six-wire system. The cables were originally installed underground in a steel conduit. However, the cables soon began to fail as a result of deteriorating insulation and were replaced with uninsulated wires on poles. As an interim measure, a two-needle system was used with three of the remaining working underground wires, which despite using only two needles had a greater number of codes. But when the line was extended to Slough in 1843, a one-needle, two-wire system was installed.

From this point the use of the electric telegraph started to grow on the new railways being built from London. The London and Blackwall Railway (another rope-hauled application) was equipped with the Cooke and Wheatstone telegraph when it opened in 1840, and many others followed. The one-needle telegraph proved highly successful on British railways, and 15,000 sets were still in use at the end of the nineteenth century. Some remained in service in the 1930s. In September 1845 the financier John Lewis Ricardo and Cooke formed the Electric Telegraph Company, the first public telegraphy company in the world. This company bought out the Cooke and Wheatstone patents and solidly established the telegraph business.

As well as the rapid expansion of the use of the telegraphs along the railways, they soon spread into the field of mass communication with the instruments being installed in post offices across the country. The era of mass personal communication had begun.

Morse System

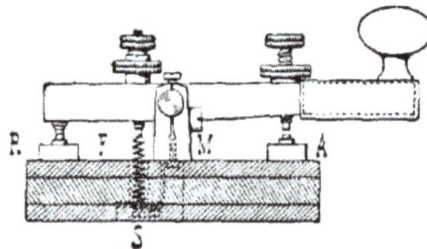

Fig. 6.

A Morse key

An electrical telegraph was independently developed and patented in the United States in 1837 by Samuel Morse. His assistant, Alfred Vail, developed the Morse code signalling alphabet with Morse. The first telegram in the United States was sent by Morse on 11 January 1838, across two miles (3 km) of wire at Speedwell Ironworks near Morristown, New Jersey, although it was only later, in 1844, that he sent the message "WHAT HATH GOD WROUGHT" from the Capitol in Washington to the old Mt. Clare Depot in Baltimore. From then on, commercial telegraphy took off in America with lines linking all the major metropolitan centres on the East Coast within the next decade. The overland telegraph connected the west coast of the continent to the east coast by 24 October 1861, bringing an end to the Pony Express.

The Morse telegraphic apparatus was officially adopted as the standard for European telegraphy in 1851. Only Great Britain with its extensive overseas empire kept the needle telegraph of Cooke and Wheatstone. In 1858, Morse introduced wired communication to Latin America when he established a telegraph system in Puerto Rico, then a Spanish Colony. The line was inaugurated on March 1, 1859, in a ceremony flanked by the Spanish and American flags.

Another early system was that of Edward Davy, who demonstrated his in Regent's Park in 1837 and was granted a patent on 4 July 1838. He also developed an electric relay.

1Telegraphy was driven by the need to reduce sending costs, either in hand-work per message or by increasing the sending rate. While many experimental systems employing moving pointers and various electrical encodings proved too complicated and unreliable, a successful advance in the sending rate was achieved through the development of telegraphese.

The first message is received by the Submarine Telegraph Company in London from Paris on the Foy-Breguet instrument in 1851

The first system that didn't require skilled technicians to operate, was Sir Charles Wheatstone's ABC system in 1840 where the letters of the alphabet were arranged around a clock-face, and the signal caused a needle to indicate the letter. This early system required the receiver to be present in real time to record the message and it reached speeds of up to 15 words a minute.

Before Telegraphy, a letter by post from London took	
days	to reach
12	New York in USA
13	Alexandria in Egypt
19	Constantinople in Ottoman Turkey
33	Bombay in India
44	Calcutta in Bengal
45	Singapore
57	Shanghai in China
73	Sydney in Australia

In 1846, Alexander Bain patented a chemical telegraph in Edinburgh. The signal current made a readable mark on a moving paper tape soaked in a mixture of ammonium nitrate and potassium ferrocyanide, which gave a blue mark when a current was passed through it.

A Baudot keyboard, 1884

David Edward Hughes invented the printing telegraph in 1855; it used a keyboard of 26 keys for the alphabet and a spinning type wheel that determined the letter being transmitted by the length of time that had elapsed since the previous transmission. The system allowed for automatic recording on the receiving end. The system was very stable and accurate and became the accepted around the world.

The next improvement was the Baudot code of 1874. French engineer Émile Baudot patented a printing telegraph in which the signals were translated automatically into typographic characters. Each character was assigned a unique code based on the sequence of just five contacts. Operators had to maintain a steady rhythm, and the usual speed of operation was 30 words per minute.

By this point reception had been automated, but the speed and accuracy of the transmission was still limited to the skill of the human operator. The first practical automated system was patented by Charles Wheatstone, the original inventor of the telegraph. The message (in Morse code) was typed onto a piece of perforated tape using a keyboard-like device called the 'Stick Punch'. The transmitter automatically ran the tape through and transmitted the message at the then exceptionally high speed of 70 words per minute.

Teleprinters

Phelps' Electro-motor Printing Telegraph from circa 1880, the last and most advanced telegraphy mechanism designed by George May Phelps

Teleprinters were invented in order to send and receive messages without the need for operators trained in the use of Morse code. A system of two teleprinters, with one operator trained to use a typewriter, replaced two trained Morse code operators. The teleprinter system improved message speed and delivery time, making it possible for messages to be flashed across a country with little manual intervention.

Early teleprinters used the ITA-1 Baudot code, a five-bit code. This yielded only thirty-two codes, so it was over-defined into two "shifts", "letters" and "figures". An explicit, unshared shift code prefaced each set of letters and figures. In 1901 Baudot's code was modified by Donald Murray and Around 1930, the CCITT introduced the International Telegraph Alphabet No. 2 (ITA2) code as an international standard.

A Siemens T100 Telex machine

By 1935, message routing was the last great barrier to full automation. Large telegraphy providers began to develop systems that used telephone-like rotary dialling to connect teletypewriters. These machines were called "Telex" (TELegraph EXchange). Telex machines first performed rotary-telephone-style pulse dialling for circuit switching, and then sent data by Baudot code. This "type A" Telex routing functionally automated message routing.

Telex began in Germany as a research and development program in 1926 that became an operational teleprinter service in 1933. The service was operated by the Reichspost (Reich postal service) and had a speed of 50 baud - approximately 66 words-per-minute.

At the rate of 45.45 (±0.5%) baud—considered speedy at the time—up to 25 telex channels could share a single long-distance telephone channel by using *voice frequency telegraphy multiplexing*, making telex the least expensive method of reliable long-distance communication.

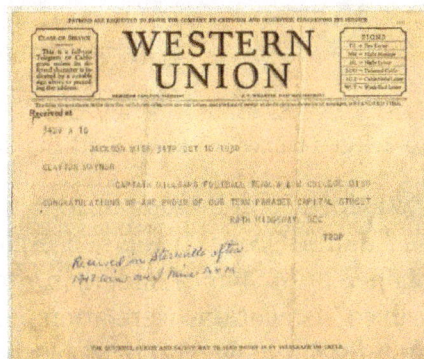

Western Union telegram circa 1930

Automatic teleprinter exchange service was introduced into Canada by CPR Telegraphs and CN Telegraph in July 1957 and in 1958, Western Union started to build a Telex network in the United States.

Beginning in 1956 telegrams begun to be transmitted over the Telex network using the ITU F.20 standard named Gentex in order to lower the costs for some European telecommunications companies by allowing the sending telegraph station to connect directly to the receiving station.

Oceanic Telegraph Cables

Soon after the first successful telegraph systems were operational, the possibility of transmitting messages across the sea by way of submarine communications cables was first mooted. One of the primary technical challenges was to insulate the submarine cable sufficiently to prevent the current from leaking out into the water. In 1842, a Scottish surgeon William Montgomerie introduced Gutta-percha, the adhesive juice of the *Palaquium gutta* tree, to Europe. Michael Faraday and Wheatstone soon discovered the merits of gutta-percha as an insulator, and in 1845, the latter suggested that it should be employed to cover the wire which was proposed to be laid from Dover to Calais. It was tried on a wire laid across the Rhine between Deutz and Cologne. In 1849, C.V. Walker, electrician to the South Eastern Railway, submerged a two-mile wire coated with gutta-percha off the coast from Folkestone, which was tested successfully.

John Watkins Brett, an engineer from Bristol, sought and obtained permission from Louis-Philippe in 1847 to establish telegraphic communication between France and England. The first undersea cable was laid in 1850 and connected London with Paris. After an initial exchange of greetings between Queen Victoria and President Napoleon, it was almost immediately severed by a French fishing vessel. The line was relaid the next year and then followed by connections to Ireland and the Low Countries.

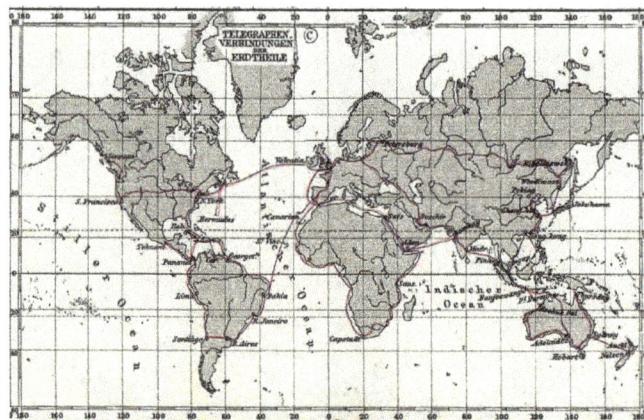

Major telegraph lines across the Earth in 1891

The Atlantic Telegraph Company was formed in London in 1856 to undertake to construct a commercial telegraph cable across the Atlantic Ocean. It was successfully completed on 18 July 1866 by the ship SS *Great Eastern*, captained by Sir James Anderson after many mishaps along the way. Earlier transatlantic submarine cables installations were attempted in 1857, 1858 and 1865. The 1857 cable only operated intermittently for a few days or weeks before it failed. The study of underwater telegraph cables accelerated interest in mathematical analysis of very long transmission lines. An overland telegraph from Britain to India was first connected in 1866 but was unreliable so a submarine telegraph cable was connected in 1870. Several telegraph companies were combined to form the *Eastern Telegraph Company* in 1872.

Australia was first linked to the rest of the world in October 1872 by a submarine telegraph cable at Darwin. This brought news reportage from the rest of the world. The telegraph across the Pacific was completed in 1902, finally encircling the world.

From the 1850s until well into the 20th century, British submarine cable systems dominated the world system. This was set out as a formal strategic goal, which became known as the All Red Line. In 1896, there were thirty cable laying ships in the world and twenty-four of them were owned by British companies. In 1892, British companies owned and operated two-thirds of the world's cables and by 1923, their share was still 42.7 percent. During World War I, Britain's telegraph communications were almost completely uninterrupted, while it was able to quickly cut Germany's cables worldwide.

Later Technology

Facsimile

Alexander Bain's facsimile machine, 1850

In 1843 Scottish inventor Alexander Bain invented a device that could be considered the first facsimile machine. He called his invention a "recording telegraph". Bain's telegraph was able to transmit images by electrical wires. Frederick Bakewell made several improvements on Bain's design and demonstrated a telefax machine. In 1855 an Italian abbot, Giovanni Caselli, also created an electric telegraph that could transmit images. Caselli called his invention "Pantelegraph". Pantelegraph was successfully tested and approved for a telegraph line between Paris and Lyon.

In 1881, English inventor Shelford Bidwell constructed the *scanning phototelegraph* that was the first telefax machine to scan any two-dimensional original, not requiring manual plotting or drawing. Around 1900, German physicist Arthur Korn invented the *Bildtelegraph*, widespread in continental Europe especially, since a widely noticed transmission of a wanted-person photograph from Paris to London in 1908, used until the wider distribution of the radiofax. Its main competitors were the *Bélinographe* by Édouard Belin first, then since the 1930s the *Hellschreiber*, invented in 1929 by German inventor Rudolf Hell, a pioneer in mechanical image scanning and transmission.

Wireless Telegraphy

Post Office Engineers inspect Marconi's equipment on Flat Holm, May 1897

The late 1880s through the 1890s saw the discovery and then development of a newly understood phenomenon into a form of wireless telegraphy, called *Hertzian wave* wireless telegraphy, radio-telegraphy, or (later) simply "radio". Between 1886 and 1888 Heinrich Rudolf Hertz published the results of his experiments where he was able to transmit electromagnetic waves (radio waves) through the air, proving James Clerk Maxwell's 1873 theory of electromagnetic radiation. Many scientists and inventors experimented with this new phenomenon but the general consensus was that these new waves (similar to light) would be just as short range as light, and therefore useless for long range communication.

At the end of 1894 the young Italian inventor Guglielmo Marconi began working on the idea of building a commercial wireless telegraphy system based on the use of Hertzian waves (radio waves), a line of inquiry that he noted other inventors did not seem to be pursuing. Building on the ideas of previous scientists and inventors Marconi re-engineered their apparatus by trial and error attempting to build a radio based wireless telegraphic system that would function the same as wired telegraphy. He would work on the system through 1895 in his lab and then in field tests making improvements to extend its range. After many breakthroughs, including applying the wired telegraphy concept of grounding the transmitter and receiver, Marconi was able, by early 1896, to transmit radio far beyond the short ranges that had been predicted. Having failed to interest the Italian government, the 22-year-old inventor brought his telegraphy system to Britain in 1896 and met William Preece, a Welshman, who was a major figure in the field and Chief Engineer of the General Post Office. A series of demonstrations for the British government followed—by March 1897, Marconi had transmitted Morse code signals over a distance of about 6 kilometres (3.7 mi) across Salisbury Plain.

Marconi watching associates raising the kite (a "Levitor" by B.F.S. Baden-Powell) used to lift the antenna at St. John's, Newfoundland, December 1901

On 13 May 1897, Marconi, assisted by George Kemp, a Cardiff Post Office engineer, transmitted the first wireless signals over water to Lavernock (near Penarth in Wales) from Flat Holm. The message sent was "ARE YOU READY". From his Fraserburgh base, he transmitted the first long-distance, cross-country wireless signal to Poldhu in Cornwall.[when?] His star rising, he was soon sending signals across The English channel (1899), from shore to ship (1899) and finally across the Atlantic (1901). A study of these demonstrations of radio, with scientists trying to figure out how a phenomenon predicted to have a short range could transmit "over the horizon", led to the discovery of a radio reflecting layer in the Earth's atmosphere in 1902, later called the ionosphere.

Radiotelegraphy proved effective for rescue work in sea disasters by enabling effective communication between ships and from ship to shore. In 1904 Marconi began the first commercial service to transmit nightly news summaries to subscribing ships, which could incorporate them into their on-board newspapers. A regular transatlantic radio-telegraph service was finally begun on 17 October 1907. Notably, Marconi's apparatus was used to help rescue efforts after the sinking of *Titanic*. Britain's postmaster-general summed up, referring to the *Titanic* disaster, "Those who have been saved, have been saved through one man, Mr. Marconi...and his marvellous invention."

Internet

Around 1965, DARPA commissioned a study of decentralized switching systems. Some of the ideas developed in this study provided inspiration for the development of the ARPANET packet switching research network, which later grew to become the public Internet.

As the PSTN became a digital network, T-carrier "synchronous" networks became commonplace in the U.S. A T1 line has a "frame" of 193 bits that repeats 8000 times per second. The first bit, called the "sync" bit, alternates between 1 and 0 to identify the start of the frames. The rest of the frame provides 8 bits for each of 24 separate voice or data channels. Customarily, a T-1 link is sent over a balanced twisted pair, isolated with transformers to prevent current flow. Europeans adopted a similar system (E-1) of 32 channels (with one channel for frame synchronisation).

Later, SONET and SDH were adapted to combine carrier channels into groups that could be sent over optic fiber. The capacity of an optic fiber is often extended with wavelength division multiplexing, rather than rerigging new fibre. Rigging several fibres in the same structures as the first fibre is usually easy and inexpensive, and many fibre installations include unused spare "dark fibre", "dark wavelengths", and unused parts of the SONET frame, so-called "virtual channels".

In 2002, the Internet was used by Kevin Warwick at the University of Reading to communicate neural signals, in purely electronic form, telegraphically between the nervous systems of two humans, potentially opening up a new form of communication combining the Internet and telegraphy.

In 2006, a well-defined communication channel used for telegraphy was established by the SONET standard OC-768, which sent about 40 gigabits per second.

The theoretical maximum capacity of an optic fiber is more than 10^{12} bits (one terabit or one trillion bits) per second. In 2006, no existing encoding system approached this theoretical limit, even with wavelength division multiplexing.

Since the Internet operates over any digital transmission medium, further evolution of telegraphic technology will be effectively concealed from users.

E-mail

E-mail was first invented for CTSS and similar time sharing systems of the era in the mid-1960s. At first, e-mail was possible only between different accounts on the same computer (typically a main-frame). ARPANET allowed different computers to be connected to allow e-mails to be relayed from computer to computer, with the first ARPANET e-mail being sent in 1971. Multics also pioneered instant messaging between computer users in the mid-1970s. With the growth of the Internet, e-mail began to be possible between any two computers with access to the Internet.

Various private networks like UUNET (founded 1987), the Well (1985), and GEnie (1985) had e-mail from the 1970s, but subscriptions were quite expensive for an individual, US$25 to US$50 per month, just for e-mail. Internet use was then largely limited to government, academia and other government contractors until the net was opened to commercial use in the 1980s.

By the early 1990s, modems made e-mail a viable alternative to Telex systems in a business en-vironment. But individual e-mail accounts were not widely available until local Internet service providers were in place, although demand grew rapidly, as e-mail was seen as the Internet's killer app. It allowed anyone to email anyone, whereas previously, different system had been walled off from each other, such that America Online subscribers could email only other America Online subscribers, Compuserve subscribers could email only other Compuserve subscribers, etc. The broad user base created by the demand for e-mail smoothed the way for the rapid acceptance of the World Wide Web in the mid-1990s. Fax machines were another technology that helped displace the telegram.

On Monday, 12 July 1999, a final telegram was sent from the National Liberty Ship Memorial, the SS Jeremiah O'Brien, in San Francisco Bay to President Bill Clinton in the White House. Officials of Globe Wireless reported that "The message was 95 words, and it took six or eight minutes to copy it." They then transmitted the message to the White House via e-mail. That event was also used to mark the final commercial U.S. ship-to-shore telegraph message transmitted from North America by Globe Wireless, a company founded in 1911. Sent from its wireless station at Half Moon Bay, California, the sign-off message was a repeat of Samuel F. B. Morse's message 155 years earlier, "What hath God wrought?"

21st Century Decline

- In Australia, Australia Post closed its telegram service on 7 March 2011. In the Victorian town of Beechworth, visitors can send telegrams to family members or friends from the Beechworth Telegraph Station.

- In Bahrain, Batelco still offers telegram services. They are thought to be more formal than an email or a fax, but less so than a letter. So should a death or anything of importance occur, telegrams would be sent.

- In Belgium, Belgacom still offers telegram services within the country and internationally. It sent 63.000 telegrams in 2010

- In Canada, Telegrams Canada still offers telegram services. AT&T Canada (previously CNCP Telecommunications) had discontinued its telegram service in 2001 and later became MTS Allstream.

- In France, Orange S.A. still offers a telegram service, although not transmitted by telegraph any more.

- In Germany, Deutsche Post delivers telegrams the next day as ordinary mail. Deutsche Post discontinued service to foreign countries on 31 December 2000. A private firm, TelegrammDirekt.de, offers delivery in Germany and service to a number of foreign countries.

- In Hungary, Magyar Posta still offers (national only) telegram services.

- In India, state-owned BSNL discontinued telegram services from 15 July 2013. Telegrams to foreign countries had been discontinued in May 2013.

- In Iran, telex services are still provided by Telecommunication Infrastructure Company of I.R.Iran.

- In Ireland, Eircom – the country's largest telecommunication company and former PTT – formally discontinued telegram service on 30 July 2002.

- In Israel, the Israel Postal Company still offers telegram services. Telegrams may be sent via the internet or by a telephone operator. Illustrated telegrams are available for special occasions.

- In Italy, Poste Italiane still offers telegram services. Around 2.5 million telegrams are sent annually, primarily for births, weddings, and funerals.

- In Japan, NTT provides a telegram (*denpou*) service used mainly for special occasions such as weddings, funerals, graduations, etc. Local offices offer telegrams printed on special decorated paper and envelopes.

- In Lithuania, telegram service was closed by the only provider Teo LT on 15 October 2007.

- In Malaysia, Telekom Malaysia has ceased its telegram service effective 1 July 2012.

- In Mexico, telegrams are still used as a low-cost service for people who cannot afford or do not have access to e-mail.

- In Nepal, Nepal Telecom closed its telegram service on 1 January 2009.

- In the Netherlands, the telegram service was sold by KPN to the Swiss-based company Unitel Telegram Services in 2001.

- In New Zealand, New Zealand Post closed its telegram service in 1999. It later reinstated the service in 2003 for use only by business customers, primarily for debt collection or other important business notices.

- In Pakistan, the Pakistan Telecommunication Company Ltd ceased telegram services on 27 January 2006.

- In the Philippines, telegram services by the government's Telecommunications Office or

Tanggapan ng Telekomunikasyon ceased on 20 September 2013. The last telegram was sent on that day at 3:15 PM.

- In Russia, Central Telegraph (subsidiary of national operator Rostelecom) still offers telegram service. "Regular" or "Urgent" telegrams can be sent to any address in Russia and other countries. So called "Stylish" telegrams printed on an artistic postcards are also available.

- In Serbia, JP Pošta Srbije Beograd, the state-owned post, provides a telegram service. It is commonly used to express condolences, official notifications of death or to congratulate anniversaries, births, graduations etc. Telegrams may be sent by using special telephone number or directly at the post office. Telegrams are delivered on the same day for recipients in territories covered by post offices with telegram delivery service and are delivered as regular mail for post offices which do not have telegram delivery service. In internal traffic, length of message is limited to 800 characters and is charged at flat rate, while in international traffic telegrams are charged by word. International delivery is possible for recipients in Croatia, Slovenia, Montenegro, Bosnia and Herzegovina, and Macedonia.

- In Slovakia, the Slovak post closed its telegram service on 1 January 2007.

- In Slovenia, Pošta Slovenije d.o.o. (Slovenian Post) provides a telegram service still commonly used for special occasions such as births, anniversaries, condolences, graduations, etc. It is considered more formal than email or SMS. Telegrams are usually printed in a typewriter font on greeting or condolences cards delivered in a specific yellow envelope. It is also possible to send gifts (e.g. chocolates, wine, plush toys, flowers) together with a message. The telegrams can be sent from local post offices, over the phone or online to addresses in Slovenia only.

- In Sweden, Telia ceased telegraph services in 2002.

- In Switzerland, Unitel Telegram Services took over telegram services from the national PTTs. Telegrams can still be sent to and from most countries.

- In Thailand, Thailand Post ceased its telegram service on 30 April 2008, at 20.00 local time.

- In the United Kingdom, the international telegram service formerly provided by British Telecom was sold in 2003 to an independent company, Telegrams Online, which promotes the use of telegrams as a retro greeting card or invitation.

- In the United States, Western Union closed its telegram service on 27 January 2006.

Social Implications

Prior to the electrical telegraph, nearly all information was limited to traveling at the speed of a human or animal. The telegraph freed communication from the constraints of space and time and revolutionized the global economy and society. By the end of the 19th century, the telegraph was becoming an increasingly common medium of communication for ordinary people. The telegraph isolated the message (information) from the physical movement of objects or the process.

Telegraphy facilitated the growth of organizations "in the railroads, consolidated financial and commodity markets, and reduced information costs within and between firms". This immense growth in the business sectors influenced society to embrace the use of telegrams.

Worldwide telegraphy changed the gathering of information for news reporting. Messages and information would now travel far and wide, and the telegraph demanded a language "stripped of the local, the regional; and colloquial", to better facilitate a worldwide media language. Media language had to be standardized, which led to the gradual disappearance of different forms of speech and styles of journalism and storytelling.

Newspaper Names

Numerous newspapers in various countries, such as *The Daily Telegraph* in Britain, *The Telegraph* in India, and *De Telegraaf* in the Netherlands, were given names which include the word "telegraph" due to their having received news by means of electric telegraphy. Some of these names are retained even though more sophisticated means are now used.

Telegram Length

The average length of a telegram in the 1900s in the US was 11.93 words, more than half of the messages were 10 words or fewer.

According to another study the mean length of the telegrams sent in the UK before 1950 was 14.6 words or 78.8 characters.

For German telegrams the mean length is 11.5 words or 72.4 characters. At the end of the 19th century the average length of a German telegram was calculated as 14.2 words.

Informatics

Informatics is the science of information and computer information systems. As an academic field it involves the practice of information processing, and the engineering of information systems. The field considers the interaction between humans and information alongside the construction of interfaces, organisations, technologies and systems. It also develops its own conceptual and theoretical foundations and utilizes foundations developed in other fields. As such, the field of informatics has great breadth and encompasses many individual specializations, including disciplines of computer science, information systems, information technology and statistics. Since the advent of computers, individuals and organizations increasingly process information digitally. This has led to the study of informatics with computational, mathematical, biological, cognitive and social aspects, including study of the social impact of information technologies.

Etymology

In 1956 the German computer scientist Karl Steinbuch coined the word *Informatik* by publishing a paper called *Informatik: Automatische Informationsverarbeitung* ("Informatics: Automatic Information Processing"). The English term *Informatics* is sometimes understood as meaning the same as computer science. The German word *Informatik* is usually translated to English as *computer science*.

The French term *informatique* was coined in 1962 by Philippe Dreyfus together with various translations—informatics (English), also proposed independently and simultaneously by Walter F. Bauer and associates who co-founded *Informatics Inc.*, and *informatica* (Italian, Spanish, Romanian, Portuguese, Dutch), referring to the application of computers to store and process information.

The term was coined as a combination of "information" and "automatic" to describe the science of automating information interactions. The morphology—*informat*-ion + -*ics*—uses "the accepted form for names of sciences, as conics, linguistics, optics, or matters of practice, as economics, politics, tactics", and so, linguistically, the meaning extends easily to encompass both the science of information and the practice of information processing.

History

The culture of library science promotes policies and procedure for managing information that fosters the relationship between library science and the development of information science to provides benefits for health informatics development; which is traced to the 1950's with the beginning of computer uses in healthcare (Nelson & Staggers p.4). Early practitioners interested in the field soon learned that; there was no formal education programs set up to educate them on the informatics science until during the late 1960's and early 1970's. Professional development begins to emerge, playing a significant role in the development of health informatics (Nelson &Staggers p.7) According to Imhoff et al., 2001. Healthcare informatics is not only the application of computer technology to problems in healthcare but covers all aspects of generation, handling, communication, storage, retrieval, management, analysis, discovery, and synthesis of data information and knowledge in the entire scope of healthcare. Furthermore, they stated that the primary goal of health informatics can be distinguished as follows: To provide solution for problems related to data, information, and knowledge processing. To study general principles of processing data information and knowledge in medicine and healthcare.

Reference Imhoff, M., Webb. A,.&Goldschmidt, A., (2001). Health Informatics. Intensive Care Med, 27: 179-186. doi:10.1007//s001340000747.

Nelson, R. & Staggers, N. Health Informatics: An Interprofessional Approach. St. Louis: Mosby, 2013. Print. (p.4,7)

This new term was adopted across Western Europe, and, except in English, developed a meaning roughly translated by the English 'computer science', or 'computing science'. Mikhailov advocated the Russian term *informatika* (1966), and the English *informatics* (1967), as names for the *theory of scientific information*, and argued for a broader meaning, including study of the use of information technology in various communities (for example, scientific) and of the interaction of technology and human organizational structures.

Informatics is the discipline of science which investigates the structure and properties (not specific content) of scientific information, as well as the regularities of scientific information activity, its theory, history, methodology and organization.

Usage has since modified this definition in three ways. First, the restriction to scientific information is removed, as in business informatics or legal informatics. Second, since most information is now digitally stored, computation is now central to informatics. Third, the representation, processing and communication of information are added as objects of investigation, since they have been recognized as fundamental to any scientific account of information. Taking *information* as the central focus of study distinguishes *informatics* from *computer science*. Informatics includes the study of biological and social mechanisms of information processing whereas computer science focuses on the digital computation. Similarly, in the study of representation and communication, informatics is indifferent to the substrate that carries information. For example, it encompasses the study of communication using gesture, speech and language, as well as digital communications and networking.

In the English-speaking world the term *informatics* was first widely used in the compound medical informatics, taken to include "the cognitive, information processing, and communication tasks of medical practice, education, and research, including information science and the technology to support these tasks". Many such compounds are now in use; they can be viewed as different areas of "*applied informatics*". Indeed, "In the U.S., however, informatics is linked with applied computing, or computing in the context of another domain."

Informatics encompasses the study of systems that represent, process, and communicate information. However, the theory of computation in the specific discipline of theoretical computer science, which evolved from Alan Turing, studies the notion of a complex system regardless of whether or not information actually exists. Since both fields process information, there is some disagreement among scientists as to field hierarchy; for example Arizona State University attempted to adopt a broader definition of informatics to even encompass cognitive science at the launch of its School of Computing and Informatics in September 2006.

A broad interpretation of *informatics*, as "the study of the structure, algorithms, behaviour, and interactions of natural and artificial computational systems," was introduced by the University of Edinburgh in 1994 when it formed the grouping that is now its School of Informatics. This meaning is now (2006) increasingly used in the United Kingdom.

The 2008 Research Assessment Exercise, of the UK Funding Councils, includes a new, *Computer Science and Informatics*, unit of assessment (UoA), whose scope is described as follows:

The UoA includes the study of methods for acquiring, storing, processing, communicating and reasoning about information, and the role of interactivity in natural and artificial systems, through the implementation, organisation and use of computer hardware, software and other resources. The subjects are characterised by the rigorous application of analysis, experimentation and design.

Academic Schools and Departments

Academic research in the informatics area can be found in a number of disciplines such as computer science, information technology, Information and Computer Science, information system, business information management and health informatics.

In France, the first degree level qualifications in Informatics (computer science) appeared in the mid-1960s.

In English-speaking countries, the first example of a degree level qualification in Informatics occurred in 1982 when Plymouth Polytechnic (now the University of Plymouth) offered a four-year BSc(Honours) degree in Computing and Informatics – with an initial intake of only 35 students. The course still runs today making it the longest available qualification in the subject.

At the Indiana University School of Informatics (Bloomington, Indianapolis and Southeast), informatics is defined as "the art, science and human dimensions of information technology" and "the study, application, and social consequences of technology." It is also defined in Informatics 101, Introduction to Informatics as "the application of information technology to the arts, sciences, and professions." These definitions are widely accepted in the United States, and differ from British usage in omitting the study of natural computation.

Texas Woman's University places its informatics degrees in its department of Mathematics and Computer Science within the College of Arts & Sciences, though it offers interdisciplinary Health Informatics degrees. Informatics is presented in a generalist framework, as evidenced by their definition of informatics ("Using technology and data analytics to derive meaningful information from data for data and decision driven practice in user centered systems"), though TWU is also known for its nursing and health informatics programs.

At the University of California, Irvine Department of Informatics, informatics is defined as "the interdisciplinary study of the design, application, use and impact of information technology. The discipline of informatics is based on the recognition that the design of this technology is not solely a technical matter, but must focus on the relationship between the technology and its use in real-world settings. That is, informatics designs solutions in context, and takes into account the social, cultural and organizational settings in which computing and information technology will be used."

At the University of Michigan, Ann Arbor Informatics interdisciplinary major, informatics is defined as "the study of information and the ways information is used by and affects human beings and social systems. The major involves coursework from the College of Literature, Science and the Arts, where the Informatics major is housed, as well as the School of Information and the College of Engineering. Key to this growing field is that it applies both technological and social perspectives to the study of information. Michigan's interdisciplinary approach to teaching Informatics gives a solid grounding in contemporary computer programming, mathematics, and statistics, combined with study of the ethical and social science aspects of complex information systems. Experts in the field help design new information technology tools for specific scientific, business, and cultural needs." Michigan offers four curricular tracks within the informatics degree to provide students with increased expertise. These four track topics include:

- *Internet Informatics*: An applied track in which students experiment with technologies behind Internet-based information systems and acquire skills to map problems to deployable Internet-based solutions. This track will replace Computational Informatics in Fall 2013.

- Data Mining & Information Analysis: Integrates the collection, analysis, and visualization of complex data and its critical role in research, business, and government to provide students with practical skills and a theoretical basis for approaching challenging data analysis problems.

- *Life Science Informatics*: Examines artificial information systems, which has helped scientists make great progress in identifying core components of organisms and ecosystems.

- Social Computing: Advances in computing have created opportunities for studying patterns of social interaction and developing systems that act as introducers, recommenders, coordinators, and record-keepers. Students, in this track, craft, evaluate, and refine social software computer applications for engaging technology in unique social contexts. This track will be phased out in Fall 2013 in favor of the new bachelor of science in information. This will be the first undergraduate degree offered by the School of Information since its founding in 1996. The School of Information already contains a Master's program, Doctorate program, and a professional master's program in conjunction with the School of Public Health. The BS in Information at the University of Michigan will be the first curriculum program of its kind in the United States, with the first graduating class to emerge in 2015. Students will be able to apply for this unique degree in 2013 for the 2014 Fall semester; the new degree will be a stem off of the most popular Social Computing track in the current Informatics interdisciplinary major in LSA. Applications will be open to upper-classmen, juniors and seniors, along with a variety of information classes available for first and second year students to gauge interest and value in the specific sector of study. The degree was approved by the University on June 11, 2012. Along with a new degree in the School of Information, there has also been the first and only chapter of an Informatics Professional Fraternity, Kappa Theta Pi, chartered in Fall 2012.

At the University of Washington, Seattle Informatics Undergraduate Program, Informatics is an undergraduate program offered by the Information School. Bachelor of Science in Informatics is described as "[a] program that focuses on computer systems from a user-centered perspective and studies the structure, behavior and interactions of natural and artificial systems that store, process and communicate information. Includes instruction in information sciences, human computer interaction, information system analysis and design, telecommunications structure and information architecture and management." Washington offers three degree options as well as a custom track.

- Human-Computer Interaction: The iSchool's work in human-computer interaction (HCI) strives to make information and computing useful, usable, and accessible to all. The Informatics HCI option allows one to blend your technical skills and expertise with a broader perspective on how design and development work impacts users. Courses explore the design, construction, and evaluation of interactive technologies for use by individuals, groups, and organizations, and the social implications of these systems. This work encompasses user interfaces, accessibility concerns, new design techniques and methods for interactive systems and collaboration. Coursework also examines the values implicit in the design and development of technology.

- Information Architecture: Information architecture (IA) is a crucial component in the development of successful Web sites, software, intranets, and online communities. Architects structure the underlying information and its presentation in a logical and intuitive way so that people can put information to use. As an Informatics major with an IA option, one will master the skills needed to organize and label information for improved navigation and search. One will build frameworks to effectively collect, store and deliver information.

One will also learn to design the databases and XML storehouses that drive complex and interactive websites, including the navigation, content layout, personalization, and transactional features of the site.

- Information Assurance and Cybersecurity: Information Assurance and Cybersecurity (IAC) is the practice of creating and managing safe and secure systems. It is crucial for organizations public and private, large and small. In the IAC option, one will be equipped with the knowledge to create, deploy, use, and manage systems that preserve individual and organizational privacy and security. This tri-campus concentration leverages the strengths of the Information School, the Computing and Software Systems program at UW Bothell, and the Institute of Technology at UW Tacoma. After a course in the technical, policy, and management foundations of IAC, one may take electives at any campus to learn such specialties as information assurance policy, secure coding, or networking and systems administration.

- Custom (Student-Designed Concentration): Students may choose to develop their own concentration, with approval from the academic adviser. Student-designed concentrations are created out of a list of approved courses and also result in the Bachelor of Science degree.

Applied Disciplines

Organizational Informatics

One of the most significant areas of application of informatics is that of organizational informatics. Organizational informatics is fundamentally interested in the application of information, information systems and ICT within organisations of various forms including private sector, public sector and voluntary sector organisations. As such, organisational informatics can be seen to be sub-category of social informatics and a super-category of business informatics.

Theories of Socio-technical Systems

Theories about Socio-Technical Systems include: Functionalism/ Transaction economics, Socio-Technical Interaction Networks/Infrastructure, Media, Information Societies, and Ethics/ Values.

Functionalism/Transaction Economics

Functionalism is defined as the impact new technologies have. Technologies are able to supply new functions or even revise preexisting functions. The main impact of these technologies is seen in the change that they provide and the new things that are possible with these changes. These changes also come with a grain of salt because of the costs that come along with these changes. Looking at the costs and benefits, transaction economics, is extremely critical in the study of the impact of a technology. The difference in costs that is given is equal to the impact of using the new technology. New technologies have made a lot of the tasks done on a daily basis a lot more cost efficient.

Socio-technical Networks/Infrastructure

Socio-technical interaction network (STIN) is a network that includes people, equipment, data, diverse resources, documents, messages, legal arrangements, enforcement mechanisms and re-source flows. STINs are embedded in all of the ICT (Information Communication Technology) that are used today. The bases of these STINs are known as infrastructure. Infrastructure is known as the basic physical and organization structures that are essential to the operation of an enterprise This infrastructure offers solutions that may occur in STINs. Infrastructure is also often not visible and for that reason it is taken for granted. However, infrastructure is extremely important.

References

- Ronalds, B.F. (2016). Sir Francis Ronalds: Father of the Electric Telegraph. London: Imperial College Press. ISBN 978-1-78326-917-4.

- Rojas, Raúl (2002). "The history of Konrad Zuse's early computing machines". In Rojas, Raúl; Hashagen, Ulf. The First Computers—History and Architectures History of Computing. MIT Press. p. 237. ISBN 0-262-68137-4.

- Engineering: Issues, Challenges and Opportunities for Development. UNESCO. 2010. pp. 127–8. ISBN 978-92-3-104156-3.

- Manual on the Use of Thermocouples in Temperature Measurement. ASTM International. 1 January 1993. p. 154. ISBN 978-0-8031-1466-1.

- Occupational Outlook Handbook, 2008–2009. U S Department of Labor, Jist Works. 1 March 2008. p. 148. ISBN 978-1-59357-513-7.

- Ronalds, B.F. (2016). Sir Francis Ronalds: Father of the Electric Telegraph. London: Imperial College Press. ISBN 978-1-78326-917-4.

- Beauchamp, K.G. (2001). History of Telegraphy: Its Technology and Application. IET. pp. 394–395. ISBN 0-85296-792-6.

- Wilson, Arthur (1994). The Living Rock: The Story of Metals Since Earliest Times and Their Impact on Civilization. p. 203. Woodhead Publishing. ISBN 978-1-85573-301-5.

- Beynon-Davies P. (2002). Information Systems: an introduction to informatics in Organisations. Palgrave, Basingstoke, UK. ISBN 0-333-96390-3

- Norman, Jeremy. "Francis Ronalds Builds the First Working Electric Telegraph (1816)". HistoryofInformation. com. Retrieved 1 May 2016.

- sector=Government, ; corporateName=Department of Economic Development, Jobs, Transport and Resources - State Government of Victoria;. "Electrical Engineer Career Information for Migrants | Victoria, Australia". www.liveinvictoria.vic.gov.au. Retrieved 2015-11-30.

- "Electrical and Electronic Engineer". Occupational Outlook Handbook, 2012-13 Edition. Bureau of Labor Statistics, U.S. Department of Labor. Retrieved November 15, 2014.

- "Computer Software Engineer". Bureau of Labor Statistics. March 19, 2010. Archived from the original on July 26, 2013. Retrieved July 20, 2012.

- "What is the difference between electrical and electronic engineering?". FAQs - Studying Electrical Engineering. Retrieved 20 March 2012.

- "Milestones:Shilling's Pioneering Contribution to Practical Telegraphy, 1828-1837". IEEE Global History Network. IEEE. Retrieved 26 July 2011.

Permissions

Index

A

Adoption, 15, 18, 115, 197, 251, 259, 275

Amplifier, 241, 243, 261, 269

Antenna, 6-7, 20-22, 27, 29-30, 36-42, 46-48, 56, 67, 77, 80-81, 90, 94, 107, 180, 201-205, 211, 215-216, 218, 232, 235-236, 282, 296

Attenuation, 5, 8, 10, 20, 22-23, 28-29, 62, 65-66, 107, 215, 238, 242-245, 262-263

Autoconfiguration Protocols, 96

B

Bandwidth-distance Product, 243

Bluetooth, 68-69, 72, 75, 77, 83, 90, 105-126, 131, 153, 156-157, 162, 175, 179, 181-183, 202, 213, 226

C

Communication Systems, 17, 21, 38, 194, 198, 201, 205, 212, 214, 218, 225, 238-240, 248, 257

Communications Satellite, 4, 226-228, 230-231, 235-236

Computer Networks, 2, 4, 8, 84, 146, 161, 166, 173, 218, 282

Computer Security, 127, 129, 153, 155, 157-158, 160-166, 186, 191

Cordless, 68, 72-73, 75, 118, 218

Current Market Demands, 258

D

Data Communications, 18, 74, 83

Decoding, 14, 104, 205, 236, 256

Dense Wdm, 244

Digital Audio Broadcasting Systems, 217, 219

Digital Radio, 194, 211-214, 217-219, 221-225, 236

Digital Television (dtv) Broadcasting Systems, 220

Dispersion, 65-66, 204, 239, 241, 243-245, 262

Dual-tone Multi-frequency Signalling, 253

Dynamic Radio Resource Management, 216

E

Electrical Telegraph, 2, 285, 288-290, 300

Electromagnetic Induction, 71, 201, 206, 266

Electromagnetic Spectrum, 73, 103

Energy Transfer, 70, 74

Extending The Useful Distance, 259

Extraterrestrial, 227, 260

F

Fiber Cable Types, 241

Fiber-optic Communication, 23, 194, 237-240, 243

Free-space Optical, 69, 71, 77, 256-258

Free-space Optical Communication, 69, 71, 77, 256-257

Frequency-division Multiplexing, 33, 104, 205, 213, 221

G

Geostationary Orbits (geo), 230

Global Area Network, 78

H

Hidden Node Problem, 80, 85

Hotspot (wi-fi), 96

I

Inter-cell Radio Resource Management, 217

Internet, 2-6, 9, 11, 13, 18-20, 57-59, 70, 74, 77-79, 82, 84, 87, 89-94, 97-99, 101, 103, 109, 116, 119, 125-126, 128, 133, 136, 138-141, 144, 149, 152-155, 157, 160, 162-163, 165-166, 169-170, 172-173, 177, 179, 181, 190, 213, 217, 219, 223, 225-226, 233, 236-238, 240, 246, 253, 285, 297-299, 304

L

Li-fi, 102-104, 125-126, 261

Low Earth Orbiting (leo) Satellites, 229

M

Medium Earth Orbit (meo), 229-230

Mobile Security, 175, 192

Mobile Telephones, 73

Modulation, 13, 16, 23, 47, 90, 94, 104, 106, 113-114, 200-203, 206, 208-210, 212, 214-215, 217-220, 241, 244, 261, 272-273, 282

Molniya Satellites, 232

Multi-radio Mesh, 94

Multipath Fading, 80

O

Operating Systems, 99, 118, 128, 131, 160, 162, 175, 184-185, 189-191, 280, 282

Optical Fiber, 5, 7, 23, 71, 199, 237-242, 244, 246-249, 257-258, 261, 264, 273, 279

Optical Telegraphy, 257

P

Photophone, 69, 194-201, 238, 257

Propagation, 12-13, 20-23, 30-31, 38, 54, 60, 62-64, 66-67, 77, 107, 184, 204, 244-245, 263

Push-button Telephone, 249, 252, 254

R

Radio, 1-4, 6-7, 9-10, 12-15, 17-23, 27, 29, 36-38, 40-42, 56, 61, 65, 67-78, 80-83, 86, 89-90, 92-94, 102-107, 110, 112-115, 117-118, 121, 125, 141, 150-151, 156, 178, 194, 197-198, 200-228, 233-236, 254-255, 261-263, 265, 268-269, 271-272, 283, 285, 296-297

Radio And Television, 3

Radio Band, 205

Radio Resource Management, 214-217, 263

Receiver, 3, 7, 20-22, 39, 69-70, 75, 80-81, 91, 131, 154, 191, 195-198, 200-212, 215, 218-219, 222, 226, 228, 232, 236, 240-241, 244, 257, 261, 263, 285, 288, 291, 296

Receiver And Demodulation, 204

Receivers, 7, 13, 42, 68-69, 74, 76, 108, 156, 198, 206-209, 212-213, 217-218, 221, 223-225, 234-236, 241-242, 247-249, 261, 273

Reception, 5, 12, 71, 80, 90, 197, 204, 208-210, 220, 235, 292

Record Speeds, 243

Regeneration, 245

Resonance, 29, 36-37, 39, 41, 201, 204

S

Satellite, 2-4, 68, 72, 74, 76, 78, 93-94, 210-213, 218-222, 224-237, 262-263, 270, 273, 278

Satellite Constellation, 230, 262

Shared Resource Problem, 80

Space Network, 23, 78

Static Radio Resource Management, 215

T

Telecommunication, 1-3, 5-6, 8-9, 17, 20-21, 24, 34, 57, 67, 91, 105, 194-195, 218, 227, 233, 249, 253, 257, 273, 278-279, 299

Telegraph, 2-3, 27, 31-32, 35, 197, 202, 225, 265, 267, 280, 285-295, 297-301, 307

Telegraph And Telephone, 2

Telephone, 1-5, 8-11, 24, 27-29, 32-34, 36, 57, 59-60, 69,

72-73, 91, 93, 176-177, 183, 194-199, 202, 207, 210, 212, 225-226, 228, 230, 234, 237-240, 249-258, 265, 270, 286, 293, 299-300

Television, 2-5, 42, 68, 72, 169, 203, 206-207, 210, 220, 223-226, 231-232, 234-235, 237-238, 240, 254-255, 269, 271-272, 279

Transmission Medium, 6-7, 57, 238, 248, 298

Transmission Windows, 245

Transmitter, 6, 20-22, 29-30, 39, 42-43, 46-47, 56, 61, 69-70, 77, 196-197, 200-202, 204-206, 208, 210, 215, 218, 225-227, 237, 240-241, 243, 268, 274, 292, 296

Transmitter And Modulation, 202

Transmitters, 6-7, 40-42, 47-48, 69, 71, 76, 80-81, 109, 204, 206, 208-209, 211, 214, 217-218, 222-223, 228, 240-242, 244, 247-248, 255-258, 261, 273-274

Two-way Digital Radio Standards, 218, 225

U

Useful Distances, 259

W

Wavelength-division Multiplexing, 239, 241, 243

Wi-fi Protected Access, 84, 127, 134, 142, 146

Wired Communication, 7, 290

Wired Equivalent Privacy, 84, 127, 134, 141, 146

Wireless, 2-3, 7, 67-99, 101-105, 107-109, 111, 113-115, 117-119, 121, 123, 125-141, 143, 145-147, 149-155, 157, 159, 161, 163, 165, 167, 169, 171, 173, 175, 177, 179, 181, 183, 185, 187, 189, 191-197, 199, 201-202, 206-207, 214-216, 218, 226-227, 238, 257-258, 261, 263, 267, 269, 282, 285, 296-298

Wireless Communication, 3, 7, 68, 73, 84, 102, 108, 129, 133, 214, 227, 258

Wireless Intrusion Prevention System, 133, 150

Wireless Lan, 68, 75, 77, 82, 85-87, 97, 118, 126, 128, 132-134, 138, 150

Wireless Man, 78

Wireless Mesh Network, 68, 78, 92

Wireless Network, 75, 78-79, 83, 86, 91, 97, 128-130, 132-133, 138, 140-141, 145-147, 215, 263, 282

Wireless Pan, 77

Wireless Security, 127-129, 131, 133, 135, 137-139, 141, 143, 145, 147, 149, 151, 153, 155, 157, 159, 161, 163, 165, 167, 169, 171, 173, 175, 177, 179, 181, 183, 185, 187, 189, 191, 193

Wireless Wan, 78, 91